T0202720

**Lecture Notes of
the Unione Matematica Italiana**

More information about this series at http://www.springer.com/series/7172

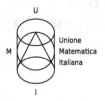

Francesco Russo

On the Geometry of Some Special Projective Varieties

Springer

Unione
Matematica
Italiana

Francesco Russo
Dipartimento di Matematica e Informatica
Università degli Studi di Catania
Catania, Italy

ISSN 1862-9113   ISSN 1862-9121   (electronic)
Lecture Notes of the Unione Matematica Italiana
ISBN 978-3-319-26764-7   ISBN 978-3-319-26765-4   (eBook)
DOI 10.1007/978-3-319-26765-4

Library of Congress Control Number: 2015958350

Mathematics Subject Classification: 14N05, 14M07, 14M10, 14M22, 14E30, 14J70, 14E05

Springer Cham Heidelberg New York Dordrecht London

Printed on acid-free paper

This Springer imprint is published by SpringerNature
The registered company is Springer International Publishing AG Switzerland.

The Unione Matematica Italiana (UMI) has established a bi-annual prize, sponsored by Springer-Verlag, to honor an excellent, original monograph presenting the latest developments in an active research area of mathematics, to which the author made important contributions in the recent years.

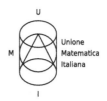

U

M Unione Matematica Italiana

I

The prize-winning monographs are published in this series.

Details about the prize can be found at:
http://umi.dm.unibo.it/en/unione-matematica-italiana-prizes/
book-prize-unione-matematicaitaliana/

This book has been awarded the 2015 Book Prize of the Unione Matematica Italiana.

The members of the scientific committee of the 2015 prize were:

Lucia Caporaso
Università Roma Tre, Italy

Ciro Ciliberto
(Presidente of the UMI)
Università degli Studi di Roma Tor Vergata, Italy

Gianni Dal Maso
Scuola Internazionale Superiore di Studi Avanzati (SISSA),
Trieste, Italy

Camillo De Lellis
University of Zurich (UZH), Switzerland

Alessandro Verra
Università Roma Tre, Italy

a Cledvane, Giulia e Luca
con tanto amore e affetto

Preface

Providing an introduction to both classical and modern techniques in projective algebraic geometry, this monograph treats the geometrical properties of varieties embedded in projective spaces, their secant and tangent lines, the behavior of tangent linear spaces, the algebro-geometric and topological obstructions to their embedding into smaller projective spaces, and the classification of extremal cases. It also provides a solution to Hartshorne's Conjecture on Complete Intersections for the class of quadratic manifolds and new short proofs of previously known results, using the modern tools of Mori Theory and of rationally connected manifolds and following the ideas and methods contained in [103, 104, 154, 156, 160].

The new approach to some of the problems considered can be summarized in the principle that, instead of studying a special embedded manifold uniruled by lines, one analyzes the original geometrical properties of the manifold of lines passing through a general point and contained in the manifold. Once this manifold of lines in its natural embedding, usually of lower codimension, is classified, one tries to reconstruct the original manifold, following a principle which also appears in other areas of geometry such as projective differential geometry and complex geometry.

These classical themes in algebraic geometry enjoyed renewed interest at the beginning of the 1980s, following some conjectures posed by Hartshorne and the discovery by Fulton and Hansen of an important connectedness theorem with new and deep applications to the geometry of algebraic varieties found by Zak, see [44, 69, 70, 198].

Catania, Italy
19 October 2015

Francesco Russo[1]

[1]Partially supported by PRIN Grant *Geometria delle varietà algebriche* and by the Research Project FIR 2014 *Aspetti geometrici e algebrici della Weak e Strong Lefschetz Property* of the UNICT. The author is a member of the GNSAGA group of INDAM.

Ringraziamenti

Sono grato a tutti i miei Maestri, non solo matematici, sparsi in giro per il mondo o racchiusi nella mia biblioteca o scoperti sul web. L'elenco sarebbe troppo vasto e poco omogeneo per poter essere inserito qui.

Un ringraziamento molto speciale va a tre amici con cui ho collaborato intensamente negli ultimi anni: Ciro (Ciliberto), Paltin (Ionescu) e Luc (Pirio). Molti dei risultati originali presentati in queste note sono frutto del nostro lavoro e/o di nostre discussioni senza le quali questo testo non avrebbe mai visto la luce. Paltin mi ha aiutato anche nella fase di correzione finale, segnalandomi tantissimi errori sia tipografici che ortografici.

Sono riconoscente al mio amico Ivan (Pan) per aver letto molto attentamente una versione preliminare, evidenziando alcune imprecisioni matematiche nell'ultimo capitolo e infiniti errori tipografici qua e là.

I due referee hanno suggerito moltissime correzioni matematiche, linguistiche, tipografiche e di esposizione consentendomi di migliorare la presentazione in vari punti, di rendere più coerente il testo e di espandere opportunamente alcune parti trattate troppo sinteticamente nelle versioni preliminari.

Il lavoro editoriale della Springer, svolto a tempo di record e in maniera altamente professionale, ha eliminato alcune espressioni linguistiche inconsuete e innumerevoli refusi tipografici presenti nelle versioni preliminari. Tutto quanto sopravviverà a questo notevole sforzo é interamente dovuto alla mia disattenzione e alla mia incapacità di scovare le rimanenti imprecisioni o gli eventuali refusi ancora annidati nel testo.

L'amore, la comprensione e l'appoggio incondizionato di Cledvane, di Giulia e di Luca hanno permesso che questo lavoro si concretizzasse in tempi ristrettissimi. Un grazie di cuore a tutti e tre.

Contents

Introduction

Algebraic geometry is a glorious topic boasting at least one and a half centuries of history and numerous famous contributors, but still full of intriguing and outstanding problems that are rather easy to state but hard to solve.

Located at the crossroad of geometry, algebra, analysis and number theory, algebraic geometry has influenced most other mathematical disciplines and recently has found applications in physics, system theory, computer science, and even technology. It would not be a gross exaggeration to say that, during the last six decades, most notorious achievements in the field of pure mathematics were connected with or used the language of algebraic geometry—this point of view can easily be justified by looking at the Proceedings of the International Congress of Mathematicians or the Mathematical Reviews databases.

This flourishing ubiquity of today's algebraic geometry has not arrived at no cost. While classical algebraic geometry—developed primarily in Italy—emphasized the geometric aspects of the theory, the progress in this field made in the last decades of the last century was largely due to a more formal and abstract approach, which yielded general results, but often failed in concrete examples. Moreover, much of the special beauty of this subject was lost in this process of generalizing and formalizing.

A new trend occurred in the last 30/40 years when geometrically meaningful and naturally formulated problems were successfully dealt with using modern tools and when beautiful interactions between different branches of algebraic geometry— such as the geometry of projective varieties and birational geometry on one side and the theory of resolutions of ideals and modules on the algebraic side—produced new interesting results.

One can consult the notable book of Robin Hartshorne *Ample Subvarieties of Algebraic Varieties*, [86], which together with *Varieties of Small Codimension in Projective Space*, [87], can be considered the starting points of this new trend. In [86] several open problems were solved, and many outstanding questions were discussed. For example, we point out the problem on set-theoretic complete intersection of curves in $\mathbb{P}^3$ (still open) or the characterization of $\mathbb{P}^N$ as the unique smooth projective variety with ample tangent bundle. The last problem was solved by Mori in [136] and cleared the path to the foundations of Mori Theory in [137]. In related fields we only mention Deligne's proof of the Weil Conjectures or later Faltings' proof of the Mordell Conjecture, which used the new machinery developed by Weil, Zariski, Serre, and above all by Grothendieck, his collaborators, and his school.

The interplay between topology and algebraic geometry continued to flourish. Lefschetz's Theorem and the Barth–Larsen Theorem also suggested that smooth projective varieties, whose codimension is small with respect to their dimension, should satisfy very strong restrictions. To get a feeling for this result, we remark that a codimension two smooth complex subvariety of $\mathbb{P}^N$, $N \geq 5$, has to be simply connected. If $N \geq 6$, there are no known examples of codimension two smooth varieties with the exception of the trivial ones, that is, the transversal intersections of two hypersurfaces.

Based on these empirical observations, Hartshorne was led to formulate his famous Complete Intersection Conjecture in 1974, see [87], whose statement is very neat and clear:

> Let $X \subset \mathbb{P}^N$ be a smooth irreducible nondegenerate projective variety.
>
> If $N < \dfrac{3}{2} \dim(X)$, equivalently if $\operatorname{codim}(X) < \dfrac{1}{2} \dim(X)$, then X is a complete intersection.

Let us quote Hartshorne: *While I am not convinced of the truth of this statement, I think it is useful to crystallize one's idea, and to have a particular problem in mind* [87]. The conjecture is sharp, as the example of $\mathbb{G}(1, 4) \subset \mathbb{P}^9$ shows.

This is not the place to remark on how many important contributions originated and still arise today from this open problem in the areas of vector bundles on projective space, of the study of defining equations of a variety, k-normality, and so on. The list of these achievements is so long that we prefer to avoid citations, being confident that every algebraic geometer has at some time met a problem or a result related to it. It is quite embarrassing that the powerful methods of modern algebraic geometry have not yet produced a solution (or a counterexample).

For a long time Hartshorne's Conjecture has been practically forgotten, and nowadays it seems not to be a primary interest for the main research groups in algebraic geometry around the world. Moreover, in the original formulation for arbitrary codimension, it remains as open as before.

Recently there were two contributions yielding a positive answer to the Complete Intersection Conjecture in the particular case of quadratic manifolds together with the classification of quadratic *Hartshorne varieties*.

We recall that a quadratic manifold is a nondegenerate manifold $X^n \subset \mathbb{P}^N$ which is the scheme-theoretic intersection of $m \geq N - n$ quadratic hypersurfaces. Also Mumford in his seminal series of lectures [140] brought attention to the fact that many interesting embedded manifolds are scheme-theoretically defined by quadratic equations, e.g., homogeneous manifolds or *special varieties*. Moreover, those having small codimension really share very special geometrical properties as later confirmed in [19, Corollary 2] and in Theorem 5.3.6 here. Let us state these two contributions, whose proofs are presented in detail in Sect. 5.2.

Theorem (Hartshorne's Conjecture for quadratic manifolds [105, Theorem 3.8, part (4)]) *Let $X^n \subset \mathbb{P}^N$ be a quadratic manifold. If $\operatorname{codim}(X) < \frac{1}{2}\dim(X)$, then X is a complete intersection.*

Theorem (Classification of quadratic Hartshorne varieties [105, Theorem 3.9]) *Let $X^n \subset \mathbb{P}^{\frac{3n}{2}}$ be a quadratic manifold. Then either $X \subset \mathbb{P}^N$ is a complete intersection or it is projectively equivalent to one of the following:*

1. $\mathbb{G}(1,4) \subset \mathbb{P}^9$
2. $S^{10} \subset \mathbb{P}^{15}$

The originality of most parts of this monograph can be summarized in the fact that the tools and methods—presented in Chaps. 2, 3, 4, and 5 and developed to solve the two previous problems—are based on a new perspective which has been able to prove in a very simple and direct way *all* the previously known results on the geometry of special varieties connected with the circle of ideas originated by Hartshorne's Conjecture. Furthermore, this new point of view allowed us to go ahead by solving the conjecture for quadratic manifolds and by obtaining the classification of quadratic Hartshorne manifolds. Last, but not least, it also enabled us to shed new light on the classification results of special varieties proved so far or to conjecture new intriguing problems, as we now illustrate briefly (see Sect. 5.3 for an expansion of these ideas).

Our main instrument of investigation in these problems is the Hilbert scheme of lines contained in a smooth projective variety $X^n \subset \mathbb{P}^N$, called from now on a manifold, and passing through a general point $x \in X$. This scheme will be denoted by $\mathscr{L}_x$ and has a natural embedding into $\mathbb{P}((t_x X)^*) = \mathbb{P}^{n-1}$.

The *principle* that $\mathscr{L}_x$ can inherit intrinsic and extrinsic geometrical properties of the manifold X has emerged recently. The first nontrivial instance of this principle is based on the theory of deformations and shows that $\mathscr{L}_x$ is smooth as soon as it is not empty, see Proposition 2.2.1. Moreover, the theory of deformations of rational curves on a projective manifold (Mori Theory), recalled here in great detail in Chap. 2, is nowadays well established and also permits us to control the dimension of each irreducible component of $\mathscr{L}_x$.

For example, the behavior of $\mathscr{L}_x \subset \mathbb{P}^{n-1}$ is easily handled for *prime Fano manifolds of index $i(X)$*, which are Fano manifolds $X^n \subset \mathbb{P}^N$ whose Picard group is generated by the hyperplane section class H and such that $-K_X = i(X)H$ for some positive integer $i(X)$. Indeed, one shows that for these manifolds, $\mathscr{L}_x$, if not empty, is equidimensional with $\dim(\mathscr{L}_x) = i(X) - 2$; see Chap. 2. In general the smoothness of $\mathscr{L}_x$ implies that, if equidimensional and of dimension of at least $\frac{n-1}{2}$, $\mathscr{L}_x$ is also irreducible. Thus on prime Fano manifolds, these conditions can be easily expressed in terms of the index. In any case, prime Fano manifolds of high index are quite rare, and all known examples are either complete intersections or quadratic. Let us also remark the notable fact that the Hilbert scheme of lines $\mathscr{L}_x \subset \mathbb{P}^{n-1}$ of a quadratic manifold $X^n \subset \mathbb{P}^N$ is also quadratic; see Corollary 2.3.6 and Proposition 2.3.8, revealing another key incarnation of the principle.

The Hilbert schemes of lines through a general point of some special homogeneous manifolds are also somehow *nested* or behave like *parts of a matryoshka*. For homogeneous manifolds or for other classes of manifolds, where the principle of reincarnation on $\mathscr{L}_x$ of a notable property of X holds, one can start an induction process which sometimes stops after only a few steps; see, e.g., [160, Theorem 2.8, Corollaries 3.1 and 3.2] and also [64].

A fundamental example of this inductive procedure is the following: if $X^n \subset \mathbb{P}^N$ is a *LQEL*-manifold of type $\delta \geq 3$, then $\mathscr{L}_{x,X} \subset \mathbb{P}^{n-1}$ is a *QEL*-manifold of type $\delta - 2$; see [160, Theorem 2.3] and Theorem 4.2.3 here. Then starting the induction with $X^n \subset \mathbb{P}^N$, a *LQEL*-manifold of type $\frac{n}{2}$, one deduces immediately $n = 2, 4, 8$, or 16. In particular, this yields the quickest proof that Severi varieties appear only in these dimensions. See Chap. 4 for the definitions of $(L)QEL$-variety and of Severi variety, the latter being introduced by Zak in [196].

The classification of Severi varieties has many different proofs nowadays (see [27, 119, 196, 198]), which are longer and less transparent than ours, but this is only half of the value provided by the new approach. Indeed, an amazing highlight is the observation that the existence of a $(L)QEL$-manifold of type $\delta = n/2$ and of dimension greater than 16 would have produced a counterexample to Hartshorne's Conjecture on Complete Intersections; see [160, Remark 3.3] and above all Sect. 5.3. Obviously the counterexample would not have been $X^n \subset \mathbb{P}^{\frac{3n}{2}+2}$ but the associated $\mathscr{L}_x \subset \mathbb{P}^{n-1}$; see Sect. 5.3.

This is a key observation which enables us to change perspective and to formulate new problems and intriguing questions by simply applying Hartshorne's Conjecture to $\mathscr{L}_x \subset \mathbb{P}^{n-1}$ and reading the output. Of course the point is to prove the conjectured statements without invoking Hartshorne's Conjecture. For those who are not convinced of the truth of the conjecture, this procedure suggests an efficient way to look for possible counterexamples; see again Sect. 5.3 for variations on these themes.

The above principle of reincarnation was also used to attack several problems in a *unified way*. The technique of studying, or even reconstructing, X from the *variety of minimal rational tangents* introduced in the work of Hwang, Mok, and others (a generalization of the Hilbert scheme of lines passing through a point) was applied to the theory of Fano manifolds (see, e.g., [64, 93–97, 135]). On the other hand, Landsberg and others investigated some possible characterizations of special homogeneous manifolds via the projective second fundamental form (see, e.g., [98, 120–122]). The Hilbert scheme of lines through a point is closely related to the base locus of the (projective) second fundamental form, a classical tool used in projective differential geometry and reconsidered in modern algebraic geometry by Griffiths and Harris (see [79] and also [107]). In this approach one tries to reconstruct a (homogeneous) variety from its second fundamental form (see, e.g., [98, 120–122]) by integrating local differential equations and obtaining global results. We note that the base locus of the second fundamental form at a general point is typically not smooth for manifolds, while smoothness is preserved by the Hilbert scheme of lines.

An important class where the two previous objects coincide is that of quadratic manifolds; see Corollary 2.3.6. Since $\mathscr{L}_x \subset \mathbb{P}^{n-1}$ coincides with the base locus scheme of the second fundamental form, it may be scheme-theoretically defined by at most $c = \operatorname{codim}(X)$ quadratic equations; see *loc. cit.* If $\mathscr{L}_{x,X}$ is also irreducible, then another *matryoshka* naturally appears in which the number of equations defining the successive manifold has decreased and, at the first step, is controlled by the codimension of X. When the process stops, one hopes to reconstruct the original quadratic manifold X or the successive pieces of the matryoshka by calculating their invariants.

From this point of view, a quadratic manifold $X^n \subset \mathbb{P}^N$ with $3n > 2N$ (equivalently with $c \leq \frac{n-1}{2}$) is a complete intersection *because* $\mathscr{L}_{x,X} \subset \mathbb{P}^{n-1}$ is a smooth irreducible nondegenerate complete intersection, defined precisely by the c quadratic equations in the second fundamental form, so that it has the *right dimension*; see [106, Theorems 4.8 and 2.4] and Theorem 5.1.3 here. If X is quadratic, we use Faltings' Criterion [56] to deduce that, when $n \geq 2c + 1$, $\mathscr{L}_x \subset \mathbb{P}^{n-1}$ is a complete intersection. It is worth remarking that Faltings' Criterion was not so effective at solving restricted forms of Hartshorne's Conjecture because it has always been applied to $X^n \subset \mathbb{P}^N$ and not to $\mathscr{L}_x$ where it reveals its power since its hypothesis in the case of quadratic manifolds reads exactly as $c \leq \frac{n-1}{2}$, which is the condition in Hartshorne's Conjecture. To classify quadratic Hartshorne varieties, we appeal to Netsvetaev's Theorem [144], which when applied to $\mathscr{L}_x$ yields that the dimension of a quadratic Hartshorne variety is only 6 or 10; see Theorem 5.1.4 and [105, Theorem 3.8]. The reconstruction is then easy due to results of Fujita in [65], respectively, of Mukai in [138].

Having illustrated the novelty of some of the methods and of some of the topics presented in the monograph, we now describe the contents in more detail, and doing this we shall also summarize the results in the last two chapters.

In the first chapter we recall the definitions of tangent cone, tangent space, and tangent star to a variety at a fixed point, and define the secant variety SX, the higher secant varieties S^kX, the tangent variety TX, and the variety of tangent stars T^*X of a variety $X \subset \mathbb{P}^N$. We consider its *join*, $S(X, Y)$, with another variety $Y \subset \mathbb{P}^N$ and prove Terracini's Lemma relating the dimension of SX, or more generally of S^kX, respectively, $S(X, Y)$, with the intersection of general tangent spaces to the involved varieties. We describe the first consequence for linear tangency at $k + 1$ general points, by defining the entry loci and by studying their first properties. We end the chapter by recalling the definition of dual variety, its first properties, the definitions of Gauss maps, and the relations with reflexivity. Quite surprisingly nowadays these notions are not part of introductory courses in algebraic geometry. For this reason we have tried to be as self-contained as possible since these concepts also play a fundamental role in the statements of some of the main problems treated in the monograph.

In the second chapter we recall without proofs the basic definitions and results leading to the construction of various Hilbert schemes and describe the infinitesimal properties of these parameter spaces. Since these tools are used systematically in the rest of the monograph, we specialized the general theory to the case of rational curves on a projective manifold providing complete proofs based on the preliminaries. Although there exist several references on this subject, e.g., [116] and [40], to which we have referred, we thought that the inclusion of these parts will make the text more self-contained and will stimulate a systematic deepening of these important techniques. Moreover, the application of these general theories to the specific case of lines might be a good opportunity to see a lot of abstract tools working effectively in the simplest possible case. Then we introduce and study in great detail the Hilbert scheme of lines passing through a (general) point of a projective variety $X \subset \mathbb{P}^N$ and contained in it. As we explained before, this instrument is applied almost everywhere in the sequel, so we have included all the details and proofs concerning the geometrical properties of $\mathscr{L}_x$ inherited from X. In particular, we analyze the singularities of $\mathscr{L}_x$ (see Proposition 2.2.1), and construct an Example 2.2.2 to illustrate some technical issues of the theory of deformations of curves on singular varieties, usually overlooked or only known to experts in the field. Moreover, the singularities of $\mathscr{L}_x$ appearing in this (singular) example point out that $\mathscr{L}_x$ is not a local invariant and it is very sensitive to global properties. The presence of singularities is related to developability, and it also seems to play an important role in other central questions discussed in Sect. 5.3.

We then introduce the second fundamental form of $X \subset \mathbb{P}^N$ at a general point, its fundamental base locus scheme B_x, and prove the inclusion as schemes $\mathscr{L}_x \subseteq B_x$ which becomes an equality of schemes for quadratic varieties or manifolds; see Corollary 2.3.6. More generally, Sect. 2.3 analyzes some relations between the equations defining scheme-theoretically $X \subset \mathbb{P}^N$ and those defining $\mathscr{L}_x \subset \mathbb{P}^{n-1}$.

The second chapter ends with an application of the notions introduced to the classical problem of the existence of projective extensions $X \subset \mathbb{P}^{N+1}$ of a manifold $Y \subset \mathbb{P}^N \subset \mathbb{P}^{N+1}$. It is well known that some special manifolds cannot be hyperplane sections of smooth varieties and that in some cases only the trivial extensions exist. These are given by cones over Y with vertex a point $p \in \mathbb{P}^{N+1} \setminus \mathbb{P}^N$ (see, e.g., [163, 166, 169, 187]). Recently the interest in the above problem (and further generalizations of it) was renewed. Complete references, many results, and a lot of interesting connections with other areas, such as deformation theory of isolated singularities, can be found in the monograph [12], especially relevant for this problem being Chaps. 1 and 5. One could also look at the survey [197] for a different point of view on this problem.

Here we present, following [161], a simple geometrical sufficient condition for non-extendability, Theorem 2.4.3, for smooth projective complex varieties uniruled by lines. The simplest version states that $Y \subset \mathbb{P}^N$ admits only trivial extensions $X \subset \mathbb{P}^{N+1}$ as soon as $\mathscr{L}_{y,X} \subset \mathbb{P}((t_y X)^*)$ admits no smooth extension (a condition easier to verify than the thesis). Indeed, one easily shows in Proposition 2.4.2 that also $\mathscr{L}_{y,X} \subset \mathbb{P}((t_y X)^*)$ is a projective extension of $\mathscr{L}_{y,Y} \subset \mathbb{P}((t_y Y)^*)$ for $y \in Y$ general, revealing another overlooked realization of the *principle* of reincarnation of geometrical properties. Then under the hypothesis of Theorem 2.4.3, one deduces the existence of a line through y and a singular point $p_y \in X$. Then $p_y = p$ does not vary with $y \in Y$ general since X has at most a finite number of singular points so that $X \subset \mathbb{P}^{N+1}$ is a cone of vertex p. The range of applications of Theorem 2.4.3 is quite wide (see Corollaries 2.4.4, 2.4.5, 2.4.7), enabling us to recover some results previously obtained differently, see, for example [12].

We were led to the analysis of the problem of extending smooth varieties by the desire to understanding geometrically why in some well-known examples the geometry of $Y \subset \mathbb{P}^N$ forces every extension to be trivial and by the curiosity of explicitly constructing the cones extending Y. Moreover, this approach reveals that Scorza's result on the non-extendability of $\mathbb{P}^a \times \mathbb{P}^b \subset \mathbb{P}^{ab+a+b}$ for $a + b \geq 3$, originally proved in [166] and rediscovered later by many authors (see, e.g., [12] and Corollary 2.4.4 here), implies the non-extendability of a lot of homogeneous varieties via the description of their Hilbert scheme of lines; see Corollary 2.4.5.

At the beginning of Chap. 3, we survey, following Fulton [68], the connectedness principle of Enriques–Severi–Zariski claiming that limits of irreducible varieties remain connected. Some deep generalizations are reported following the circle of ideas which led to the Connectedness Theorem of Fulton–Hansen and to some new results related to it. We include a proof of the Fulton–Hansen Theorem according to Deligne, and we describe its consequences for the geometry of embedded projective varieties proved by Zak: Theorem on Tangency, finiteness of the Gauss map, dimension of the dual variety, and hyperplane sections of low codimensional varieties. These are modern and deep generalizations of many classical theorems in projective geometry which usually dealt with *general* objects.

In Sect. 3.3 we define some classical tangential invariants of projective varieties, connect them to tangential projections, and prove Scorza's Lemma (see Theorem 3.3.3), which gives a sufficient condition for a manifold to have quadratic (secant) entry locus. Then in Sect. 3.4 we prove the well-known characterization of the Veronese surface, due to Severi in [176], as the unique nondegenerate surface in $\mathbb{P}^N$, $N \geq 5$, not a cone and with $\dim(SX) = 4$. The novelty of our proof of this result and of its generalization to arbitrary dimension in Theorem 3.4.4 relies on a complete parallel with Mori's proof of the characterization of the projective space as the unique manifold with ample tangent bundle; see Sect. 3.4 for the details. This opened the way to some striking generalizations of Severi's Theorem presented in Chap. 6, following [154]; see Theorem 6.3.2 here.

Chapter 4 contains the definition of *QEL*, respectively, *LQEL*, manifolds as those $X \subset \mathbb{P}^N$ whose general entry locus is a quadric hypersurface of dimension equal to the secant defect of X, respectively, the union of quadric hypersurfaces of dimension equal to the secant defect of X. We present the main results of the theory of *LQEL*-manifolds introduced in [160], leading to the Divisibility Theorem for the secant defect of *LQEL*-manifolds in Theorem 4.2.8, which is based on an inductive argument due to the fact that the *LQEL* property passes to $\mathscr{L}_x \subset \mathbb{P}^{n-1}$; see

Theorem 4.2.3. The main applications concern the classification of *LQEL*-manifolds of type $\delta \geq \frac{n}{2}$, which also include Severi varieties.

The rest of the chapter surveys the classification of conic-connected manifolds obtained in [104] and some of the results of Ionescu and Russo [103, 106] on dual defective manifolds. In particular, an astonishing simple proof of the famous Landman Parity Theorem for dual defective manifolds appears as an application of the tools developed in Chap. 2. This famous result was recently reproved in [106] in this way and previously in [51, Theorem 2.4] using completely different methods. This section also contains a self-contained proof of the classification of manifolds with small dual presented in Theorem 4.4.9 and based on [51, Theorem 4.5], reproducing the proof in [103, Theorem 4.4]. We would like to remark that the theories of *LQEL* and *CC* manifolds recently found a lot of new applications to several problems; see, for example, [9, 63, 64, 181, 182].

Chapter 5 begins with the statements of Hartshorne's Conjecture on Complete Intersections recalled above and of Hartshorne's Conjecture on Linear Normality. In Theorem 5.1.6 we present Zak's proof of the Linear Normality Conjecture, and we formalize the definition of Severi variety from the perspective of this result. In Sect. 5.2 we prove Faltings' and Netsvetaev's Criterions for Complete Intersections needed for the proof of Hartshorne's Conjecture for quadratic manifolds, presented in Theorem 5.1.3 and originally proved in [105, Theorem 3.8]. We end the section with the classification of quadratic Hartshorne manifolds, first obtained in [105, Theorem 3.9] and reproduced in Theorem 5.1.4 here. Section 5.3 concerns various speculations originated by Hartshorne's Conjecture and contains the formulation of various open problems arising by the principles described above. The chapter ends with a well-known explicit reconstruction of Severi varieties of dimension 2, 4, 8, and 16 using the variety $\mathscr{L}_x$, which essentially follows the approach of Zak in [198].

Chapter 6 deals with the properties of projective varieties $X^{r+1} \subset \mathbb{P}^N$ of dimension $r+1$ which are n-covered by irreducible curves of degree $\delta \geq n-1 \geq 1$, that is, varieties such that through n-general points there passes an irreducible curve of degree δ contained in X. From now this class of projective varieties will be denoted by $X^{r+1}(n, \delta)$.

The previous condition depends on the embedding, on the number $n \geq 2$, and on the degree $\delta \geq 1$, and natural constraints for the existence of such varieties immediately appear. It has been recently realized in [158] that the study of these varieties is also related to the Castelnuovo bound for the geometric genus of an irreducible projective variety and also to the theory of Castelnuovo varieties, see Sect. 6.2.3.

We shall present the sharp Pirio–Trépreau bound for the embedding dimension N in terms of r, n, δ (see [158] and Theorem 6.2.3 here), which is obtained geometrically via the iteration of projections from general osculating spaces to $X^{r+1}(n, \delta) \subset \mathbb{P}^N$ determined by the irreducible curves of degree δ which n-cover the variety. The varieties extremal for the previous bound are subject to even stronger restrictions—e.g., they are rational and through n-general points there passes a unique rational normal curve of degree δ—see Theorems 6.3.2 and 6.3.3. Moreover, Theorem 6.3.3 generalizes the previous embedded results to a bound for $h^0(\mathscr{O}_X(D))$

with D a Cartier divisor on a proper irreducible variety X of dimension $r + 1$, n-covered by irreducible curves C such that $(D \cdot C) = \delta \geq n - 1$. Furthermore, equality occurs in the bound if and only if $\phi_{|D|}$ maps X birationally onto an extremal $X^{r+1}(n, \delta)$.

Another consequence of the previous bound is that under the same hypotheses, we have $D^{r+1} \leq \delta^{r+1}/(n-1)^r$ if D is nef; see Theorem 6.3.4. This is a generalization of a result usually attributed to Fano in the case $n = 2$ (see for example [116, Proposition V.2.9]), which is useful to prove the boundedness of families of varieties of a certain fixed type such as Fano manifolds; see [116].

A lot of examples of extremal $X^{r+1}(n, \delta)$ (for arbitrary $n \geq 2$, $r \geq 1$, and $\delta \geq n - 1$) have been described in [158] via the theory of Castelnuovo varieties, and their construction will be briefly recalled in Sect. 6.2.3. The main result of Pirio and Trépreau [158] ensures that these examples *of Castelnuovo type* are the only extremal varieties except possibly when $n > 2$, $r > 1$, and $\delta = 2n - 3$.

The first open case, that is, the classification of extremal varieties $X = X^{r+1}(3, 3) \subset \mathbb{P}^{2r+3}$ not of Castelnuovo type, is considered in Sect. 6.4, where it is proved that these varieties are in one-to-one correspondence, modulo projective transformations, with quadro-quadric Cremona transformations on $\mathbb{P}^r$, [154, Theorem 5.2], and Theorem 6.4.5 here. One of the key steps to establish this unknown and somehow unexpected connection is to prove a priori the equality $B_x = \mathscr{L}_x$ as schemes for a general $x \in X$. This has some striking consequences, such as the fact that a quadro-quadric Cremona transformation is, modulo projective transformations acting on the domain and on the codomain, an involution; see Corollary 6.4.6. To the best of our knowledge, this result has not been conjectured before. From this one easily deduces the classification of smooth extremal varieties $X^{r+1}(3, 3) \subset \mathbb{P}^{2r+3}$ showing that there are two infinite series: smooth rational normal scrolls and $\mathbb{P}^1 \times Q^r$ Segre embedded and four isolated examples appearing for $r = 5, 8, 14$, and 26 whose $\mathscr{L}_x \subset \mathbb{P}^r$ is one of the four Severi varieties. Related results and generalizations are also contained in [34].

In Sect. 6.5 we included a self-contained presentation of the basics of the theory of power associative algebras and of their subclass of Jordan algebras, defining the notion of rank of these algebras and generalizing to this setting the usual Laplace formulas for inversion of a square matrix. In this way we present a new point of view on birational involutions with special emphasis on those of type $(2, 2)$ defining the notion of birational involution associated to a power associative algebra. This approach culminates in Theorem 6.5.22 according to which every quadro-quadric Cremona transformation is linearly equivalent to the *cofactor or adjoint map* of a suitable rank three Jordan algebra; see [34] for details.

We end the chapter by surveying the recent results in [156] showing that extremal $X^{r+1}(3, 3)$ and quadro-quadric Cremona transformations are also in bijection with the isotopy classes of rank three complex Jordan algebras; see Sect. 6.6 for precise formulations of these equivalences leading to the so-called *XJC-correspondence*, defined in [156]. In this correspondence X stands for varieties, J for Jordan algebras of rank three, and C for quadro-quadric Cremona transformations. Among other things, it shows that smoothness of X corresponds to semi-simple Jordan algebras and to *semi-special quadro-quadric* Cremona transformations. The *XJC*-correspondence has many applications and allowed us to introduce the notion of simplicity and semi-simplicity for quadro-quadric Cremona transformations or for extremal $X^{r+1}(3,3)$; see [155]. Moreover, the well-established classification of rank three Jordan algebras in dimension at most 6 enabled us to obtain in [157] the complete classification of quadro-quadric Cremona transformations of $\mathbb{P}^r$ for $r \leq 5$ (see [157]), strongly generalizing the previous contributions via a completely new point of view on the subject.

The last chapter concerns an algebraic problem formulated by O. Hesse in [90] to the effect that the determinant of the Hessian matrix of a homogeneous polynomial vanishes identically if and only if, modulo a linear change of coordinates, the polynomial depends on fewer variables. Geometrically this means that the associated projective hypersurface is a *cone*.

Of course cones have vanishing Hessian determinant, and Hesse claimed twice in [90] and in [91] that a hypersurface $X = V(f) \subset \mathbb{P}^N$ is a cone if hess$_X = 0$. Clearly the claim is true if $\deg(f) = 2$ so that the first relevant case is that of cubic hypersurfaces. One immediately sees that $V(x_0 x_3^2 + x_1 x_3 x_4 + x_2 x_4^2) \subset \mathbb{P}^4$ is a cubic hypersurface with vanishing Hessian but not a cone (e.g., because the first partial derivatives of the equation are linearly independent).

Actually the question is quite subtle because, as was firstly pointed out by Gordan and Noether in [78], the claim is true for $N \leq 3$ and in general false for $N \geq 4$; see Sects. 7.3.1 and 7.4.2. The cases $N = 1, 2$ are easily handled, but beginning from $N = 3$ the problem is related to nontrivial characterizations of cones among developable hypersurfaces or, from a differential point of view, to characterize algebraic cones (or algebraic cylinders in the affine setting) among hypersurfaces with zero Gaussian curvature at every regular point; see Sect. 7.2.2.

We shall provide here a complete and detailed account of Gordan–Noether Theory (see Sects. 7.3, 7.3.2, and 7.4); as well as its wide range of applications in different areas of mathematics including differential geometry (see Sect. 7.2.2) and commutative algebra (see Sect. 7.2.4). The approach via the Gordan–Noether Identity to various questions follows the original treatment in [78] as revisited also in [127] and in [193]. In particular, we present applications to the Cremona equivalence of hypersurfaces with vanishing Hessian with cones or to the classification of hypersurfaces with vanishing Hessian in $\mathbb{P}^4$; see Sect. 7.4.2. The chapter ends with a survey of the classical work of Perazzo dealing with the classification of cubic hypersurfaces with vanishing Hessian for $N \leq 6$ (see [151]), which we revisit here in Sect. 7.6, following the recent treatment in [77].

This complete account of hypersurfaces with vanishing Hessian seems to be the most exhaustive and updated survey written so far, and it might be helpful to people interested in these fascinating and unexplored classical problems, which should deserve some attention, also due to their connections with other areas of mathematics partially described at the beginning of the chapter.

Chapter 1
Tangent Cones, Tangent Spaces, Tangent Stars: Secant, Tangent, Tangent Star and Dual Varieties of an Algebraic Variety

1.1 Tangent Cones and Tangent Spaces of an Algebraic Variety and Their Associated Varieties

Let X be an algebraic variety, or more generally a scheme of finite type, over a fixed algebraically closed field K. Let $x \in X$ be a closed point. We briefly recall the definitions of *tangent cone to X at x* and of *tangent space to X at x*. For more details one can consult [142] or [178].

Definition 1.1.1 (Tangent Cone at a Point) Let $U \subset X$ be an open affine neighborhood of x, let $i : U \to \mathbb{A}^N$ be a closed immersion and let U be defined by the ideal $I \subset K[X_1, \ldots, X_N]$. There is no loss of generality in supposing $i(x) = (0, \ldots, 0) \in \mathbb{A}^N$. Given $f \in K[X_1, \ldots, X_N]$ with $f(0, \ldots, 0) = 0$, we can define the *leading form* (or *initial form/term*) f^{in} of f as the non-zero homogeneous polynomial of lowest degree in its expression as a sum of homogenous polynomials in the variables X_i's. Let

$$I^{\mathrm{in}} = \{\text{the ideal generated by the leading form (or initial term)} \, f^{\mathrm{in}} \text{ of all } f \in I\}.$$

Then

$$C_x X := \mathrm{Spec}(\frac{K[X_1, \ldots, X_N]}{I^{\mathrm{in}}}), \tag{1.1}$$

is called *the affine tangent cone to X at x*.

The previous definition does not depend on the choice of $U \subset X$ or on the choice of $i : U \to \mathbb{A}^N$. Indeed, letting $(\mathscr{O}_x, m_x)$ be the local ring of regular functions of X

© Springer International Publishing Switzerland 2016
F. Russo, *On the Geometry of Some Special Projective Varieties*,
Lecture Notes of the Unione Matematica Italiana 18,
DOI 10.1007/978-3-319-26765-4_1

at x, one immediately verifies that

$$\frac{k[X_1, \ldots, X_N]}{I^{\text{in}}} \simeq \text{gr}(\mathcal{O}_x) := \bigoplus_{n \geq 0} \frac{m_x^n}{m_x^{n+1}},$$

yielding an intrinsic expression for the K-algebra defining C_xX.

The previous isomorphism assures that we can calculate C_xX by choosing an arbitrary open neighborhood U of x and moreover that the definition is *local* on X. It should be observed that C_xX is a scheme, which can be neither irreducible nor reduced as shown by easy examples, e.g. a cubic plane curve with a node, respectively a cubic plane curve with a cusp. We now obtain a geometrical interpretation of this cone and find some of its properties.

Since C_xX is locally defined by homogeneous forms, it can be naturally projectivized and thought of as a subscheme of $\mathbb{P}^{N-1} = \mathbb{P}(\mathbb{A}^N)$. If we consider the blow-up of $x \in U \subset \mathbb{A}^N$, $\pi : \text{Bl}_x U \to U$, then $\text{Bl}_x U$ is naturally a subscheme of $U \times \mathbb{P}^{N-1} \subset \mathbb{A}^N \times \mathbb{P}^{N-1}$ and the exceptional divisor $E := \pi^{-1}(x)$ is naturally a subscheme of $x \times \mathbb{P}^{N-1}$. With these identifications one shows that $E \simeq \mathbb{P}(C_xX) \subset \mathbb{P}^{N-1}$ as schemes, see [142, p. 225]. In particular, if X is equidimensional at x, then C_xX is an equidimensional scheme of dimension $\dim(X)$ because E is a Cartier divisor on $\text{Bl}_x U$.

If $X \subset \mathbb{P}^N$ is quasi-projective, we define *the projective tangent cone to X at x*, indicated by $\mathbf{C}_xX$, as the closure of $C_xX \subset \mathbb{A}^N$ in $\mathbb{P}^N$, where $x \in U = \mathbb{A}^N \cap X$ is a suitable chosen affine neighborhood.

We now recall the definition of tangent space to X at $x \in X$.

Definition 1.1.2 (Tangent Space at a Point; Tangent Variety to a Variety) Let the notation be as in the previous definition and let $\mathbf{0} = (0, \ldots, 0) \in \mathbb{A}^N$. Given $f \in K[X_1, \ldots, X_N]$ with $f(\mathbf{0}) = 0$, we can define the *linear term f^{lin}* of f as the degree one homogeneous polynomial in its expression as a sum of homogenous polynomials in the variables X_i's. If the degree one term is zero we define $f^{\text{lin}} = 0$. In other words,

$$f^{\text{lin}} = \sum_{i=1}^{N} \frac{\partial f}{\partial X_i}(\mathbf{0})X_i.$$

Let

$$I^{\text{lin}} = \{\text{the ideal generated by the linear terms } f^{\text{lin}} \text{ of all } f \in I\}.$$

Then

$$t_xX := \text{Spec}\left(\frac{K[X_1, \ldots, X_N]}{I^{\text{lin}}}\right) \tag{1.2}$$

is called *the affine tangent space to X at x*.

Geometrically it is the locus of *tangent lines to X at x*, where a line through x is tangent to X at x if it is tangent to the hypersurfaces $V(f) = 0, f \in I$, i.e. if the multiplicity of intersection of the line with $V(f)$ at $(0, \ldots, 0)$ is greater than one for all $f \in I$. In particular, this locus is a linear subspace of $\mathbb{A}^N$, being an intersection of linear subspaces.

Since $I^{\text{lin}} \subseteq I^{\text{in}}$, we deduce the inclusion of schemes

$$C_x X \subseteq t_x X;$$

and that $t_x X$ is a *linear* subspace of $\mathbb{A}^N$ containing $C_x X$ as a subscheme (and not only as a set). In particular, for every $x \in X$ $\dim(t_x X) \geq \dim(X)$ holds.

We recall that a point $x \in X$ is non-singular if and only if $C_x X = t_x X$. Since $t_x X$ is reduced and irreducible and since $C_x X$ is of dimension $\dim(X)$, we have that $x \in X$ is non-singular if and only if $\dim(t_x X) = \dim(X)$.

Once again there is an intrinsic definition of $t_x X$ since

$$\frac{K[X_1, \ldots, X_N]}{I^{\text{lin}}} \simeq S(\frac{m_x}{m_x^2}),$$

where $S(m_x/m_x^2)$ is the symmetric algebra of the K-vector space m_x/m_x^2.

The natural surjection

$$S(\frac{m_x}{m_x^2}) \to \text{gr}(\mathcal{O}_x) = \bigoplus_{n \geq 0} \frac{m_x^n}{m_x^{n+1}}$$

shows that $t_x X$ is the linear span of $C_x X$ inside $\mathbb{A}^N$, that is the smallest linear space of $\mathbb{A}^N$ containing $C_x X$ as a subscheme (and not only as a set).

If $X \subset \mathbb{P}^N$ is a quasi-projective variety, we define *the projective tangent space to X at x*, indicated by $T_x X$, as the closure of $t_x X \subset \mathbb{A}^N$ in $\mathbb{P}^N$, where $x \in U = \mathbb{A}^N \cap X$ is a suitable chosen affine neighborhood. Then $T_x X$ is a linear projective space which varies with $x \in X$ and clearly $C_x X \subseteq T_x X$ as schemes. We also set, for a (quasi)-projective variety $X \subset \mathbb{P}^N$,

$$TX = \bigcup_{x \in X} T_x X,$$

the variety of tangents, or *the tangent variety of X*.

At a non-singular point $x \in X \subset \mathbb{P}^N$, the equality $C_x X = T_x X$ says that every tangent line to X at x is the *limit* of a secant line $< x, y >$ with $y \in X$ approaching x.

An interesting question is to investigate what are the limits of the secant lines $< x_1, x_2 >$, $x_i \in X$, $x_1 \neq x_2$, when the x_i's, $i = 1, 2$, approach a fixed $x \in X$. As shown in Example 1.1.6 below, for a non-singular point $x \in X$, every tangent line to X at x arises in this way, but for singular points this is not the case. These limits generate a cone, the tangent star cone to X at x, which contains but does not usually coincide with $\mathbf{C}_x X$. From now on we restrict ourselves to the projective setting and we will not treat local questions related to tangent star cones considered for example in [179]. Firstly we introduce the notion of secant variety to a variety $X \subset \mathbb{P}^N$.

Definition 1.1.3 (Secant Varieties to a Variety) For simplicity let us suppose that $X \subset \mathbb{P}^N$ is a closed irreducible subvariety.

Let

$$S_X^0 := \{((x_1, x_2), z) \; : \; z \in < x_1, x_2 >\} \subset ((X \times X) \setminus \Delta_X) \times \mathbb{P}^N.$$

The set is closed in $(X \times X) \setminus \Delta_X) \times \mathbb{P}^N$ so that, taken with the reduced scheme structure, it is a quasi-projective variety. We remark that, by definition, it is a $\mathbb{P}^1$-bundle over $(X \times X) \setminus \Delta_X$, which is irreducible, so that S_X^0 is an irreducible quasi-projective variety of dimension $\dim(S_X^0) = 2\dim(X) + 1$.

Let S_X be its closure in $X \times X \times \mathbb{P}^N$. Then S_X is an irreducible projective variety of dimension $2\dim(X) + 1$, called *the abstract secant variety to X*. Let us consider the projections of S_X onto the factors $X \times X$ and $\mathbb{P}^N$,

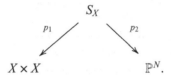

The secant variety to X, SX, is the scheme-theoretic image of S_X in $\mathbb{P}^N$, i.e.

$$SX = p_2(S_X) = \overline{\bigcup_{x_1 \neq x_2, x_i \in X} < x_1, x_2 >} \subseteq \mathbb{P}^N.$$

Thus $SX \subseteq \mathbb{P}^N$ is an irreducible projective variety of dimension

$$s(X) = \dim(SX) \leq \min\{2\dim(X) + 1, \; N\},$$

which geometrically is *the variety swept out by the secant lines to X*.

Let now $k \geq 1$ be a fixed integer. We can generalize the construction to the case of $(k + 1)$-secant $\mathbb{P}^k$, i.e. to the variety swept out by the linear spaces generated by $k + 1$ independent points on X.

Define

$$(S_X^k)^0 \subset \underbrace{X \times \ldots \times X}_{k+1} \times \mathbb{P}^N$$

as the locally closed irreducible set

$$(S_X^k)^0 := \{((x_0, \ldots, x_k), z) \; : \; \dim(< x_0, \ldots, x_k >) = k \; , \; z \in < x_0, \ldots, x_k >\}.$$

Let S_X^k, *the abstract k-secant variety of X,* be

$$S^k X := \overline{(S_X^k)^0} \subset \underbrace{X \times \ldots \times X}_{k+1} \times \mathbb{P}^N.$$

The closed set S_X^k is irreducible and of dimension $(k+1)\dim(X) + k$. Consider the projections of S_X^k onto the factors $\underbrace{X \times \ldots \times X}_{k+1}$ and $\mathbb{P}^N$,

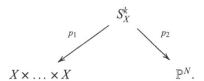

The k-secant variety to X, $S^k X$, is the scheme-theoretic image of S_X^k in $\mathbb{P}^N$, i.e.

$$S^k X = p_2(S_X^k) = \overline{\bigcup_{x_i \in X,\, \dim(<x_0,\ldots,x_k>)=k} < x_0, \ldots, x_k >} \subseteq \mathbb{P}^N.$$

It is an irreducible algebraic variety of dimension

$$s_k(X) = \dim(S^k X) \leq \min\{N,\ (k+1)\dim(X) + k\}.$$

Remark 1.1.4 Secant varieties have recently become very popular in connection with tensor decompositions and other problems coming from numerical analysis due to the work of Landsberg, Ottaviani, Chiantini and other authors. The notation commonly used by these specialists slightly differs from the one used here and usually k is replaced by $k+1$. This recalls that we are dealing with $k+1$ general points on X and they frequently use the symbol $\sigma^{k+1}(X)$ to indicate $S^k X$. The index k in $S^k X$ reminds us that it is the locus of the $\mathbb{P}^k$'s generated by $k+1$ general points on X.

Some authors define k-defective secant varieties as those for which $s_k(X) < \min\{N,\ (k+1)\dim(X) + k\}$ while others as those for which $s_k(X) < (k+1)\dim(X) + k$. We shall mainly be interested in the case $k = 1$ here and we shall define secant defective varieties as those for which $s_1(X) = \dim(SX) < 2\dim(X) + 1$, following the approach of Zak in [198] but not that of many other references cited in the bibliography. So, to avoid confusion, one should check the definition used in each reference consulted.

We are now in a position to define the last cone attached to a point $x \in X$. This notion was introduced by Johnson in [109] and further studied extensively by Zak. Algebraic properties of tangent star cones and of the algebras related to them are investigated in [179].

Definition 1.1.5 (Tangent Star at a Point; Variety of Tangent Stars, [109]) Let $X \subset \mathbb{P}^N$ be an irreducible projective variety.

The abstract variety of tangent stars to X, T_X^*, is defined by the following Cartesian diagram

$$
\begin{array}{ccc}
T_X^* & \lhook\joinrel\longrightarrow & S_X \\
{\scriptstyle p}\Big\downarrow & & \Big\downarrow{\scriptstyle p_1} \\
\Delta_X & \lhook\joinrel\longrightarrow & X \times X \,.
\end{array}
$$

The tangent star to X at x, T_x^*X, is defined by

$$T_x^*X := p_2(p_1^{-1}((x,x))) \subseteq \mathbb{P}^N.$$

It is a scheme whose underling set $(T_x^*X)_{\mathrm{red}}$ can be described geometrically as follows:

$$(T_x^*X)_{\mathrm{red}} = \overline{\bigcup_{(x_1,x_2)\in X\times X\backslash \Delta_X} \lim_{x_i\to x} <x_1,x_2>} \subset \mathbb{P}^N.$$

The variety of tangent stars to X is by definition

$$T^*X = p_2(T_X^*) \subseteq \mathbb{P}^N,$$

so that by construction

$$T^*X \subseteq SX.$$

Moreover, letting only one point vary, we deduce

$$\mathbf{C}_xX \subseteq T_x^*X.$$

It is also clear that the limit of a secant line is a tangent line, i.e. that

$$T_x^*X \subseteq T_xX.$$

From the previous definitions and properties we deduce that for an arbitrary point $x \in X \subset \mathbb{P}^N$, the following inclusions hold:

$$\mathbf{C}_xX \subseteq T_x^*X \subseteq T_xX.$$

Moreover, we recall that a point $x \in X$ is non-singular if and only if $\mathbf{C}_x X = T_x^* X = T_x X$. At singular points strict inequalities can hold, i.e. at singular points there could exist tangent lines which are not the limit of secant lines, as we shall immediately see in the next example.

Example 1.1.6 (Singular Points for Which $\mathbf{C}_x X \subsetneq T_x^* X \subsetneq T_x X$) Let $Y \subset \mathbb{P}^N \subset \mathbb{P}^{N+1}$ be an irreducible, non-degenerate variety in $\mathbb{P}^N$. Consider a point $p \in \mathbb{P}^{N+1} \setminus \mathbb{P}^N$ and let $X := S(p, Y)$ be *the cone over Y of vertex p*, i.e.

$$S(p, Y) = \bigcup_{y \in Y} <p, y> .$$

Then X is an irreducible, non-degenerate variety in $\mathbb{P}^{N+1}$. In fact, modulo a projective transformation, the variety X is defined by the same equations of Y, now thought of as homogeneous polynomials with one more variable. In particular, $\dim(X) = \dim(Y) + 1$.

The line $<p, y>$ is contained in X for every $y \in Y$, so that $X \subset T_p X$ and

$$T_p X = \mathbb{P}^{N+1} = \langle X \rangle. \tag{1.3}$$

For an algebraic variety $Z \subset \mathbb{P}^M$ we indicate by $\langle Z \rangle \subseteq \mathbb{P}^M$ the *linear span of Z*, that is the smallest linear subspace of $\mathbb{P}^M$ containing Z.

It follows from the definition of tangent cone to a variety that

$$\mathbf{C}_p S(p, Y) = S(p, Y).$$

We also have that

$$S(p, SY) = SX. \tag{1.4}$$

Indeed, by projecting from p onto $\mathbb{P}^N$, it is clear that a general secant line to X projects onto a secant line to Y, proving $SX \subseteq S(p, SY)$. On the contrary if we get a general point $q \in S(p, SY)$, by definition it projects onto a general point $q' \in SY$, which belongs to a secant line $<p_1', p_2'>$, $p_i' \in Y$. The plane $<p, p_1', p_2'>$ contains the point q, while the lines $<p, p_i'>$, $i = 1, 2$, are contained in X by definition of cone; hence through q there pass infinitely many secant lines to X, yielding $S(p, SY) \subseteq SX$. The claim is proved.

The previous argument proves the following general fact:

$$T_p^* S(p, Y) = S(p, SY).$$

Indeed, by definition $T_p^* X \subseteq SX = S(p, SY)$ as schemes. On the other hand, by fixing two general points $p_1, p_2 \in X$, $p_1 \neq p_2$, $p_i \neq p$, the plane $<p, p_1, p_2>$ is contained in $T_p^* X$ as is easily seen by varying the velocity of approaching p of two points $q_i \in < p, p_i >$. By the generality of the points p_i we get the inclusion $SX \subseteq T_p^* X$ as schemes and the proof of the claim.

As an immediate application one constructs examples of irreducible singular varieties X with a point $p \in \text{Sing}(X)$ for which

$$\mathbf{C}_p X \subsetneq T_p^* X \subsetneq T_p X.$$

One can take as $Y \subset \mathbb{P}^4 \subset \mathbb{P}^5$ an irreducible, smooth, non-degenerate curve in $\mathbb{P}^4$ and consider the cone X over Y of vertex $p \in \mathbb{P}^5 \setminus \mathbb{P}^4$. Then $\mathbf{C}_p X = S(p, Y) = X$, $T_p^* X = S(p, SY) = SX$ is a hypersurface in $\mathbb{P}^5$, because SY is a hypersurface in $\mathbb{P}^4$, while $T_p X = \mathbb{P}^5$. Every variety Y such that $SY \subsetneq \mathbb{P}^N$ will produce analogous examples.

1.2   Join of Varieties

We generalize to arbitrary irreducible varieties $X, Y \subset \mathbb{P}^N$ the notion of *cone* and the definition of the *join* of linear spaces.

Let us recall that if $L_i \simeq \mathbb{P}^{N_i} \subseteq \mathbb{P}^N$, $i = 1, 2$, is a linear subspace, then

$$< L_1, L_2 > := \bigcup_{x_i \in L_i \,, \, x_1 \neq x_2} < x_1, x_2 >,$$

is a linear space called *the join of L_1 and L_2*. It is the smallest linear subspace of $\mathbb{P}^N$ containing L_1 and L_2 so that $< L_1, L_2 > = \langle L_1 \cup L_2 \rangle$. From Grassmann's Formula we deduce

$$\dim(< L_1, L_2 >) = \dim(L_1) + \dim(L_2) - \dim(L_1 \cap L_2), \tag{1.5}$$

where $\dim(\emptyset) = -1$ by definition. Grassmann's Formula shows that the dimension of the join depends on the intersection of the two linear spaces.

On the other hand, if $X \subset \mathbb{P}^N \subset \mathbb{P}^{N+1}$ is an irreducible subvariety and if $p \in \mathbb{P}^{N+1} \setminus \mathbb{P}^N$ is an arbitrary point, if we define as before

$$S(p, X) = \bigcup_{x \in X} < p, x >,$$

the cone of vertex p over X, then for every $z \in < p, x >$, $z \neq p$, we have by construction

$$T_z S(p, X) = < p, T_x X > = < T_p p, T_x X >, \tag{1.6}$$

showing the well-known fact that the tangent space to a cone is constant along its rulings.

As we shall see in the next section, once we have defined the join of two varieties we can *linearize* the problem looking at the tangent spaces and calculating the dimension of the *join* by studying the affine cones over the varieties, exactly as in the proof of the formula (1.5). The dimension of the join of two varieties will still depend on the intersection of the corresponding tangent spaces, a result known as Terracini's Lemma, [185]. Moreover, a kind of property similar to the second tautological inequality in (1.6) will hold generically, at least in characteristic zero, see Theorem 1.4.1.

Definition 1.2.1 (Join of Varieties; Relative Secant, Tangent Star and Tangent Varieties) Let $X, Y \subset \mathbb{P}^N$ be closed irreducible subvarieties. Let

$$S^0_{X,Y} := \{((x, y, z), x \neq y : z \in < x, y >\} \subset X \times Y \times \mathbb{P}^N.$$

The set is locally closed so that, taken with the reduced scheme structure, it is a quasi-projective irreducible variety of dimension $\dim(S^0_{X,Y}) = \dim(X) + \dim(Y) + 1$. Let $S_{X,Y}$ be its closure in $X \times Y \times \mathbb{P}^N$. Then $S_{X,Y}$ is an irreducible projective variety of dimension $\dim(X) + \dim(Y) + 1$, called *the abstract join of X and Y*. Let us consider the projections of $S_{X,Y}$ onto the factors $X \times Y$ and $\mathbb{P}^N$,

$$(1.7)$$

The join of X and Y, $S(X, Y)$, is the scheme-theoretic image of $S_{X,Y}$ in $\mathbb{P}^N$, i.e.

$$S(X, Y) = p_2(S_{X,Y}) = \overline{\bigcup_{x \neq y, x \in X, y \in Y} < x, y >} \subseteq \mathbb{P}^N.$$

Thus $S(X, Y) \subseteq \mathbb{P}^N$ is an irreducible projective variety of dimension

$$s(X, Y) = \dim(S(X, Y)) \leq \dim(X) + \dim(Y) + 1,$$

swept out by lines joining points of X with points of Y.

With this notation, after setting $X = S^0(X)$, we have $S(X, X) = SX$ and

$$S(X, S^{k-1}X) = S^k X = S(S^l X, S^h X),$$

if $h \geq 0, l \geq 0, h + l = k - 1$. Moreover, for arbitrary irreducible varieties X, Y and Z, we have

$$S(X, S(Y, Z)) = S(S(X, Y), Z),$$

that is *join* is an associative operation on the set of irreducible varieties of a fixed projective space.

When $Y \subseteq X \subset \mathbb{P}^N$ is an irreducible closed subvariety, the variety $S(Y, X)$ is usually called *the relative secant variety of X with respect to Y*. Analogously,

$$T(Y, X) = \bigcup_{y \in Y} T_y X.$$

In this case, by taking $\Delta_Y \subset Y \times X$ and by looking at (1.7), we can define $T^*_{Y,X} := p_1^{-1}(\Delta_Y) \subseteq S_{Y,X}$ to be *the abstract relative tangent star variety* and finally

$$T^*(Y, X) := p_2(T^*_{Y,X}) \subseteq S(X, Y) \tag{1.8}$$

to be *the relative tangent star variety*. Letting

$$T^*_y(Y, X) = p_2(p_1^{-1}(y \times y)),$$

then

$$(T^*_y(Y, X))_{\mathrm{red}} = \overline{\bigcup_{(y_1, x_1) \in Y \times X \setminus \Delta_Y} \lim_{\substack{y_1 \to y \\ x_1 \to y}} < y_1, x_1 >} \subset \mathbb{P}^N,$$

and $T^*(Y, X) = \bigcup_{y \in Y} T^*_y(Y, X)$. With these definitions we have $T^*_y(y, X) = \mathbf{C}_y X$ and $T^*_y(X, X) = T^*_y X$ for every $y \in X$. In particular,

$$\mathbf{C}_y X = T^*_y(y, X) \subseteq T^*_y(X, X) = T^*_y X,$$

a fact previously proved with an intuitive *dynamical* (or synthetic) geometric proof.

We provide some immediate applications of the definition of join to properties of $S^k X$ and to characterizations of linear spaces. Let us recall that a variety $X \subset \mathbb{P}^N$ is said to be *non-degenerate* if $\langle X \rangle = \mathbb{P}^N$.

Proposition 1.2.2 ([148]) *Let* $X, Y \subset \mathbb{P}^N$ *be closed irreducible subvarieties. Then:*

1. for every $x \in X$,

$$Y \subseteq S(x, Y) \subseteq T_x S(x, Y) \subseteq T_x S(X, Y)$$

and in particular

$$< x, \langle Y \rangle > \subseteq T_x S(x, Y);$$

2. if $S^k X = S^{k+1} X$ for some $k \geq 0$, then $S^k X = \mathbb{P}^{s_k(X)} \subseteq \mathbb{P}^N$;
3. if $\dim(S^{k+1} X) = \dim(S^k X) + 1$ for some $k \geq 0$, then $S^{k+1} X = \mathbb{P}^{s_{k+1}(X)}$ and $S^k X$ is a hypersurface in $\mathbb{P}^{s_{k+1}(X)}$;
4. let $k \geq 0$ be an integer. If $S^{k+1} X \subset \mathbb{P}^N$ is not a linear space, then $S^k X \subseteq$ $\operatorname{Sing}(S^{k+1} X)$.

Proof By definition of join we get the inclusion $S(x, Y) \subseteq S(X, Y)$ yielding $T_x S(x, Y) \subseteq T_x S(X, Y)$ for every $x \in X$. Moreover, for every $y \in Y$, $y \neq x$, the line $< x, y >$ is contained in $S(x, Y)$ and passes through x so that it is contained in $T_x S(x, Y)$ and part 1) easily follows.

Let $z \in S^k X$ be a smooth point of $S^k X$. From part 1) we get

$$X \subseteq T_z S(S^k X, X) = T_z S^{k+1} X = T_z S^k X = \mathbb{P}^{s_k(X)}.$$

Thus $S^k X \subseteq \langle X \rangle \subseteq T_z S^k X = \mathbb{P}^{s_k(X)}$ so that $S^k X = \langle X \rangle = \mathbb{P}^{s_k(X)}$.

To prove part 3), take a general point $z \in S^{k+1} X \setminus S^k X$. For general $x \in X$ we get $S^k X \subsetneq S(x, S^k X) \subseteq S(X, S^k X) = S^{k+1} X$. Thus for general $x \in X$ we get $S(x, S^k X) = S^{k+1} X$ since $s_{k+1}(X) = s_k(X) + 1$. In particular, $z \in S(x, S^k X)$ for $x \in X$ general, i.e. for $z \in S^{k+1} X$ general there exists a $y \in S^k X$ such that $z \in< x, y >\subset S^{k+1} X$. Thus a general point $x \in X$ is contained in $T_z S^{k+1} X$ yielding

$$S^{k+1} X \subseteq \langle X \rangle \subseteq T_z S^{k+1} X$$

and $S^{k+1} X = \langle X \rangle = \mathbb{P}^{s_{k+1}(X)}$.

To prove 4) let us remark that $T_z S^{k+1} X \subseteq \langle S^{k+1} X \rangle = \langle X \rangle$. Take $z \in S^k X$ and observe that, via the last part of (1), we have

$$\langle S^{k+1} X \rangle = \langle X \rangle = S(z, \langle X \rangle) \subseteq T_z S^{k+1} X \subseteq \langle S^{k+1} X \rangle,$$

so that $T_z S^{k+1} X = \langle S^{k+1} X \rangle \supseteq S^{k+1} X$. By hypothesis the last inclusion is strict, yielding $\dim(T_z S^{k+1} X) > \dim(S^{k+1} X)$, and z is a singular point of $S^{k+1} X$ as claimed. $\square$

Thus to a non-degenerate irreducible closed subvariety $X \subset \mathbb{P}^N$ we can associate an ascending filtration of irreducible projective varieties, whose inclusions are strict by Proposition 1.2.2,

$$X = S^0 X \subsetneq SX \subsetneq S^2 X \subsetneq \ldots \subsetneq S^{k_0} X = \mathbb{P}^N. \tag{1.9}$$

Therefore the integer $k_0 = k_0(X) \geq 1$ is well defined as the least integer $k \geq 1$ such that $S^k X = \mathbb{P}^N$.

The above immediate consequences of the definitions also give the next result, which was classically very well known, see for example [147, footnote p. 635]. It has been considered as an open problem by Atiyah in [10, p. 424] and reproved repeatedly in the modern literature. Via an argument of Atiyah, the next geometrical property of secant varieties yields a proof of C. Segre's and Nagata's Theorem on the minimal section of a geometrically ruled surface, see [123] for details.

Corollary 1.2.3 ([147]) *Let $C \subset \mathbb{P}^N$ be an irreducible non-degenerate projective curve. Then $s_k(C) = \min\{2k + 1, N\}$.*

Let $X \subset \mathbb{P}^N$ be an irreducible non-degenerate projective variety of dimension $n \geq 1$ and let $k < k_0$. Then:

1. *$s_k(X) \geq n + 2k$ for every $k < k_0$.*
2. *If $s_j(X) = n + 2j$ for some $j \geq 1$, $j < k_0$, then $s_k(X) = n + 2k$ for every $k \leq j$. In particular, if $s_k(X) = n + 2k$ for some $k \geq 1$, $k < k_0$, then $s(X) = n + 2$ (and $SX \subsetneq \mathbb{P}^N$).*

Proof For $k = 0$ the assertion is true since $s_0(C) = \dim(C) = 1$ by definition and we can argue by induction. Suppose $S^k C \subsetneq \mathbb{P}^N$. By Proposition 1.2.2 $s_k(C) \geq s_{k-1}(C) + 2$ and the description $S^k(C) = S(C, S^{k-1}C)$ yields $s_k(C) \leq s_{k-1}(C) + 2$. Combining the two inequalities we get $s_k(C) = s_{k-1}(C) + 2 = 2(k - 1) + 1 + 2 = 2k + 1$, as claimed.

To prove the second part we argue as above. Thus, for $k < k_0$, $s_k(X) \geq s_{k-1}(X) + 2 = n + 2(k - 1) + 2 = n + 2k$. Moreover, $s_j(X) = n + 2j$ yields $s_k(X) = n + 2k$ for every $k \leq j$, proving all the remaining claims.   □

1.3 Linear Projections

We now define and study linear projections with the terminology just introduced, generalizing the dimension formula (1.5) to the case of arbitrary cones, at least in characteristic zero.

Definition 1.3.1 (Linear Projections and Linear Cones) Let $L = \mathbb{P}^l \subset \mathbb{P}^N$ be a fixed linear space, $l \geq 0$, and let $M = \mathbb{P}^{N-l-1}$ be a linear space skew to L, i.e. $L \cap M = \emptyset$ (note that this implies $< L, M > = \mathbb{P}^N$). Let $X \subseteq \mathbb{P}^N$ be a closed irreducible variety not contained in L and let

$$\pi_L : X \dashrightarrow \mathbb{P}^{N-l-1} = M$$

be the rational map defined on $X \setminus (L \cap X)$ by

$$\pi_L(x) = < L, x > \cap M.$$

The map is well defined by Grassmann's formula (1.5). Let $X' = \overline{\pi_L(X)} \subset \mathbb{P}^{N-l-1}$ be the closure of the image of X by π_L. The whole process can be described using the terminology of joins. Indeed, we have

$$X' = S(L, X) \cap M,$$

that is X' is the intersection of M with the *cone over X of vertex L* and clearly $S(L, X) = S(L, X')$.

The projective differential of π_L at a point $x \in X$ where π_L is defined coincides with the projection of T_xX from L. Thus, for $x \in X \setminus (L \cap X)$, we have

$$d_{\pi_L}(T_xX) =< L, T_xX > \cap M \subseteq T_{\pi_L(x)}X',$$

a fact which can also be easily verified by passing to a suitable chosen (local) coordinate system.

Suppose $L \cap X = \emptyset$, then we claim that $\pi_L : X \to X'$ is a finite morphism, which implies $\dim(X) = \dim(X')$. Being a morphism between projective varieties, it is sufficient to show that it has finite fibers by Stein Factorization. By definition, for $x' \in X'$,

$$\pi_L^{-1}(x') =< L, x' > \cap X \subset < L, x' >= \mathbb{P}^{l+1}.$$

If there existed an irreducible curve $C \subset < L, x' > \cap X \subset < L, x' >$, then $\emptyset \neq L \cap C \subseteq L \cap X$, contrary to our assumption.

In particular, for an arbitrary L, the dimension of X' does not depend on the choice of the position of M, except for the requirement $L \cap M = \emptyset$.

The relation $S(L, X) = S(L, X')$ allows us to calculate the dimension of the irreducible variety $S(L, X)$ for an arbitrary L. Exactly as in (1.6) for $z \in S(L, X) \setminus L$, we have

$$z \in < L, x >=< L, \pi_L(z) >=< L, x' >,$$

with $x \in X$ and $\pi_L(z) = \pi_L(x) = x' \in X'$. Since $S(L, X')$ is, modulo a projective transformation, the variety defined by the same homogeneous polynomials of X' now thought of as polynomials in $N + 1$ variables, we have

$$T_zS(L, X) =< L, T_{\pi_L(z)}X' > \supseteq < L, T_xX > . \tag{1.10}$$

Taking $z \in S(L, X)$ general and recalling that $L \cap M = \emptyset$ we deduce:

$$\dim(S(L, X)) = \dim(< L, T_{\pi_L(z)}X' >) = \dim(X') + l + 1. \tag{1.11}$$

Suppose until the end of the subsection that $\mathrm{char}(K) = 0$. By generic smoothness, the differential map is surjective so that $T_{\pi_L(x)}X' = \pi_L(T_xX)$ for $x \in X$ general.

In this case $\pi_L(x) = x' \in X'$ will be general on X' and finally

$$\dim(X') = \dim(T_{x'}X') = \dim(\pi_L(T_xX)) = \dim(X) - \dim(L \cap T_xX) - 1,$$

which combined with (1.11) gives the following generalization of (1.5):

$$\dim(S(L, X)) = \dim(L) + \dim(X) - \dim(L \cap T_xX), \qquad (1.12)$$

$$x \in X \text{ general point.}$$

Moreover, we get the following refinement of (1.10)

$$T_zS(L, X) = < L, T_xX >, \qquad (1.13)$$

$$x \in X , \ z \in< L, x > \text{ general points.}$$

The definition $S(L, X)$ of a cone of vertex L over X generalized the notion of a cone over a variety lying in a space skew with respect to the vertex introduced before. Then by projecting the variety X from the vertex L, we find the description of $S(L, X)$ as a usual cone $S(L, X')$ with $L \cap \langle X' \rangle = \emptyset$.

Now we investigate under which condition a variety is a *cone*, i.e. there exists a *vertex* $L \simeq \mathbb{P}^l \subseteq X$ such that $X = S(L, X) = S(L, X')$ with X' a section of $S(X, L)$ with a general $\mathbb{P}^{N-l-1}$ skew with the *vertex* L. Clearly the *vertex* is not uniquely determined if we do not require some maximality condition. Let us introduce the precise definitions.

Definition 1.3.2 (Cone; Vertex of a Variety) Let $X \subset \mathbb{P}^N$ be a closed (irreducible) subvariety. The variety X is a *cone* if there exists a $p \in X$ such that $S(p, X) = X$. Geometrically this means that given $x \in X$, $x \neq p$, the line $< p, x >$ is contained in X. In particular, $p \in \bigcap_{x \in X} T_xX$.

This motivates the definition of the vertex of a variety. Given $X \subset \mathbb{P}^N$ an irreducible closed subvariety, *the vertex of X*, Vert(X), is the set

$$\text{Vert}(X) = \{ x \in X : S(x, X) = X \}.$$

In particular, a variety X is a cone if and only if Vert$(X) \neq \emptyset$. Moreover, by definition $S(X, Y) = X$ if and only if $Y \subseteq \text{Vert}(X)$.

We list some obvious consequences of the previous definition.

Proposition 1.3.3 Let $X \subset \mathbb{P}^N$ be a closed irreducible variety of dimension $\dim(X) = n$. Then:

1. Vert(X), if not empty, is a linear subspace $\mathbb{P}^l$ and moreover

$$\text{Vert}(X) \subseteq \bigcap_{x \in X} T_xX;$$

2. *if* $\operatorname{codim}(\operatorname{Vert}(X), X) \leq 1$, *then* $\operatorname{Vert}(X) = X = \mathbb{P}^n \subset \mathbb{P}^N$;
3. *if* $\dim(S(X, Y)) = \dim(X) + 1$, *then* $Y \subseteq \operatorname{Vert}(S(X, Y))$;
4. *if* $\operatorname{char}(K) = 0$,

$$\operatorname{Vert}(X) = \bigcap_{x \in X} T_x X.$$

In particular,

$$\bigcap_{x \in X} T_x X \subseteq X,$$

a non-obvious fact, which is false in positive characteristic (cf. Exercise 1.5.19);
5. *suppose* $\operatorname{char}(K) = 0$ *and* $\emptyset \neq \operatorname{Vert}(X) \subsetneq X$, *then*

$$X = S(\operatorname{Vert}(X), X')$$

is a cone, where X' *is the projection of* X *from* $\operatorname{Vert}(X)$ *onto a* $\mathbb{P}^{N-l-1}$ *skew to* $\operatorname{Vert}(X)$ *and* $\dim(X') = n - l - 1$.

Proof To prove 1) it is sufficient to show that, given two points $x_1, x_2 \in \operatorname{Vert}(X)$, the line $< x_1, x_2 >$ is contained in $\operatorname{Vert}(X)$, forcing $\operatorname{Vert}(X)$ to be irreducible and linear by Proposition 1.2.2 part 2). Given $y \in < x_1, x_2 > \setminus \{x_1, x_2\}$ and $x \in X \setminus \operatorname{Vert}(X)$, it is sufficient to prove that $< y, x > \subset X$. By definition the lines $< x_i, x >$ are contained in X and by varying the point $q \in < x_2, x > \subset X$ and by joining it with x_1 we see that the line $< x_1, q >$ is contained in X for every such q. Therefore the plane

$$\Pi_x = < x_1, x_2, x >$$

is contained in X. Since y and x belong to Π_x, the claim follows.
  If $\operatorname{Vert}(X) = X$, then $X = \mathbb{P}^n$ by part 1). If there exists a

$$W = \mathbb{P}^{n-1} \subseteq \operatorname{Vert}(X) = \mathbb{P}^l \subseteq X,$$

we can take $x \in X \setminus W$. Therefore $S(x, W) = \mathbb{P}^n$ and $\mathbb{P}^n = S(W, x) \subseteq X$ forces $X = \mathbb{P}^n$.
  To prove 3) take $y \in Y \setminus \operatorname{Vert}(X)$ and observe that for dimensional reasons $S(y, X) = S(Y, X)$ and $S(y, S(X, Y)) = S(y, S(y, X)) = S(y, X) = S(Y, X)$ gives the desired conclusion.
  Set $L = \bigcap_{x \in X} T_x X$ and assume $\operatorname{char}(K) = 0$. From (1.12) we get $\dim(S(L, X)) = \dim(X)$, yielding $X = S(L, X)$ and $L \subseteq \operatorname{Vert}(X)$, proving part 4). Part 5) now follows in a straightforward way. $\square$

  Later we will use the following result.

Corollary 1.3.4 *Let $X \subset \mathbb{P}^N$ be an irreducible non-degenerate variety of dimension $n = \dim(X)$. Assume char$(K)=0$, $N \geq n + 3$ and $\dim(SX) = n + 2$. If through the general point $x \in X$ there passes a line l_x contained in X, then X is a cone.*

Proof Let $x \in X$ be a general point. Then $x \notin \mathrm{Vert}(X)$ and $x \notin \mathrm{Vert}(SX)$ since X is non-degenerate. Moreover, $X \subsetneq S(l_x, X) \subseteq SX$. If $\dim(S(l_x, X)) = n + 2$, then $S(l_x, X) = SX$. Since $S(l_x, SX) = S(l_x, S(l_x, X)) = S(l_x, X) = SX$, we would deduce $x \in l_x \subseteq \mathrm{Vert}(SX)$. In conclusion, l_x is not contained in $\mathrm{Vert}(SX)$ and $\dim(S(l_x, X)) = n + 1$.

Thus the general tangent space to X, $T_y X$, will cut l_x in a point $p_{x,y} := l_x \cap T_y X$. If this point varies with y, then join of two general tangent spaces $T_{y_1} X$ and $T_{y_2} X$ would contain l_x, yielding

$$< l_x, < T_{y_1} X, T_{y_2} X >>=< T_{y_1} X, T_{y_2} X > .$$

This would force $S(l_x, SX) = SX$, i.e. $l_x \subseteq \mathrm{Vert}(SX)$ and $x \in \mathrm{Vert}(SX)$ contrary to our assumption. So the point remains fixed when varying y. Therefore $p_{x,y} = p \in \cap_{y \in X} T_y X = \mathrm{Vert}(X)$ and X is a cone by Proposition 1.3.3. $\square$

We end this section by relating the projections of a variety to the dimension of its secant or tangent varieties.

If $L = \mathbb{P}^l \subset \mathbb{P}^N$ is a linear space and if $\pi_L : \mathbb{P}^N \setminus L \to \mathbb{P}^{N-l-1}$ is the projection onto a skew complementary linear space, then π_L restricts to a finite morphism $\pi_L : X \to \mathbb{P}^{N-l-1}$, as soon as $L \cap X = \emptyset$. Assuming in principle that studying varieties whose codimension is small with respect to the dimension is easier (this is true only from some points of view), we can ask when this finite morphism is one-to-one, or a closed embedding. Let us examine these conditions.

Proposition 1.3.5 *Let the notation be as above. Then:*

1. *the morphism $\pi_L : X \to \mathbb{P}^{N-l-1}$ is one-to-one if and only if $L \cap SX = \emptyset$;*
2. *the morphism $\pi_L : X \to \mathbb{P}^{N-l-1}$ is unramified if and only if $L \cap TX = \emptyset$;*
3. *the morphism $\pi_L : X \to \mathbb{P}^{N-l-1}$ is a closed embedding if and only if $L \cap SX = L \cap TX = \emptyset$.*

Proof The morphism $\pi_L : X \to X' \subseteq \mathbb{P}^{N-l-1}$ is one-to-one if and only it there exists no secant line to X cutting the center of projection: $< L, x >=< L, y >$ if and only if $< x, y > \cap L \neq \emptyset$. It is ramified at a point $x \in X$ if and only if $T_x X \cap L = \emptyset$ by looking at the projective differential of π_L. A finite morphism is a closed embedding if and only if it is one-to-one and unramified. $\square$

We have to state the following well-known result, which only takes into account that for smooth varieties the equality $TX = T^*X$ yields $TX \subseteq SX$.

Corollary 1.3.6 *Let $X \subset \mathbb{P}^N$ be a smooth irreducible closed projective variety. If $N > \dim(SX)$, then X can be isomorphically projected into $\mathbb{P}^{N-1}$. In particular, if $N > 2\dim(X) + 1$, then X can be isomorphically projected into $\mathbb{P}^{N-1}$.*

One could ask what is the meaning of $L \cap T^*X = \emptyset$. This means that π_L (or $d(\pi_L)$) restricted to T_x^*X is finite for every $x \in X$. This is the notion of J-unramified morphism, where J stands for Johnson [109], and it can be expressed in terms of affine tangent stars, see [198]. We take the above condition as the definition of J-unramified projection. In particular, if $L \cap SX = \emptyset$, then π_L is one-to-one and J-unramified and it is said to be *a J-embedding*. If the projection $\pi_L : X \to X' \subset \mathbb{P}^{N-l-1}$ is a J-embedding, then $\mathrm{Sing}(\pi_L(X)) = \pi_L(\mathrm{Sing}(X))$ so that X' does not acquire singularities from the projection.

It is clearly weaker than the usual notion of embedding and it is well behaved to study the projections of singular varieties. For example, take $C \subset \mathbb{P}^4 \subset \mathbb{P}^5$ a smooth non-degenerate curve in $\mathbb{P}^4$ and let $p \in \mathbb{P}^5 \setminus \mathbb{P}^4$. If $X = S(p, C)$ is the cone over C, then $T_pX = \mathbb{P}^5$, see (1.3), and X cannot be projected isomorphically in $\mathbb{P}^4$. Since $SX = S(p, SC)$, see (1.4), is a hypersurface in $\mathbb{P}^5$, there exists a point $q \in \mathbb{P}^5 \setminus X$ such that $\pi_q : X \to X'$ is a J-embedding and $X' = S(\pi_q(p), C)$ is a cone over C of vertex $\pi_q(p) = p'$. In this example the morphism π_q is one-to-one and unramified outside the vertex of the cones and maps the tangent star at p, $T_p^*X = S(p, SC)$, m-to-one onto $\mathbb{P}^4$, where $m = \deg(S(p, SC)) = \deg(SC) = \binom{d-1}{2} - g$, $d = \deg(C)$, g the genus of C, see also Exercise 1.5.20.

The conditions $L \cap S(Y, X) = \emptyset$, respectively $L \cap T^*(Y, X) = \emptyset$ or $L \cap T(Y, X) = \emptyset$, with $Y \subseteq X$, mean that π_L is one-to-one in a neighborhood of Y, respectively is J-unramified in a neighborhood of Y or unramified in a neighborhood of Y.

1.4 Terracini's Lemma and Its First Applications

By definition the secant variety SX of $X \subset \mathbb{P}^N$ is the join of X with itself and it is not clear a priori how to calculate its dimension, see for example Exercise 1.5.11. More generally one would like to compute the dimension of $S(X, Y)$ for two arbitrary varieties $X, Y \subset \mathbb{P}^N$.

In fact, the circle of ideas, which allowed Terracini to solve the problem of calculating the dimension of SX, or more generally of S^kX, originated precisely from the study of examples like the ones considered in Exercise 1.5.11 and from the pioneering work of Scorza on secant defective varieties:

> Alcune considerazioni che qui non riporto perché non mi é ancora riuscito di presentarle in maniera compiuta, mi fan pensare che le V_3 di S_r $(r \geq 7)$ le cui corde non riempiono una V_7 rientrano tra le V_3 a spazi tangenti mutuamente secantisi,
> G. Scorza, 1909, [165, p. 265].

Let Terracini explain this process to us by quoting the beginning of Terracini [185]:

> É noto, [46], che la sola V_2, non cono, di S_r, i cui S_2 tangenti si incontrano a due a due, é, se $r \geq 5$, la superficie di VERONESE; e che questa superficie, [176], é pure caratterizzata dall' essere, in un tale S_r, la sola, non cono, le cui corde riempiono una V_4. Recentemente lo SCORZA, [165, p. 265], disse di aver ragione di credere, sebbene non gli fosse venuto

fatto di darne una dimostrazione, che le V_3 di S_7, o di uno spazio piú ampio, le cui corde non riempiono una V_7 *rientrino* tra le V_3 a spazi tangenti mutuamente secantisi. Ora si puó dimostrare, piu' precisamente, che queste categorie di V_3 coincidono, anzi, piu' in generale, che: Se una V_k di S_r ($r > 2k$) gode di una delle due proprietá, che le corde riempiano una varietá di dimensione $2k - i$ ($i \geq 0$), o che due qualsiansi S_k tangenti si seghino in uno S_i, gode pure dell' altra. Questo teorema, a sua volta, non é se non un caso particolare di un teorema piú generale che ora dimostreremo, teorema che pone in relazione l' eventuale abbassamento di dimensione della varietá degli S_h ($h+1$)-seganti di una V_k immersa in uno spazio di dimensione $r \geq (h + 1)k + h$, coll' esistenza di $h + 1$ qualsiansi suoi S_k tangenti in uno spazio minore dell' ordinario.

A. Terracini, 1911, [185].

To compute the dimension of $S(X, Y)$ in a simple way and to determine the relation between $T_z S(X, Y)$, $T_x X$ and $T_y Y$, where $z \in\, <x, y>$, $z \neq x$, $z \neq y$, $x \neq y$, we recall the definition of an affine cone over a projective variety $X \subset \mathbb{P}^N$.

Let $\pi : \mathbb{A}^{N+1} \setminus \mathbf{0} \to \mathbb{P}^N$ be the canonical projection. If $X \subset \mathbb{P}^N$ is a closed subvariety, we indicate by $C_0(X)$ *the affine cone over X*, i.e.

$$C_0(X) = \pi^{-1}(X) \cup \mathbf{0} \subset \mathbb{A}^{N+1}$$

is the affine variety cut out by the homogeneous polynomials in $N + 1$ variables defining X in $\mathbb{P}^N$. If $\mathbf{x} \neq \mathbf{0}$ is a point such that $\pi(\mathbf{x}) = x \in X$, then

$$\pi(t_{\mathbf{x}} C_0(X)) = T_x X.$$

Moreover, if $L_i = \pi(U_i)$, $i = 1, 2$, U_i a vector subspace of $\mathbb{A}^{N+1}$, then by definition

$$< L_1, L_2 > = \langle L_1 \cup L_2 \rangle = \pi(U_1 + U_2),$$

where $+ : \mathbb{A}^{N+1} \times \mathbb{A}^{N+1} \to \mathbb{A}^{N+1}$ is the vector space operation. Therefore, regarded as a morphism of algebraic varieties, the differential of the sum coincides with the operation, that is

$$d_{(\mathbf{x},\mathbf{y})} : t_{(\mathbf{x},\mathbf{y})}(\mathbb{A}^{N+1} \times \mathbb{A}^{N+1}) = t_{\mathbf{x}}\mathbb{A}^{N+1} \times t_{\mathbf{y}}\mathbb{A}^{N+1} \to t_{\mathbf{x}+\mathbf{y}}\mathbb{A}^{N+1}$$
$$(\mathbf{u}, \mathbf{v}) \qquad \to \qquad \mathbf{u} + \mathbf{v}.$$

With the above notation and definitions we deduce

$$\overline{C_0(X) + C_0(Y)} = C_0(S(X, Y)). \tag{1.14}$$

We are now in position to prove the so-called Terracini Lemma. The original proof of Terracini relies on the study of the differential of the second projection morphism $p_2 : S_{X,Y} \to S(X, Y)$. Here we follow Ådlandsvik, [1], to avoid the difficulty, if any, of writing the tangent space at a point $(x, y, z) \in S_{X,Y}^{\circ}$. When writing $z \in\, <x, y>$, we always suppose $x \neq y$.

Theorem 1.4.1 (Terracini's Lemma) *Let $X, Y \subset \mathbb{P}^N$ be irreducible subvarieties. Then:*

1. for every $x \in X$, for every $y \in Y$, $x \neq y$, and for every $z \in <x, y>$,

$$< T_x X, T_y Y > \subseteq T_z S(X, Y);$$

2. if $\text{char}(K) = 0$, there exists an open subset U of $S(X, Y)$ such that

$$< T_x X, T_y Y > = T_z S(X, Y)$$

for every $z \in U$, $x \in X$, $y \in Y$, $z \in \langle x, y \rangle$. In particular

$$\dim(S(X, Y)) = \dim(X) + \dim(Y) - \dim(T_x X \cap T_y Y)$$

for $x \in X$ and $y \in Y$ general points.

Proof The first part follows from Eq. (1.14) and from the interpretation of the differential of the affine sum. The second part from generic smoothness applied to the affine cones over X, Y and $S(X, Y)$. $\qquad\square$

Since we have quoted the original form given by Terracini, let us state it as an obvious corollary.

Corollary 1.4.2 ([185]) *Let $X \subset \mathbb{P}^N$ be an irreducible subvariety of $\mathbb{P}^N$. Then:*

1. for every $x_0, \ldots, x_k \in X$ and for every $z \in \langle x_0, \ldots, x_k \rangle$,

$$< T_{x_0} X, \ldots, T_{x_k} X > \subseteq T_z S^k X;$$

2. if $\text{char}(K) = 0$, there exists an open subset U of $S^k X$ such that

$$< T_{x_0} X, \ldots, T_{x_k} X > = T_z S^k X$$

for every $z \in U$, $x_i \in X$, $i = 0, \ldots, k$, $z \in <x_0, \ldots, x_k>$. In particular,

$$\dim(SX) = 2\dim(X) - \dim(T_x X \cap T_y X)$$

for $x, y \in X$ general points.

The first application we give is the so-called *Trisecant Lemma*. Let us recall that a line $L \subset \mathbb{P}^N$ is said to be a *trisecant line* to $X \subset \mathbb{P}^N$ if $\text{length}(L \cap X) \geq 3$.

Proposition 1.4.3 (Trisecant Lemma) *Let $X \subset \mathbb{P}^N$ be a non-degenerate, irreducible closed subvariety. Suppose $\text{char}(K)=0$ and $\text{codim}(X) > k \geq 1$.*

Then a general $(k+1)$-secant $\mathbb{P}^k$, $< x_0, \ldots, x_k > = L = \mathbb{P}^k$, is not $(k+2)$-secant, i.e. $L \cap X = \{x_0, \ldots, x_k\}$ as schemes.

In particular, if $\text{codim}(X) > 1$, the projection of X from a general point on it, $\pi_x : X \dashrightarrow X' \subset \mathbb{P}^{N-1}$, is a birational map.

Proof We claim that it is sufficient to prove the result for $k = 1$. Indeed, X is not linear so that by taking a general $x \in X$ and projecting X from this point we get a non-degenerate, irreducible subvariety $X' = \pi_x(X) \subset \mathbb{P}^{N-1}$ with codim$(X') =$ codim$(X) - 1 > k - 1$. If the general $L = \langle x_0, \ldots, x_k \rangle$ as above were $k + 2$-secant, by taking $x = x_k$, the linear space $\langle \pi_x(x_0), \ldots, \pi_x(x_{k-1}) \rangle = \mathbb{P}^{k-1} = L'$ would be a general k-secant $\mathbb{P}^{k-1}$, which is also $(k + 1) = ((k - 1) + 2)$-secant. So we can assume $k = 1$ and we set $n = \dim(X)$.

Take $x \in X \setminus \text{Vert}(X)$. Then a general secant line through x, $L = \langle x, y \rangle$, is neither tangent to X at x nor at y. If L is a trisecant line then necessarily there exists a $u \in (L \cap X) \setminus \{x, y\}$. Consider the projection of X from x. Since $x \notin \text{Vert}(X)$, letting $X' = \pi_x(X) \subset \mathbb{P}^{N-1}$, we have $\dim(X') = \dim(X)$ and moreover $\pi_x(y) = \pi_x(u) = x'$ is a general smooth point of X'. By generic smoothness

$$< x, T_{x'}X' > = < x, T_yX > = < x, T_uX >$$

so that T_yX and T_uX are hyperplanes in $< x, T_{x'}X' > = \mathbb{P}^{n+1}$, yielding

$$\dim(T_yX \cap T_uX) = n - 1.$$

Taking $z \in < x, y > = < y, u >$ general, we have a point in the set U specified in Corollary 1.4.2 so that $\dim(SX) = \dim(T_zSX) = \dim(< T_yX, T_uX >) = n + 1$. This implies codim$(X) = 1$ by Proposition 1.2.2 part 3). The last part follows from the fact that a generically one-to one morphism is birational if char$(K)=0$, being generically étale. $\square$

As a second application we reinterpret Terracini's Lemma as the tangency of the tangent space to higher secant varieties at a general point along the locus described on X by the secant spaces passing through the point.

Definition 1.4.4 (Tangency Along a Subvariety)   Let $Y \subset X$ be a closed (irreducible) subvariety of X and let $L = \mathbb{P}^l \subset \mathbb{P}^N$, $l \geq \dim(X)$, be a linear subspace. The linear space L is said to be *tangent to X along Y* if for every $y \in Y$

$$T_yX \subseteq L,$$

i.e. if and only if $T(Y, X) \subseteq L$.

The linear space L is said to be *J-tangent to X along Y* if for every $y \in Y$

$$T_y^*X \subseteq L.$$

The linear space L is said to be *J-tangent to X with respect to Y* if for every $y \in Y$

$$T_y^*(Y, X) \subseteq L,$$

i.e. if and only if $T^*(Y, X) \subseteq L$.

Clearly if L is tangent to X along Y, it is also J-tangent to X along Y and if L is J-tangent to X along Y it is also J-tangent to X with respect to Y.

In the case $L = \mathbb{P}^{N-1}$, the scheme-theoretic intersection $L \cap X = D$ is a divisor, i.e. a subscheme of pure dimension $\dim(X) - 1$. By definition, for every $y \in D$, we have $T_y D = T_y X \cap L$ so that, if X is a smooth variety, $L = \mathbb{P}^{N-1}$ is tangent to X exactly along $\mathrm{Sing}(D) = \{y \in D : \dim(T_y D) > \dim(D)\}$.

We define the important notions of entry loci and k-secant defect and we study their first properties.

Definition 1.4.5 (Entry Loci) Let $X \subset \mathbb{P}^N$ be a closed irreducible non-degenerate subvariety. Let us recall the diagram defining the higher secant varieties $S^k X$ as the join of X with $S^{k-1}X$:

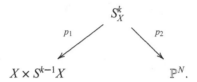

Let us define $\phi : X \times S^{k-1}X \to X$ to be the projection onto the first factor of this product.

Then, for $z \in S^k X$, *the k-entry locus of X with respect to z* is the scheme-theoretic image

$$\Sigma_z^k = \Sigma_z^k(X) := \phi(p_1(p_2^{-1}(z))). \tag{1.15}$$

In the sequel mostly the 1-entry locus of a variety will be considered, which will simply be called the *entry locus*, and the suffix 1 will be omitted using the notation $\Sigma_z(X) := \Sigma_z^1(X)$. Sometimes we shall simply use Σ_z without referring to X.

Geometrically, the support of Σ_z^k is the locus described on X by the $(k+1)$-secant $\mathbb{P}^k$ of X passing through $z \in S^k X$. If $z \in S^k X$ is general, then through z there passes an ordinary $(k+1)$-secant $\mathbb{P}^k$, i.e. given by $k+1$ distinct points on X. Thus for a general $z \in S^k X$ we can describe the support of Σ_z^k in this way

$$(\Sigma_z^k)_{\mathrm{red}} = \overline{\{x \in X : \exists x_1, \ldots, x_k \text{ distinct and } z \in< x, x_1, \ldots, x_k >\}}.$$

Moreover, by the Theorem on the dimension of the fibers of a morphism, for general $z \in S^k X$, the support of Σ_z^k is equidimensional and every irreducible component contains ordinary $\mathbb{P}^k$'s since necessarily $\mathrm{codim}(X) > k$, see Proposition 1.4.3. If $\mathrm{char}(K)=0$ and if X is smooth, then for general $z \in S^k X$ the scheme $p_1^{-1}(z)$ is smooth so that Σ_z^k is reduced.

To recover the scheme structure of Σ_z^k geometrically, one could define Π_z as the locus of $(k+1)$-secant $\mathbb{P}^k$'s through z and define $\Sigma_z^k = \Pi_z \cap X$ as schemes. For example, if through $z \in SX$ there passes a unique tangent line L to X, then in this way we get $\Pi_z = L$ and $\Sigma_z = L \cap X$ consists of the point of tangency with the double structure taking into account the multiplicity of intersection.

Let us study the dimension of Σ_z^k for $z \in S^k X$ general. First, let us remark that if $x \in \Sigma_z^k$ is a general point in an irreducible component, $z \in S^k X$ general, then, as sets,

$$\phi^{-1}(x) = \{y \in S^{k-1}X : z \in< x, y >\} = < z, x > \cap S^{k-1}X \neq \emptyset$$

and $\dim(\phi^{-1}(x)) = 0$ because $z \in S^k X \setminus S^{k-1}X$ by the generality of z.

Definition 1.4.6 (k-Secant Defect δ_k) We define *the k-secant defect of X, $1 \leq k \leq k_0(X)$, $\delta_k(X)$, as the integer*

$$\delta_k(X) = \dim(\Sigma_z^k) = \dim(p_1(p_2^{-1}(z))) = s_{k-1}(X) + \dim(X) + 1 - s_k(X), \quad (1.16)$$

where $z \in S^k X$ is a general point.

For $k = 1$, as advertised above we usually put $\Sigma_z = \Sigma_z^1$, for $z \in SX \setminus X$, and we define $\delta(X) = \delta_1(X) = 2\dim(X) + 1 - \dim(SX)$; for $k = 0$, $\delta_0(X) = 0$ by definition.

Sometimes we shall omit X and simply use $\delta = \delta(X)$ to indicate the secant defect. Let us reinterpret Terracini's Lemma from the point of view of these definitions.

Corollary 1.4.7 (Tangency Along the Entry Loci) *Let $X \subset \mathbb{P}^N$ be an irreducible non-degenerate closed subvariety. Let $k < k_0(X)$, i.e. $S^k X \subsetneq \mathbb{P}^N$, and let $z \in S^k X$ be a general point. Then:*

1. *the linear space $T_z S^k X$ is tangent to X along $(\Sigma_z^k)_{\text{red}} \setminus \text{Sing}(X)$;*
2. $\delta_k(X) < \dim(X)$;
3. $\delta_{k_0}(X) = \dim(X)$ *if and only if $s_{k_0-1}(X) = N - 1$, i.e. if and only if $S^{k_0-1}X$ is a hypersurface;*
4.

$$s_k(X) = (k+1)(n+1) - 1 - \sum_{i=1}^{k} \delta_i(X) = \sum_{i=0}^{k} (\dim(X) - \delta_i(X) + 1);$$

5. *(cf. 1.2.3) if X is a curve, $s_k(X) = 2k + 1$, that is $\delta_k(X) = 0$ for every $k < k_0(X)$.*

Proof Part 1) is clearly a restatement of part 1) of Corollary 1.4.2 when we take into account the geometrical properties of Σ_z^k, $z \in S^k X$ general, described in the definition of entry loci and the fact that the locus of tangency of a linear space is closed in $X \setminus \text{Sing}(X)$, see also Definition 1.5.9. Recall that if char(K)=0, the scheme Σ_z^k is reduced.

If $\dim(\Sigma_z^k) = \delta_k(X) = \dim(X)$, then a general tangent space to $S^k X$ would be tangent along X and X would be degenerate.

With regard to (3), we remark that $\delta_{k_0}(X) = s_{k_0-1}(X) + \dim(X) + 1 - N$ so that $\dim(X) - \delta_{k_0}(X) = N - 1 - s_{k_0-1}(X)$.

Part 4) is an easy computation by induction, while part 5) follows from part 4) since for a curve $\delta_k(X) < \dim(X)$ yields $\delta_k(X) = 0$. □

Remark 1.4.8 The statement of part 1) cannot be improved. Take for example a cone $X \subset \mathbb{P}^5$ of vertex a point $p \in \mathbb{P}^5 \setminus \mathbb{P}^4$ over a smooth non-degenerate projective curve $C \subset \mathbb{P}^4$. If $z \in S(p, SC) = SX$ is general and if $z \in < x, y >$, $x, y \in X$, it is not difficult to see that $\Sigma_z(X) = < p, x > \cup < p, y >$. The hyperplane $T_z SX$ is tangent to X at x and at y by Terracini's Lemma so that it is tangent to X along the rulings $< p, x >$ and $< p, y >$ minus the point p. Since $T_p X = \mathbb{P}^5$, the hyperplane $T_z SX$ is not tangent to X at p (neither J-tangent to X at p).

Remark 1.4.9 A phenomenon studied classically firstly by Scorza—see [163, 164, 166]—and then by Terracini in [186] is the case in which imposing tangency of a hyperplane at $k + 1$ general points, $k \geq 0$, of a variety $X \subset \mathbb{P}^N$ forces tangency along a positive dimensional variety, even if $\delta_k(X) = 0$.

Indeed, Terracini's Lemma says that if $\delta_k(X) > 0$, $k < k_0(X)$, then a hyperplane tangent at $k+1$ points is tangent along the corresponding entry locus. The interesting and exceptional behavior occurs for varieties with $\delta_k(X) = 0$. The first examples are the tangent developable to a non-degenerate curve or cones of arbitrary dimension. Indeed, they have $\delta_0 = 0$ as every variety but by imposing tangency at a general point, we get tangency along the ruling passing through the point.

Varieties for which a hyperplane tangent at $k + 1$, $k \geq 0$, general points is tangent along a positive dimensional subvariety are called k–*weakly defective varieties*, according to Chiantini and Ciliberto, [28]. We observe that for $k < k_0$ a k-defective variety, which for us means $\delta_k > 0$, is k-weakly defective but the converse is not true, as recalled above.

In [28] many interesting properties of k-weakly defective varieties are investigated and a refined Terracini Lemma is proved, also putting in modern terms the classification of k-weakly defective irreducible surfaces obtained classically by Scorza in [164] and Terracini in [186]. Let us remark that, as shown in [28], for every $k \geq 1$ there exist smooth varieties of dimension greater than one which are k-weakly defective but have $\delta_k(X) = 0$ or $s_k(X) = \min\{N, (k + 1)n + k\}$. We shall come back to these definitions and phenomena in Sect. 3.3.

As a further application, we study the dimension of the projection of a variety from linear subspaces generated by general tangent spaces. Terracini's Lemma says that in this case we are indeed projecting from a general tangent space to the related higher secant variety.

As we have seen, when the center of projection L cuts the variety it is difficult to control the dimension of the image of X under projection because we do not know a priori how a general tangent space intersects L. In the case of $L = T_z S^{k-1} X$ this information is encoded in the dimension of $S^k X$ or equivalently in the defect $\delta_k(X)$. Projections from tangent spaces, or more generally from $T_z S^k X$, were a classical tool of investigation—see for example [22–24, 55, 163, 166]—and were recently used to study classical and modern problems, see [28, 33, 36, 104].

Proposition 1.4.10 (Projections from Tangent Spaces) *Let $X \subset \mathbb{P}^N$ be an irreducible, non-degenerate closed subvariety. Let $n = \dim(X)$ and suppose $\mathrm{char}(K)=0$ and $N \geq s_k$, $k \geq 1$, where $s_k = s_k(X)$. Set $\delta_k = \delta_k(X)$.*

Let $x_1, \ldots, x_k \in X$ be k general points, let $L = < T_{x_1}, \ldots, T_{x_k} >$ and let

$$\pi_k = \pi_L : X \to X' \subset \mathbb{P}^{N-s_{k-1}(X)-1}.$$

Then $\dim(L) = s_{k-1}(X) = s_{k-1}$ and, letting $X'_k = \pi_k(X) \subset \mathbb{P}^{N-s_{k-1}-1}$, we have:

1. *$\dim(X'_k) = s_k - s_{k-1} - 1 = n - \delta_k$;*
2. *suppose $N \geq (k+1)n + k$ and $s_{k-1} = kn + k - 1$, i.e. that $\delta_{k-1} = 0$. Then $s_k = (k+1)n + k$ (or equivalently $\delta_k = 0$) if and only if $\dim(X'_k) = n$. In particular, if $N = (k+1)n + k$ and if $s_{k-1} = kn + k - 1$, then $S^k X = \mathbb{P}^{(k+1)n+k}$ if and only if $\pi_k : X \dashrightarrow \mathbb{P}^n$ is dominant.*

Proof If $z \in < x_1, \ldots, x_k >$ is a general point, then z is a general point of $S^{k-1}X$ and by Terracini's Lemma $s_{k-1} = \dim(T_z S^{k-1}X) = \dim(< T_{x_1}, \ldots, T_{x_k} >)$. By (1.11) we get

$$\dim(X'_k) = \dim(S(T_z S^{k-1}X, X)) - s_{k-1} - 1 = s_k - s_{k-1} - 1 = n - \delta_k.$$

The other claims are only reformulations of part 1).   □

1.5   Dual Varieties and Contact Loci of General Tangent Linear Spaces

Let $X \subset \mathbb{P}^N$ be a projective, irreducible non-degenerate variety of dimension $n = \dim(X)$. Let $X_{\mathrm{reg}} = X \setminus \mathrm{Sing}(X)$ be the locus of non-singular points of X. By definition

$$X_{\mathrm{reg}} = \{x \in X : \dim(T_x X) = n\}.$$

If we take a hyperplane section of X, $Y = X \cap H$, where $H = \mathbb{P}^{N-1}$ is an arbitrary hyperplane, then for every $y \in Y$ we get

$$T_y Y = T_y X \cap H, \tag{1.17}$$

as schemes.

Since Y is a pure dimensional scheme of dimension $n - 1$, we deduce that

$$\mathrm{Sing}(Y) \setminus (\mathrm{Sing}(X) \cap H) = \{y \in Y \setminus (\mathrm{Sing}(X) \cap Y) : T_y X \subseteq H\},$$

which is an open subset in the locus of points of X at which H is tangent to X.

Therefore to construct a hyperplane section with a non-empty open set of non-singular points, we need to exhibit a hyperplane H which is not tangent to X at all the points in which it intersects X_{reg}.

There naturally arises the necessity of patching together all the *bad* hyperplanes and then showing that there always exists a hyperplane section of X, non-singular at least outside $\text{Sing}(X)$. Since hyperplanes can be naturally parametrized by points in the dual projective space $\mathbb{P}^{N*}$, we can define a subvariety of $\mathbb{P}^{N*}$ parametrizing hyperplane sections which are also singular outside $\text{Sing}(X)$. This locus is the so-called *dual variety*.

Definition 1.5.1 (Conormal Variety and Dual Variety) Let $X \subset \mathbb{P}^N$ be as above and let

$$\mathscr{P}_X := \overline{\{(x, [H]) \in X_{\text{reg}} \times \mathbb{P}^{N*} : T_x X \subseteq H\}} \subset X \times \mathbb{P}^{N*}$$

be the *conormal variety of X*.

Let us consider the projections of $\mathscr{P}_X$ onto the factors X and $\mathbb{P}^{N*}$,

(1.18)

The dual variety to X, indicated by X^*, is the scheme-theoretic image of $\mathscr{P}_X$ in $\mathbb{P}^{N*}$, that is

$$X^* := p_2(\mathscr{P}_X) \subseteq \mathbb{P}^{N*}.$$

The set $\mathscr{P}_X$ is easily seen to be a closed subset of $X \times \mathbb{P}^{N*}$. For $x \in X_{\text{reg}}$, we have $p_1^{-1}(x) \simeq (T_x X)^* = \mathbb{P}^{N-n-1} \subset \mathbb{P}^{N*}$. Then the set $\mathscr{P}_X$ is irreducible since

$$p_1^{-1}(X_{\text{reg}}) \to X_{\text{reg}}$$

is a $\mathbb{P}^{N-n-1}$-bundle. Therefore $\dim(\mathscr{P}_X) = N - 1$, yielding $\dim(X^*) \leq N - 1$.

Definition 1.5.2 (Dual Defect) The *dual defect of X*, $\text{def}(X)$, is defined as

$$\text{def}(X) = N - 1 - \dim(X^*) \geq 0.$$

To justify the name of conormal variety for $\mathscr{P}_X$ and to get some practice with the definitions, we refer to Exercise 1.5.17.

A famous result of Landman, known as the Landman Parity Theorem, asserts that the dual defect of a smooth manifold $X^n \subset \mathbb{P}^N$ has the same parity as

$\dim(X) = n$, see Proposition 4.4.7 for a simple self-contained proof and also the discussion/references in [141].

Definition 1.5.3 (Reflexive Variety) A variety is said to be *reflexive* if the natural isomorphism between $\mathbb{P}^N$ and $\mathbb{P}^{N^{**}} = (\mathbb{P}^{N^*})^*$ induces an isomorphism between

$$\mathscr{P}_X \subset X \times \mathbb{P}^{N^*} \subset \mathbb{P}^N \times \mathbb{P}^{N^*}$$

and

$$\mathscr{P}_{X^*} \subset X^* \times \mathbb{P}^{N^{**}} \subset \mathbb{P}^{N^*} \times \mathbb{P}^{N^{**}} \simeq \mathbb{P}^{N^*} \times \mathbb{P}^N.$$

For a reflexive variety, the natural identification between $\mathbb{P}^N$ and $\mathbb{P}^{N^{**}}$ induces the equality $X = X^{**} := (X^*)^*$. See part e) of Example 1.5.19 for a well-known example, considered firstly by Wallace in [190], of a non-reflexive smooth plane curve $X \subset \mathbb{P}^2$ defined over a field of positive characteristic such that $X = X^{**}$.

Let us take $[H] \in X^*$. By definition

$$C_H := C_H(X) = p_2^{-1}(H) = \overline{\{x \in X_{\mathrm{reg}} : T_xX \subseteq H\}}$$

is precisely the closure of the non-singular points of X where H is tangent to X, the so-called *contact locus of H on X*.

By definition C_H is not empty so that $H \cap X$ is singular outside $\mathrm{Sing}(X)$ for $H \in X^*$. On the contrary if $[H] \notin X^*$, the hyperplane section $H \cap X$ can be singular only along $\mathrm{Sing}(X)$. This is the classical and well-known *Bertini's Theorem*, which we now state explicitly, see [18].

Theorem 1.5.4 (Bertini's Theorem on Hyperplane Sections) *Let $X \subset \mathbb{P}^N$ be a projective, irreducible non-degenerate variety of dimension $n = \dim(X)$. Then for every $H \in (\mathbb{P}^N)^* \setminus X^*$ the divisor $H \cap X$ is non-singular outside $\mathrm{Sing}(X)$.*

In particular, if X has at most a finite number of singular points $p_1, \ldots, p_m$, then for every $H \notin X^* \cup (p_1)^* \cup \ldots \cup (p_m)^*$, the hyperplane section $H \cap X$ is a non-singular subscheme of pure codimension 1.

Later we shall see in Chap. 3 that for $n \geq 2$ every hyperplane section of an irreducible variety of dimension n is connected. Thus for non-singular varieties with $n = \dim(X) \geq 2$, the hyperplane sections with hyperplanes $H \notin X^*$, being connected and non-singular, are also irreducible so that they are irreducible non-singular algebraic varieties.

As we have seen, the dual varieties encode information about the tangency of hyperplanes. Terracini's Lemma says that linear spaces containing tangent spaces to higher secant varieties are tangent along $(\Sigma_z^k)_{\mathrm{red}} \setminus \mathrm{Sing}(X)$, see Corollary 1.4.7. Thus the maximal dimension of the fibers of $p_2 : \mathscr{P}_X \to X^* \subset \mathbb{P}^{N^*}$ is an upper bound for $\delta_k(X)$ as soon as $S^k X \subsetneq \mathbb{P}^N$, as we shall immediately see. More refined versions with the higher Gauss maps γ_m, see below, can be formulated but in those cases the condition expressed by the numbers $\varepsilon_m(X)$, which can be defined as below, is harder to control.

Theorem 1.5.5 (Dual Variety and Higher Secant Varieties) *Let $X \subset \mathbb{P}^N$ be an irreducible non-degenerate projective variety. Let $p_2 : \mathscr{P}_X \to X^* \subset \mathbb{P}^{N^*}$ be as above and let*

$$\varepsilon(X) = \max\{\dim(p_2^{-1}(H)), H \in X^*\}.$$

If $S^k X \subsetneq \mathbb{P}^N$, then $\delta_k(X) \leq \varepsilon(X)$. In particular, if $p_2 : \mathscr{P}_X \to X^*$ is a finite morphism, then $\dim(S^k X) = \min\{(k+1)n + k, N\}$.

Proof Let $z \in S^k X$ be a general point. There exists an $x \in \Sigma_z^k(X) \cap X_{\mathrm{reg}}$ and moreover $T_z S^k X$ is contained in a hyperplane H. Then

$$p_1(p_2^{-1}(H)) \supseteq \mathrm{Sing}(X \cap H) \setminus (\mathrm{Sing}(X) \cap H),$$

more precisely $\mathrm{Sing}(X \cap H) \setminus (\mathrm{Sing}(X) \cap H)$, contains the irreducible component of $\Sigma_z^k(X) \setminus (\mathrm{Sing}(X) \cap \Sigma_z^k(X))$ passing through x by Corollary 1.4.7. Then $p_1(p_2^{-1}(H))$ has dimension at least $\delta_k(X) = \dim(\Sigma_z^k(X))$ and the conclusion follows. $\qquad\square$

For a smooth variety $X \subset \mathbb{P}^N$ the condition that $p_2 : \mathscr{P}_X \to X^*$ is a finite morphism is equivalent to the ampleness of the locally free sheaf $\mathscr{N}_{X/\mathbb{P}^N}(-1)$, where $\mathscr{N}_{X/\mathbb{P}^N}$ is the normal bundle of X in $\mathbb{P}^N$, see Exercise 1.5.17.

Corollary 1.5.6 (cf. Corollaries 1.2.3 and 1.4.7) *Let $X \subset \mathbb{P}^N$ be either an irreducible non-degenerate curve or a smooth non-degenerate complete intersection. Then*

$$\dim(S^k X) = \min\{(k+1)n + k, N\}.$$

Proof By Exercise 1.5.17 we know that for a smooth non-degenerate complete intersection $p_2 : \mathscr{P}_X \to X^*$ is a finite morphism, a property which is immediate for projective curves. Thus the conclusions follow immediately from Theorem 1.5.5.

$\qquad\square$

For reflexive varieties $X \subset \mathbb{P}^N$ a general contact locus $C_H(X) \subset X$ is a linear space of dimension $\mathrm{def}(X)$. This is nothing but an interpretation of the isomorphism $X \simeq (X^*)^*$.

One should be careful in the interpretation of the result: it does not mean that the hyperplane remains tangent along the whole "contact locus", see remark 1.4.8 and adapt it to the more general situation of a ruling of a cone. This is true only for non-singular varieties. In particular, reflexive varieties of positive dual defect contain positive dimensional families of linear spaces, imposing strong restrictions on their existence. We formalize this discussion by stating it for further reference in the text.

Proposition 1.5.7 *Let $X \subset \mathbb{P}^N$ be a reflexive variety. Then for $[H] \in X^*_{\text{reg}}$,*

$$p_1(p_2^{-1}([H])) = \overline{\{x \in X_{\text{reg}} \; : \; T_x X \subset H\}} = (T_{[H]}X^*)^\perp = \mathbb{P}^{\text{def}(X)}.$$

The next result has an elementary and direct proof. It is considered a classical theorem, known at least to C. Segre. According to Kleiman it was also discovered in some form by Monge, see [114, 115]. The modern treatment pointing out the pathologies occurring in positive characteristic and stating it explicitly seems to be due to Wallace in [190, 191].

In Exercise 1.5.19 we shall show, following Wallace, several well-known examples of smooth irreducible non-degenerate varieties $X \subset \mathbb{P}^N$, whose associated map $p_2 : \mathscr{P}_X \to X^*$ is not generically smooth so that X cannot be reflexive. As we now prove, the condition of generic smoothness of p_2 is indeed equivalent to reflexivity.

Theorem 1.5.8 (Reflexivity Theorem, [190]) *Let $X \subset \mathbb{P}^N$ be an irreducible non-degenerate projective variety. Then X is reflexive if and only if $p_2 : \mathscr{P}_X \to X^*$ is a generically smooth morphism.*

In particular, if char$(K)=0$, *all projective irreducible varieties are reflexive.*

Proof Suppose that $\mathscr{P}_X = \mathscr{P}_{X^*}$. Then p_2 is generically smooth since the general fiber is smooth, being isomorphic as a scheme to a projective space of dimension def(X), see also Proposition 1.5.7. Let $\mathbf{x} = (x_0 : \ldots : x_N)$ be homogeneous coordinates on $\mathbb{P}^N$ and let $\mathbf{a} = (a_0 : \ldots : a_N)$ be homogeneous coordinates on $\mathbb{P}^{N^*}$. Then the tautological incidence relation

$$I = \{(p, [H]) \; : \; p \in H\} \subset \mathbb{P}^N \times \mathbb{P}^{N^*}$$

is a divisor whose bi-homogeneous equation is

$$\sum_{i=0}^N a_i x_i = 0.$$

Let $(p, [H]) \in \mathscr{P}_X$ be a general point. Without loss of generality we can suppose that $p = (1 : \tilde{p}_1 : \ldots : \tilde{p}_N) \in \mathbb{A}^N := \mathbb{P}^N \setminus V(x_0)$ and that $[H] = (1 : \tilde{h}_1 : \ldots : \tilde{h}_N) \in (A_0^N)' := \mathbb{P}^{N^*} \setminus V(a_0)$.

Let $y_j = x_j/x_0$ and let $b_j = a_j/a_0$ for $j = 1, \ldots N$ be the associated affine coordinates, identify $\mathbb{A}^{2N}_{0,0}$ with $\mathbb{A}^N_0 \times (A_0^N)'$ and let $I_{0,0} = I \cap \mathbb{A}^{2N}_{0,0}$. Then

$$t_{(p,[H])}I \simeq t_{(p,[H])}I_{0,0} \subset \mathbb{A}^{2N}_{0,0}$$

is the hyperplane of equation

$$\sum_{j=1}^N \tilde{h}_i(y_i - \tilde{p}_j) + \sum_{j=1}^N \tilde{p}_j(b_j - \tilde{h}_j) = 0. \tag{1.19}$$

Clearly $\mathscr{P}_X \subset I$ and from (1.18) we get

$$t_{(p,[H])}\mathscr{P}_X \subset t_p X \times t_{[H]} X^*.$$

We deduce

$$t_{(p,[H])}\mathscr{P}_X \simeq t_{(p,[H])}(\mathscr{P}_X \cap \mathbb{A}_{0,0}^{2N}) \subset [t_p(X \cap \mathbb{A}_0^N) \times t_{[H]}(X^* \cap (\mathbb{A}_0^N)')] \cap t_{(p,[H])}I_{0,0}.$$
$$(1.20)$$

Moreover, the differential map of $p_2 : \mathscr{P}_X \to X^*$ at the point $(p, [H])$ is the restriction to $t_{(p,[H])}\mathscr{P}_X$ of the projection of $t_p X \times t_{[H]} X^*$ onto $t_{[H]} X^*$.

Suppose $p_2 : \mathscr{P}_X \to X^*$ is generically smooth. Then the differential map of p_2 at $(p, [H])$ maps $t_{(p,[H])}\mathscr{P}_X$ onto $t_{[H]} X^*$. Since $\mathscr{P}_X$ and $\mathscr{P}_{X^*}$ are irreducible varieties of dimension $N - 1$, to prove the equality $\mathscr{P}_X = \mathscr{P}_{X^*}$ it will be sufficient to show that a general point $(p, [H]) \in \mathscr{P}_X$ belongs to $\mathscr{P}_{X^*}$.

Let $s \in t_{[H]} X^*$ be an arbitrary point. By hypothesis, there exists a $q \in t_p X$ such that $(q, s) \in t_{(p,[H])}\mathscr{P}_X$. By definition of dual variety we have $T_p X \subset H$ so that, imposing $q = (\tilde{q}_1, \ldots, \tilde{q}_N) \in H \cap \mathbb{A}_0^N$, we deduce

$$\sum_{j=1}^{N} \tilde{h}_j(\tilde{q}_j - \tilde{p}_j) = 0. \tag{1.21}$$

From $(q, s) \in t_{(p,[H])}I$, from (1.19) and from (1.21), letting $s = (\tilde{s}_1, \ldots, \tilde{s}_N) \in (\mathbb{A}_0^N)'$, we get

$$0 = \sum_{j=1}^{N} \tilde{p}_j(\tilde{s}_j - \tilde{h}_j) + \sum_{j=1}^{N} \tilde{h}_j(\tilde{q}_j - \tilde{p}_j) = \sum_{j=1}^{N} \tilde{p}_j(\tilde{s}_j - \tilde{h}_j).$$

This means that s belongs to the hyperplane $L \subset \mathbb{A}_0^N$ of equation

$$\sum_{j=1}^{N} \tilde{p}_j(b_j - \tilde{h}_j) = 0,$$

which is the equation of the affine part in $(\mathbb{A}_0^N)'$ of the hyperplane represented in $\mathbb{P}^{N^{**}} = \mathbb{P}^N$ by the point p. Since s was arbitrary we get $T_{[H]} X^* \subset [p]$, proving $(p, [H]) \in \mathscr{P}_{X^*}$ by definition of conormal variety. $\square$

Another natural and similar problem is to ask whether a general tangent space to a (smooth) variety X is tangent to X at more than one point. During the discussion we will always suppose char$(K) = 0$ to avoid the intriguing and strange pathologies in positive characteristic, which are usually variations on the themes described in Exercise 1.5.19.

By Proposition 1.5.7 a general tangent space to an irreducible non-degenerate curve, not a line, is tangent only at one point. On the other hand if X is a cone over a curve, we know that a general tangent space is tangent precisely along the ruling passing through the point. From this point of view the unique common feature of these two classes of irreducible varieties is the linearity of the locus of points at which a general tangent linear space remains tangent.

Definition 1.5.9 (Gauss Maps) Let $X \subset \mathbb{P}^N$ be an irreducible projective variety of dimension $n = \dim(X) \geq 1$, let $m \geq n$ and let $\mathbb{G}(m, N)$ be the Grassmannian parametrizing linear subspaces of dimension m in $\mathbb{P}^N$. Let

$$\mathscr{P}_X^m := \overline{\{((x, [L]) \in X_{\mathrm{reg}} \times \mathbb{G}(m, N) \; : \; T_x X \subseteq L\}} \subset X \times \mathbb{G}(m, N).$$

Let us consider the projections of $\mathscr{P}_X^m$ onto the factors X and $\mathbb{G}(m, N)$,

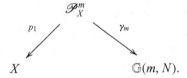

The variety of m-dimensional tangent subspaces to X, X_m^*, is the scheme-theoretic image of $\mathscr{P}_X^m$ in $\mathbb{G}(m, N)$, that is

$$X_m^* := \gamma_m(\mathscr{P}_X^m) \subset \mathbb{G}(m, N).$$

For $m = N - 1$, we recover the dual variety and its definition, while for $m = n$, we get the usual Gauss map $\mathscr{G}_X : X \dashrightarrow \mathbb{G}(n, N)$ which associates to a point $x \in X_{\mathrm{reg}}$ the point $[T_x X] \in \mathbb{G}(n, N)$, that is $\mathscr{G}_X(x) := \gamma_n(x) = [T_x X]$ for $x \in X_{\mathrm{reg}}$.

If $X = V(f) \subset \mathbb{P}^N$ is a hypersurface, then $n = N - 1$ and clearly the Gauss map

$$\mathscr{G}_X : X \dashrightarrow \mathbb{P}^{N^*} = \mathbb{G}(N - 1, N)$$

associates to a smooth point p of X its tangent hyperplane. Thus in coordinates the Gauss map is given by the formula

$$\mathscr{G}_X(p) = (\frac{\partial f}{\partial X_0}(p) : \ldots : \frac{\partial f}{\partial X_N}(p)).$$

The following result is another interesting consequence of reflexivity, whose proof follows easily from Proposition 1.5.7. For smooth varieties several notable improvements will be proved in Corollary 3.2.4 as applications of Zak's Theorem on Tangency.

Theorem 1.5.10 (Linearity of General Contact Loci for Reflexive Varieties) *Let* $X \subset \mathbb{P}^N$ *be an irreducible projective non-degenerate reflexive variety of dimension* $n = \dim(X) \geq 1.$

Then the general fiber of the morphism $\gamma_m : \mathscr{P}_X^m \to X_m^*$ *is a linear space of dimension* $\dim(\mathscr{P}_X^m) - \dim(X_m^*)$.

In particular, the closure of a general fiber of $\mathscr{G}_X : X \dashrightarrow X_n^* \subset \mathbb{G}(n, N)$ *is a linear space of dimension* $n - \dim(\mathscr{G}_X(X))$ *so that a general linear tangent space is tangent along an open subset of a linear space of dimension* $n - \dim(\mathscr{G}_X(X))$.

Proof For $m = N - 1$ the statement is proved in Proposition 1.5.7. Suppose $m < N - 1$ and let $[L] \in X_m^*$ be a general point. By definition L is tangent to X at all points of

$$p_1(\gamma_m^{-1}([L])) \cap X_{\mathrm{reg}}$$

and

$$p_1(\gamma_m^{-1}([L])) \cap X_{\mathrm{reg}} = \bigcap_{H \supset L} p_1(\gamma_{N-1}^{-1}([H])) \cap X_{\mathrm{reg}}.$$

By definition $[H] \in X^*$ and in the previous expression we can assume, without loss of generality, that $[H] \in X_{\mathrm{reg}}^*$ so that

$$p_1(\gamma_{N-1}^{-1}([H])) = (T_{[H]}X^*)^{\perp} = \mathbb{P}^{\mathrm{def}(X)}$$

by Proposition 1.5.7. It follows that

$$p_1(\gamma_m^{-1}([L])) = \overline{p_1(\gamma_m^{-1}([L])) \cap X_{\mathrm{reg}}} = \bigcap_{H \supset L} (T_{[H]}X^*)^{\perp}$$

is a linear space, being an intersection of linear spaces. $\square$

Exercises

1.5.11 Let K be a(n algebraically closed) field of characteristic different from 2 and let $X = \nu_2(\mathbb{P}^2) \subset \mathbb{P}^5$ be the 2-Veronese surface in $\mathbb{P}^5$. We shall identify $\mathbb{P}^5$ with

$$\mathbb{P}(\{A \in M_{3 \times 3}((K) : A = A^t\}),$$

so that $X = \{[A] : \mathrm{rk}(A) = 1\}$, as is well known. Prove the following facts:

(a) $SX = TX = V(\det(A)) \subset \mathbb{P}^5$ so that SX is a cubic hypersurface and $\delta(X) = 1$.
(b) If $x_1, x_2 \in X$, then $T_{x_1}X \cap T_{x_2}X \neq \emptyset$ and more precisely it consists of a point for every $x_1, x_2 \in X$, $x_1 \neq x_2$.
(c) $\mathrm{Sing}(SX) = X$.

1.5.12 Let K be a(n algebraically closed) field and let $X = \mathbb{P}^2 \times \mathbb{P}^2 \subset \mathbb{P}^8$ be the Segre embedding of $\mathbb{P}^2 \times \mathbb{P}^2$ in $\mathbb{P}^8$. Identifying $\mathbb{P}^8$ with

$$\mathbb{P}(M_{3\times 3}(K)),$$

we have $X = \{[A] : \mathrm{rk}(A) = 1\}$. Prove that:

(a) $SX = TX = V(\det(A)) \subset \mathbb{P}^8$ is the cubic hypersurface given by the cubic polynomial $\det(A)$;
(b) if $x_1, x_2 \in X$, then $T_{x_1}X$ and $T_{x_2}X$ intersect in a line and $\delta(X) = 2$;
(c) if H is a general hyperplane in $\mathbb{P}^8$, then $Y := X \cap H$ is a smooth, irreducible, non-degenerate threefold $Y \subset \mathbb{P}^7$ such that $SY = SX \cap H \subsetneq H$ so that $\delta(Y) = 1$;
(d) given $y_1, y_2 \in Y$, then $T_{y_1}Y \cap T_{y_2}Y \neq \emptyset$ (and it consists of a point if $y_1, y_2 \in Y$ are general);
(e) $\mathrm{Sing}(SX) = X$.
(f) Let $p \in \mathbb{P}^9 \backslash \mathbb{P}^8$, let $Z = S(p, X) \subset \mathbb{P}^9$ and let $X' = Z \cap W$, with $W \subset \mathbb{P}^9$ a general hypersurface of degree $d \geq 2$ not passing through p. Then X' is a smooth, irreducible, non-degenerate fourfold such that $SX' = SZ = S(p, SX)$;
(g) Prove that $\dim(SX') = 8$, yielding $\delta(X') = 1$. Deduce from Terracini's Lemma that two general tangent spaces to X' intersect in a point, yielding $\delta(X') = 1$.

Use the fact that Z is a cone over X to prove directly that two general tangent spaces to X' intersect in a point.

1.5.13 Generalize part f) of the previous problem and find the relation between $SX \subset \mathbb{P}^N$ and $SX' \subset \mathbb{P}^{N+1}$ for $X' \subset \mathbb{P}^{N+1}$ a general intersection of

$$Z = S(p, X) \subset \mathbb{P}^{N+1}$$

with a general hypersurface $W \subset \mathbb{P}^{N+1}$ of degree $d \geq 2$ not passing through the point $p \in \mathbb{P}^{N+1} \backslash \mathbb{P}^N$, which belongs to the vertex of Z.

1.5.14 Prove Edge's result in [49], see also [159], to the effect that smooth irreducible divisors of type $(0, 2)$, $(1, 2)$ and $(2, 1)$ on the Segre varieties $Y = \mathbb{P}^1 \times \mathbb{P}^n \subset \mathbb{P}^{2n+1}$, $n \geq 2$, have *one apparent double point*, that is there exists a unique secant line to the variety passing through a general point of $\mathbb{P}^{2n+1}$, using the following steps.

(a) Prove first that the only smooth curves, not necessarily irreducible, on a smooth quadric in $\mathbb{P}^3$ having one apparent double point are of the above types.
(b) For every $p \notin Y := \mathbb{P}^1 \times \mathbb{P}^n$, the entry locus $\Sigma_p(Y)$ has the form $\mathbb{P}^1 \times \mathbb{P}^1_p$ for some $\mathbb{P}^1_p \subset \mathbb{P}^n$ and spans a linear $\mathbb{P}^3_p$.
(c) If $X_{a,b}$ is a divisor of type (a, b) on Y, the secant lines of $X_{a,b}$ passing through p are exactly the secant lines of $X_{a,b} \cap \mathbb{P}^3_p$ passing through p.
(d) For a general $p \in \mathbb{P}^{2n+1}$ and for a general $X_{a,b}$ the variety $X_{a,b} \cap \mathbb{P}^3_p$ is a reduced, not necessarily irreducible curve and it is a divisor of type (a, b) on $\mathbb{P}^1 \times \mathbb{P}^1_p$.

(e) Deduce that the divisor $X_{a,b}$ has one apparent double point if and only if $(a, b) \in \{(1, 2), (2, 1), (2, 0), (0, 2)\}$. Observe that $X_{2,0} = \mathbb{P}^n \sqcup \mathbb{P}^n$ is reducible.

(f) Prove that the irreducible divisors of type $(2, 1)$ are the rational normal scrolls of *minimal degree* in $\mathbb{P}^{2n+1}$ by showing that $\deg(X_{2,1}) = n + 2 = \text{codim}(X_{2,1}) + 1$.

(g) The irreducible divisors of type $(0, 2)$ are isomorphic to $\mathbb{P}^1 \times Q^{n-1}$, where $Q^{n-1} \subset \mathbb{P}^n$ is a quadric hypersurface.

1.5.15 Let $X = \mathbb{P}^{m_1} \times \mathbb{P}^{m_2} \subset \mathbb{P}^{m_1 m_2 + m_1 + m_2}$ be the Segre variety and let $p \in\ <x_1, x_2>\in SX$ be a general point with $x_i = (a_i, b_i) \in X$, $x_1 \neq x_2$ general. Show that the two-dimensional quadric $Q_p =\ < a_1, a_2 > \times < b_1, b_2 >$ is contained in Σ_p. Prove that $Q_p = \Sigma_p$ and deduce $\delta(X) = 2$.

1.5.16 Prove the following facts, under the assumption $\text{char}(K) = 0$:

(a) If $X \subsetneq \mathbb{P}^M \subsetneq \mathbb{P}^N$ is a degenerate variety the dual variety $X^* \subset \mathbb{P}^{N*}$ is a cone of vertex $(\mathbb{P}^M)^\perp = \mathbb{P}^{N-M-1} \subset \mathbb{P}^{N*}$ over the dual variety of X in $\mathbb{P}^M$.

(b) Suppose $X = S(L, X')$ is a cone of vertex $L = \mathbb{P}^l$, $l \geq 0$, over a variety $X' \subset M = \mathbb{P}^{N-l-1}$, $M \cap L = \emptyset$. Then $X^* \subset (\mathbb{P}^l)^\perp = \mathbb{P}^{N-l-1} \subset (\mathbb{P}^N)^*$ is degenerate. Is there any relation between X^* and the dual of X' in M?

(c) Suppose $X \subset \mathbb{P}^N$ is a cone. Prove that $X^* \subset \mathbb{P}^{N*}$ is degenerate. Conclude that $X \subset \mathbb{P}^N$ is degenerate if and only if $X^* \subset \mathbb{P}^{N*}$ is a cone; and, dually, that $X \subset \mathbb{P}^N$ is a cone if and only if $X^* \subset \mathbb{P}^{N*}$ is degenerate.

(d) Assume now $p = \text{char}(K) > 0$ and show that for the irreducible curve $X \subset \mathbb{P}^2$ of equation $x_0^{p-1} x_2 - x_1^p = 0$ the dual curve is a line.

1.5.17 Let $X \subset \mathbb{P}^N$ be a non-singular variety. Then, using Grothendieck's notation, prove that:

(a)

$$\mathscr{P}_X \simeq \mathbb{P}(N_{X/\mathbb{P}^N}(-1)),$$

where $N_{X/\mathbb{P}^N}(-1)$ is the twist of the normal bundle of X in $\mathbb{P}^N$, $N_{X/\mathbb{P}^N}$, by $\mathscr{O}_{\mathbb{P}^N}(-1)$ and that

$$p_2 : \mathscr{P}_X \to X^* \subset \mathbb{P}^{N*}$$

is given by a sublinear system of $|\mathscr{O}_{\mathbb{P}(N_{X/\mathbb{P}^N}(-1))}(1))|$ (cf. [45, Exposé XVII]).

(b) Deduce that for a smooth non-degenerate variety $X \subset \mathbb{P}^N$ the locally free sheaf $N_{X/\mathbb{P}^N}(-1)$ is ample, that is $\mathscr{O}_{\mathbb{P}(N_{X/\mathbb{P}^N}(-1))}(1)$ is ample, if and only if every $[H] \in X^*$ is tangent to X in at most a finite number of points, see [124, Proposition 6.3.5, 6.3.6, 6.3.8]. In particular, for such varieties $\text{def}(X) = 0$, $\delta_k(X) = 0$ for every $k < k_0(X)$ and a general tangent linear space $\mathbb{P}^m$, $m \geq \dim(X)$, is tangent to X only at one point.

(c) Let $X^n \subset \mathbb{P}^N$ be a smooth non-degenerate complete intersection and let $c = \mathrm{codim}(X) = N - \dim(X)$. If $I(X) = < f_1, \ldots, f_c >$ with f_i a homogeneous polynomial of degree $d_i \geq 2$, then

$$N_{X/\mathbb{P}^N}(-1) \simeq \bigoplus_{i=1}^{c} \mathcal{O}_X(d_i - 1)$$

is an ample locally free sheaf. Deduce that $\mathrm{def}(X) = 0$, see [70, Remark 7.5], that $\delta_k(X) = 0$ for $k < k_0(X)$ and that a general tangent linear space $\mathbb{P}^m$, $m \geq \dim(X)$, is tangent to X only at one point.

1.5.18 Let V be a complex vector space of dimension $N + 1$ such that $\mathbb{P}(V) = \mathbb{P}^N$ and let $X \subset \mathbb{P}^N$ be a smooth non-degenerate variety of dimension $n \geq 1$.

Consider the restriction of the Euler sequence on $\mathbb{P}^N$ to X

$$0 \to \Omega^1_{\mathbb{P}^N|X} \to V \otimes \mathcal{O}_X(-1) \to \mathcal{O}_X \to 0,$$

and the exact sequence on X

$$0 \to N^*_{X/\mathbb{P}^N} \to \Omega^1_{\mathbb{P}^N|X} \to \Omega^1_X \to 0.$$

From these exact sequences define

$$0 \to N^*_{X/\mathbb{P}^N}(1) \to V \otimes \mathcal{O}_X \to P^1_X \to 0, \tag{1.22}$$

and

$$0 \to \Omega^1_X(1) \to P^1_X \to \mathcal{O}_X(1) \to 0, \tag{1.23}$$

where P^1_X is the so-called *first jet bundle of* $\mathcal{O}_X(1)$. Prove the following facts:

(a) For every closed point $x \in X$ we have $\mathbb{P}(p^1_X \otimes k(x)) = T_x X \subset \mathbb{P}(V)$.
(b) The surjection in (1.22) gives an embedding over X of $\mathbb{P}(p_X) \to X$ into $\mathbb{P}(V) \times X \to X$ in such a way that the restriction π_1 of the projection $\mathbb{P}^N \times X \to \mathbb{P}^N$ to $\mathbb{P}(p^1_X)$ maps $\mathbb{P}(p^1_X)$ onto $TX = \bigcup_{x \in X} T_x X$.
(d) Deduce from (1.22) that $\det(P^1_X) \simeq \det(N_{X/\mathbb{P}^N}(-1)) \simeq \omega_X \otimes \mathcal{O}_X(n+1)$ and use this result for $n = 1$ to calculate/compare $\deg(TX)$ and $\deg(X^*)$.

1.5.19 Let $C \subset \mathbb{P}^N$, $N \geq 2$, be an irreducible non-degenerate curve.

(a) Assume $\mathrm{char}(K) = 0$ and prove that a general tangent linear space $\mathbb{P}^m$ with $m \geq 1$ is tangent to C only at one point. Deduce in particular that a general tangent line to C is tangent to C only at one point.
(b) (cf. [114, 115, 190]) Let $p = \mathrm{char}(K) > 0$, let $X \subset \mathbb{P}^2$ be the irreducible curve of equation $x_0^p x_2 + x_0 x_2^p - x_1^{p+1} = 0$ and let $(u_0 : u_1 : u_2)$ be the dual coordinates

on $\mathbb{P}^{2^*}$. Show that $X^* \subset \mathbb{P}^{2^*}$ has equation $u_0^p u_2 + u_0 u_2^p - u_1^{p+1} = 0$ and that the Gauss map $\mathscr{G}_X : X \to \mathbb{P}^{2^*}$, which is naturally identified with $p_2 : \mathscr{P}_X \to \mathbb{P}^{2^*}$, is bijective but not generically smooth. Deduce that $X^{**} = X$ and that X is not reflexive.

(c) Let $p = \operatorname{char}(K) > 0$, let $N \geq 2$ and let $X \subset \mathbb{P}^N$ be the irreducible smooth hypersurface of equation

$$\sum_{i=0}^{N} x_i^{p+1} = 0.$$

Calculate X^* and deduce that $X^{**} = X$. Show directly that X is not reflexive by observing that, if $p_0 = (\alpha_0 : \ldots : \alpha_N) \in X$, then the tangent hyperplane to X at $q_0 = (\alpha_0^p : \ldots, \alpha_N^p) = \mathscr{G}_X(p_0)$ is not tangent to X at p_0.

(d) (cf. [150]) Let $p \geq 3$ be the characteristic of the base field K and let $X \subset \mathbb{P}^2$ be the irreducible curve of equation $x_0 x_1^{2p} + x_1 x_2^{2p} + x_2 x_0^{2p} = 0$. Show that $X^* \subset \mathbb{P}^{2^*}$ has degree $2(2p + 1)$ and that $X^{**} \neq X$.

1.5.20 Let $C \subset \mathbb{P}^N$ be a smooth non-degenerate projective curve of degree $d = \deg(X)$ and genus $g = g(C)$. Calculate geometrically as a function of d and g the following:

(a) $\deg(TC)$ for $N \geq 3$;
(b) $\deg(SC)$ for $N \geq 4$;
(c) $\deg(C^*)$ for $N \geq 2$.

1.5.21 Prove the following facts.

(a) Let $X = \mathbb{P}^1 \times \mathbb{P}^n \subset \mathbb{P}^{2n+1}$, $n \geq 1$, be the Segre embedding of $\mathbb{P}^1 \times \mathbb{P}^n$. Identify $\mathbb{P}^{2n+1}$ with $\mathbb{P}(M_{2 \times (n+1)}(K))$ and show that $(\mathbb{P}^1 \times \mathbb{P}^n)^* \simeq \mathbb{P}^1 \times \mathbb{P}^n$. Deduce that $\operatorname{def}((\mathbb{P}^1 \times \mathbb{P}^n)) = n - 1$.
(b) Show that if $X = v_2(\mathbb{P}^2) \subset \mathbb{P}^5$, or if $X = \mathbb{P}^2 \times \mathbb{P}^2 \subset \mathbb{P}^8$, then $X^* \simeq SX$.
(c) Let $C = v_3(\mathbb{P}^1) \subset \mathbb{P}^3$ be a twisted cubic. Prove that $TC \simeq C^*$.

1.5.22 (cf. [37, Proposition 1.1]) Let $X \subset \mathbb{P}^N$ be an irreducible non-degenerate variety, let $L = \mathbb{P}^s$, $s \geq 0$, be a linear space, let $\pi_L : \mathbb{P}^N \dashrightarrow \mathbb{P}^{N-1-s} = (L^\perp)^*$ be the projection with center L and let $X_L = \overline{\pi_L(X)}$. Suppose $n = \dim(X) < N - s - 1$. Prove the following facts:

(a) $X_L^* \subseteq X^* \cap L^\perp$;
(b) If $X^* \cap L^\perp$ is irreducible and reduced and if $\dim(X^* \cap L^\perp) = \dim(X_L^*)$, then $X_L^* = X^* \cap L^\perp$ as schemes. Thus in particular for $L = p \in \mathbb{P}^N$ a general point, we have $(\pi_p(X))^* = X^* \cap p^\perp$.
(c) Let $X = v_3(\mathbb{P}^1) \subset \mathbb{P}^3$ and let $L = p \in X$. Then $X_L \subset \mathbb{P}^2$ is a smooth conic so that $X_L^* \subset \mathbb{P}^{2^*}$ is a smooth conic. Let $H = p^\perp$. Then $H \cap X^*$ is a plane quartic curve containing X_L^*. Determine $H \cap X^*$ as a scheme.
(d) If $X = S(1, 3) \subset \mathbb{P}^5$ and if $L \subset S(1, 3)$ is the directrix line, then $X_L \subset \mathbb{P}^3 = (L^\perp)^*$ is a twisted cubic while $\mathbb{P}^3 = L^\perp \subset X^*$, yielding the strict inclusion $X_L^* \subsetneq X^* \cap L^\perp$.

1.5.23 Let $X, Y \subset \mathbb{P}^N$ be closed, irreducible subvarieties and let $L = \mathbb{P}^l$ be a linear subspace of dimension $l \geq 0$, which does not contain either X or Y. Let $\pi_L : \mathbb{P}^N \dashrightarrow \mathbb{P}^{N-l-1}$ be the projection from L. Then:

$$\pi_L(S(X, Y)) = S(\pi_L(X), \pi_L(Y)).$$

In particular, if L does not contain X, then for any non–negative integer k one has:

$$\pi(S^k(X)) = S^k(\pi_L(X)).$$

1.5.24 Let $X, Y \subset \mathbb{P}^N$ be closed irreducible subvarieties and assume char$(K)=0$. Suppose $S(X, Y) \supsetneq X$ and $S(X, Y) \supsetneq Y$ to avoid trivialities.

If $z \in S(X, Y)$ is a general point, if $x \in \Sigma_z(X)$ is a general point and if

$$< z, x > \cap Y = y \in \Sigma_z(Y),$$

then y is a smooth point of $\Sigma_z(Y)$,

$$T_x \Sigma_z(X) = T_x X \cap < x, T_y \Sigma_z(Y) >= T_x X \cap < x, T_y Y >,$$

$$T_y \Sigma_z(Y) = T_y Y \cap < y, T_x \Sigma_z(X) >= T_y Y \cap < y, T_x X >$$

and

$$T_x X \cap T_y Y = T_x \Sigma_z(X) \cap T_y \Sigma_z(Y).$$

In particular, for $z \in SX$ a general point, X not linear, and for $x \in \Sigma_z(X)$ a general point, we have that, if $< x, z > \cap X = y \in \Sigma_z(X)$, then y is a smooth point of $\Sigma_z(X)$,

$$T_x \Sigma_z(X) = T_x X \cap < x, T_y \Sigma_z(X) >= T_x X \cap < x, T_y Y > .$$

Hint for Problems of Chap. 1

1.5.11 (a) Observe that the linear combination of two symmetric matrixes of rank 1 has rank at most 2 and that every symmetric matrix of rank 2 can be written as the linear combination of two symmetric matrixes of rank 1. Then deduce that SX is the locus of matrices of rank at most 2, proving the claim.

1.5.17 (a) Restrict the Euler sequence of $\mathbb{P}^N$ twisted by $\mathcal{O}(-1)$ to X and use the standard conormal sequence. Interpret these sequences in terms of the associated projective bundles and of the incidence correspondence defining $\mathscr{P}_X$.

1.5.20 In cases (a) and (c) apply Riemann–Hurwitz formula to the projection of C onto $\mathbb{P}^1$ from a general $\mathbb{P}^{N-1}$. Compare also with the last part of Exercise 1.5.18.

In case (b) project from a general $L = \mathbb{P}^{N-3}$ onto a plane and compute the number of nodes of the projection of C onto $\mathbb{P}^2$.

1.5.21 For (a), (b) and (c) one can use the fact that they are homogeneous varieties and that there are few orbits (in case (a) and (b) the rank of the matrix determined by the point; in case (c) the number of roots of the associated cubic form in two variables).

A different direct geometrical proof consists in analyzing the most singular hyperplane sections of the variety considered inside the dual space and describing a variety isomorphic to the original X. Then the secant (or tangent) variety to this variety is exactly X^*.

For example, in case (b) they correspond to quadratic embeddings of conics and so there are only three types and conics of rank 1 naturally correspond to hyperplane sections consisting of a double conic; in case (c) the singular plane sections of a twisted cubic are only of two types, one of them consisting of a length three scheme and reconstructing C inside C^*. In this case we can associate to each point $p \in C$ the unique plane $H \subset \mathbb{P}^3$ such that $H \cap C = 3p$. The curve $\overline{C}$, isomorphic to C, and parametrizing the osculating planes, is such that $T\overline{C} = C^*$.

Another approach in case (b) and (c) could be via the Gauss map of the dual variety.

1.5.22 For (a) and (b) one can consult for example [37, Proposition 1.1]. For (c) prove that $p^\perp \cap X^*$ is a quartic plane curve consisting of the conic X_L^* and the line $(T_p C)^\perp$ with multiplicity two.

1.5.23 It is clear that $\pi(S(X,Y)) \subseteq S(X',Y')$. Let $x' \in X', y' \in Y'$ be general points. Then there are $x \in X, y \in Y$ such that $\pi(x) = x', \pi(y) = y'$, yielding $\pi(< x,y >) = < x', y' >$, and hence $S(X',Y') \subseteq \pi(S(X,Y))$. The last part of the statement follows by an induction on k.

1.5.24 Let us remark that by assumption and by the generality of z and of x, we can suppose that $y \notin T_x X$ and that $x \notin T_y Y$.

Take $S(z, \Sigma_z(X)) = S(z, \Sigma_z(Y))$. Then $\dim(S(z, \Sigma_z(X))) = \dim(\Sigma_z(X)) + 1$. If $u \in < z, x > = < z, y >$ is a general point, then

$$T_u S(z, \Sigma_z(X)) = < z, T_x \Sigma_z(X) > = \mathbb{P}^{\dim(S(z, \Sigma_z(X)))}$$

because $z \notin T_x X$. In particular, u is a smooth point of $S(z, \Sigma_z(X))$. By Terracini's Lemma, we get $T_u S(z, \Sigma_z(X)) \supseteq < z, T_y \Sigma_z(Y) >$, which together with $z \notin T_y Y$ yields $\dim(T_y \Sigma_z(Y)) = \dim(\Sigma_z(Y))$ so that $y \in \Sigma_z(Y)$ is a smooth point. Moreover,

$$T_x \Sigma_z(X) \subseteq T_u S(z, \Sigma_z(Y)) = < z, T_y \Sigma_z(Y) > = < x, T_y \Sigma_z(Y) > \subseteq < x, T_y Y > .$$

Since $T_x \Sigma_z(X) \subseteq T_x X$, to conclude it is enough to observe that

$$\dim(T_x X \cap < x, T_y Y >) = \dim(X) + \dim(Y) + 1 - \dim(< T_x X, T_y X >)$$
$$= \dim(\Sigma_z(X)) = \dim(T_x \Sigma_z(X)).$$

The other claims follow from the symmetry between x and y or are straightforward.

Chapter 2
The Hilbert Scheme of Lines Contained in a Variety and Passing Through a General Point

We recall some definitions and results concerning the parameter spaces of rational curves on smooth algebraic varieties. These notions will be immediately applied to some explicit problems in the next sections with special emphasis on the study of lines contained in a projective variety and passing through a (general) point.

2.1 Basics of Deformation Theory of (Smooth) Rational Curves on Smooth Projective Varieties

The standard reference for most of the technical results presented in this section is the book [116]. A more geometrical and slightly simpler introduction to the subject (but containing almost all the results needed here) is presented in [40] (see also [41]). Other recent and very interesting sources for the theory of Hilbert Schemes and for Deformation Theory of more general varieties are the books [175] and [89].

We begin with the basic definitions.

Definition 2.1.1 (Hilbert Scheme of Subschemes with a Fixed Hilbert Polynomial) Let $P = P(x) = \sum_{i=0}^{d} a_i \binom{x+d}{i} \in \mathbb{Q}[x]$ be a numerical polynomial of degree $d \geq 0$. By definition $a_i \in \mathbb{Z}$ for every $i = 0, \ldots d$ and $a_d \neq 0$.

Let T be a fixed scheme, which in most of the applications will be $\mathrm{Spec}(K)$ with K an algebraically closed field. Let X be a scheme projective over T and let $\mathscr{O}(1)$ be a ϕ-ample line bundle, where $\phi : X \to T$ is the structural morphism.

For every $t \in T$ we shall indicate by $P_{X_t}(x)$, or simply by P_{X_t}, the Hilbert polynomial of X_t relative to $\mathscr{O}(1)_{|X_t}$. Let

$$Hilb_T^P(X) : \{T - Schemes\}^o \to Sets$$

© Springer International Publishing Switzerland 2016
F. Russo, *On the Geometry of Some Special Projective Varieties*,
Lecture Notes of the Unione Matematica Italiana 18,
DOI 10.1007/978-3-319-26765-4_2

be the functor defined for every T–scheme S by

$$Hilb_T^P(X)(S) = \{\chi \subset S \times_T X, \text{proper and flat over } S \text{ such that } P_{\chi_s} = P \; \forall s \in S\}.$$

When $T = \text{Spec}(K)$ we set $Hilb^P(X) = Hilb_{\text{Spec}(K)}^P(X)$.

If the functor $Hilb_T^P(X)$ is representable, we indicate by $\text{Hilb}_T^P(X)$ the T-scheme representing it which will be called the *Hilbert scheme of closed T-subvarieties of X with Hilbert polynomial $P = P(t)$*.

If $\text{Hilb}_T^P(X)$ exists, there is a *universal family* $\chi^P \subset X \times_T \text{Hilb}_T^P(X)$ proper and flat over $\text{Hilb}_T^P(X)$ such that for every T-scheme S and for every $\chi \in Hilb_T^P(X)(S)$ there exists a unique T-morphism $f : S \to \text{Hilb}^P(X)$ such that

$$\chi = S \times_{\text{Hilb}^P(X)} \chi^P \subset S \times_T X,$$

via the base change induced by f.

Finally, we can define *the Hilbert scheme of X* as

$$\text{Hilb}_T(X) = \bigsqcup_P \text{Hilb}_T^P(X).$$

For every $W \in Hilb_T^P(X)(T)$, we have a unique morphism $f : T \to \text{Hilb}^P(X)$ such that

$$W = T \times_{\text{Hilb}^P(X)} \chi^P \subset T \times_T X \simeq X,$$

via the base change induced by f. If $T = \text{Spec}(K)$, let

$$[W] = f(\text{Spec}(K)) \in \text{Hilb}^P(X)(K).$$

In this case, the above tautological nonsense means precisely that the restrictions of the obvious projections to the factors define a tautological diagram

$$
\begin{array}{ccc}
 & \chi^P & \\
\pi \downarrow & & \searrow \phi \\
\text{Hilb}^P(X) & & X,
\end{array}
\tag{2.1}
$$

such that $\phi(\pi^{-1}([W])) = W$ as schemes.

The first important and rather difficult result in the theory of Hilbert schemes is the following Existence Theorem, see for example [116, I.1.4] or [175, Theorem 4.3.4].

Theorem 2.1.2 (Existence of Hilbert Schemes) *For every scheme X projective over T and for every numerical polynomial $P(t)$, the Hilbert scheme $\mathrm{Hilb}_T^P(X)$ exists and is projective over T. In particular, $\mathrm{Hilb}^P(X)$ is projective over K for every scheme X projective over K, so that under these hypotheses it has a finite number of irreducible components.*

It is well known that in general $\mathrm{Hilb}^P(X)$ is neither irreducible nor reduced so that it is important to analyze its scheme structure, locally and globally.

Let $g : \mathrm{Spec}(K[\epsilon]/(\epsilon^2)) \to \mathrm{Hilb}^P(X)$ such that $g(\mathrm{Spec}(K)) = [W]$, where K is the residue field at the unique closed point of $\mathrm{Spec}(K[\epsilon]/(\epsilon^2))$. Then it is well known that such g's correspond to tangent vectors to $\mathrm{Hilb}^P(X)$ at $[W]$, i.e.

$$t_{[W]} \mathrm{Hilb}^P(X) = Hilb^P(X)(\mathrm{Spec}(K[\epsilon]/(\epsilon^2))).$$

Let $\mathscr{I}_W/\mathscr{I}_W^2$ be the *conormal bundle of W in X* and let $N_{W/X} = (\mathscr{I}_W/\mathscr{I}_W^2)^*$ be *the normal bundle of W in X.* A fundamental result in the infinitesimal study of Hilbert schemes is the following, see, for example, [175, Theorem 4.3.5].

Theorem 2.1.3 (Infinitesimal Properties of Hilbert Schemes) *Let X be a scheme projective over K. Then:*

1.

$$t_{[W]} \mathrm{Hilb}^P(X) = H^0(N_{W/X}).$$

In particular, $\dim_{[W]}(\mathrm{Hilb}^P(X)) \le h^0(N_{W/X})$.

2.

$$\dim_{[W]}(\mathrm{Hilb}^P(X)) \ge h^0(N_{W/X}) - h^1(N_{W/X}).$$

3. If $W \subset X$ is a regular embedding and if $h^1(N_{W/X}) = 0$, then

$$\dim_{[W]}(\mathrm{Hilb}^P(X)) = h^0(N_{W/X})$$

and $\mathrm{Hilb}^P(X)$ is smooth at $[W]$. In particular, there exists a unique irreducible component of $\mathrm{Hilb}^P(X)$ containing $[W]$.

Definition 2.1.4 (Hilbert Scheme of Closed Subschemes Intersecting a Fixed Subscheme) For a closed subscheme $Z \subset X$ of a scheme X projective over K, we can define the functor

$$Hilb^{P,Z}(X) : \{K - Schemes\}^o \to Sets$$

which to every K-scheme S associates the set

$$Hilb^{P,Z}(X)(S) = \{\chi \subset S \times_K X\}$$

of flat families of closed subschemes of X parametrized by S, having Hilbert polynomial $P = P(t)$ and such that $Z \cap \chi_s \neq \emptyset$ for every $s \in S(K)$.

The functor $Hilb^{P,Z}(X)$ is representable by a closed subscheme

$$\mathrm{Hilb}^{P,Z}(X) \subseteq \mathrm{Hilb}^P(X),$$

called the *Hilbert scheme of subvarieties of X with Hilbert polynomial $P = P(t)$ and intersecting Z.*

If Z consists of a finite number of points (this will always be the case in which we shall use this scheme), then we shall call it the *Hilbert scheme of subvarieties of X with Hilbert polynomial $P = P(t)$ and passing through Z.*

For $Z = x \in X$ a point we set $\mathrm{Hilb}^P_x(X) = \mathrm{Hilb}^{P,x}(X)$, respectively $\mathrm{Hilb}_x(X) = \mathrm{Hilb}^x(X)$, being sure that no confusion will arise from this abuse of notation and that everyone will agree that it is not the case that $T = x$!

As in the general case, there exists a *universal family* $\chi^P_Z \subset X \times \mathrm{Hilb}^{P,Z}(X)$, flat over $\mathrm{Hilb}^{P,Z}(X)$, of closed subschemes of X such that for every scheme S and for every $\chi \in Hilb^{P,Z}(X)(S)$ there exists a unique morphism $f : S \to \mathrm{Hilb}^{P,Z}(X)$ such that this base change induces the equality

$$\chi = S \times_{\mathrm{Hilb}^{P,Z}(X)} \chi^P_Z \subset S \times X.$$

For $\mathrm{Hilb}^{P,Z}(X)$ there are results analogous to those of Theorem 2.1.2 and of Theorem 2.1.3, which will not be recalled here since we shall describe them for rational curves on a smooth variety from a different point of view we now introduce.

Definition 2.1.5 (Hilbert Scheme of Morphisms) Let Y be a projective scheme over the fixed algebraically closed field K and let X be a scheme quasi-projective over K. Fix an ample line bundle $H = \mathcal{O}(1)$ on X, a numerical polynomial $P(t) \in \mathbb{Q}[t]$ and, for every morphism $f : S \times_K Y \to S \times_K X$, let $f_s : Y \to X$ be the morphism induced via base extension by a point $s \in S(K)$. Let

$$Hom^P(Y, X) : \{K - Schemes\}^o \to Sets$$

be the contravariant functor defined for every K-scheme S and for every $s \in S(K)$ by

$$Hom^P(Y, X)(S) = \{f : S \times_K Y \to S \times_K X : \chi(f_s^*(\mathcal{O}(m))) = P(m) \; \forall m \gg 0\}.$$

Then $Hom^P(Y, X)$ is called the *functor of morphisms from Y to X with Hilbert polynomial P.*

If the functor $Hom^P(Y, X)$ is representable, we shall indicate by $\mathrm{Hom}^P(Y, X)$ the scheme representing it and we shall call it the *scheme of morphisms from Y to X with Hilbert polynomial $P = P(t)$.*

If $\text{Hom}^P(Y,X)$ exists, there is a *universal morphism*

$$\text{ev} : Y \times \text{Hom}^P(Y,X) \to X$$

such that for every K-scheme S and for every $g : Y \times_K S \to X \times_K S \in Hom^P(Y,X)(S)$ there exists a unique K-morphism $\tilde{g} : S \to \text{Hom}^P(Y,X)$ such that

$$g = \text{ev} \circ (\mathbb{I}_Y \times \tilde{g}).$$

Then we define the *scheme of morphisms from Y to X*:

$$\text{Hom}(Y,X) = \bigsqcup_P \text{Hom}^P(Y,X).$$

For every morphism $f : Y \to X \in \text{Hom}^P(Y,X)(K)$ having Hilbert polynomial P, we have a unique morphism $\tilde{f} : \text{Spec}(K) \to \text{Hom}^P(Y,X)$ such that $f = \text{ev} \circ (\mathbb{I}_Y \times \tilde{f})$. Let $[f] = \tilde{f}(\text{Spec}(K)) \in \text{Hom}^P(Y,X)(K)$. The above tautological nonsense means precisely that the restriction of ev to $Y \times_K [f]$ is exactly $f : Y \to X$, i.e. that for every $y \in Y$ we have

$$\text{ev}(y,[f]) = f(y).$$

The Existence of Hilbert Schemes implies the existence of $\text{Hom}(Y,X)$ under the above hypotheses, see especially [81] and also [116, I.1.10], [175, Sect. 4.6.6].

Theorem 2.1.6 (Existence of Hilbert Scheme of Morphisms) *For every scheme Y projective over a field K, for every scheme X quasi-projective over K and for every numerical polynomial $P(t)$, the scheme $\text{Hom}^P(Y,X)$ exists and is quasi-projective over K.*

Proof (Sketch) Suppose first that X is projective over K. Let S be a K-scheme and let $f : Y \times_K S \to X \times_K S \in Hom(Y,X)(S)$. If we consider f as an S-morphism, $Y \times_K S$ and $X \times_K S$ as S-schemes, we can define

$$\Gamma = (\mathbb{I}_Y, f) : Y \times_K S \to (Y \times_K S) \times_S (X \times_K S) \simeq Y \times_K X \times_K S.$$

Then Γ is a closed immersion so that $\Gamma(Y \times_K S) \simeq Y \times_K S$ is proper and flat over S and $\Gamma(Y \times_K S) \in \text{Hilb}(Y \times_K X)(S)$. This gives a morphism

$$\Gamma : Hom(Y,X) \to Hilb(Y \times_k X),$$

and hence a morphism $\Gamma : \text{Hom}(Y,X) \to \text{Hilb}(Y \times_K X)$, which is an open immersion of schemes. From the projectivity over K of every $\text{Hilb}_X^P(Y \times_K X)$, see Theorem 2.1.2, observing that $Y \times_K X$ is projective over K, one deduces that $\text{Hom}^P(Y,X)$ is quasi-projective over K for every numerical polynomial $P(t)$.

If X is quasi-projective over K, then there exists an open immersion i : $X \to \overline{X}$ with $\overline{X}$ projective over K. For every K-scheme S there is a morphism $Hom(Y, X)(S) \to Hom(Y, \overline{X})$ defined by sending $f : Y \times_K S \to X \times_K S$ to $(i, \mathbb{I}_S) \circ f : Y \times_K S \to \overline{X} \times_K S$. One then deduces the existence of a morphism $j : \mathrm{Hom}^P(Y, X) \to Hom(Y, \overline{X})$ and proves that j is an open immersion. Finally, one proves the quasi-projectivity over K of $\mathrm{Hom}^P(Y, X)$. For more details one can consult [81].   □

Let $g : \mathrm{Spec}(K[\epsilon]/(\epsilon^2)) \to \mathrm{Hom}^P(Y, X)$ such that $g(\mathrm{Spec}(K)) = [f]$, where K is the residue field at the unique closed point of $\mathrm{Spec}(K[\epsilon]/(\epsilon^2))$. Then such g's correspond to tangent vectors to $\mathrm{Hom}^P(Y, X)$ at $[f]$, i.e.

$$t_{[f]} \mathrm{Hom}^P(Y, X) = Hom^P(Y, X)(\mathrm{Spec}(K[\epsilon]/(\epsilon^2))).$$

Let $\Omega_{X/k}$ be the *sheaf of Kähler differentials of X*. We have the following infinitesimal results for $\mathrm{Hom}(Y, X)$, see [40, Sect. 2.2] and [116, I.2.16].

Theorem 2.1.7 (Infinitesimal Properties of the Hilbert Scheme of Morphisms) *Let the notation be as above. Then:*

1.

$$t_{[f]} \mathrm{Hom}(Y, X) = H^0(\mathscr{H}om(f^*(\Omega_{X/K}), \mathscr{O}_Y)).$$

In particular, if X is smooth along $f(Y)$, then

$$\dim_{[f]}(\mathrm{Hom}(Y, X)) \leq h^0(f^*(T_X)),$$

where $T_X = \mathscr{H}om(\Omega_{X/K}, \mathscr{O}_X)$ is the tangent sheaf of X.
2. If X is projective and non-singular along $f(Y)$, then

$$\dim_{[f]}(\mathrm{Hom}(Y, X)) \geq h^0(f^*(T_X)) - h^1(f^*(T_X)).$$

In particular, if $h^1(f^*(T_X)) = 0$, then $\dim_{[f]}(\mathrm{Hom}(Y, X)) = h^0(f^*(T_X))$, $\mathrm{Hom}(Y, X)$ is smooth at $[f]$ and there exists a unique irreducible component of $\mathrm{Hom}(Y, X)$ containing $[f]$.

Suppose that $C \subset X$ is a smooth rational curve and that $X \subset \mathbb{P}^N$ is a smooth projective variety. Then

$$N_{C/X} \simeq \bigoplus_{i=1}^{n-1} \mathscr{O}_{\mathbb{P}^1}(a_i)$$

since $N_{C/X}$ is a locally free sheaf of rank $n - 1 = \mathrm{codim}(C, X)$ on $C \simeq \mathbb{P}^1$. If $f : C \to X$ is the embedding of C into X, then

$$f^*(T_X) = T_{X|C} = \bigoplus_{i=1}^{n} \mathcal{O}_{\mathbb{P}^1}(b_i)$$

and everything is easily computable. Moreover, we have the exact sequence

$$0 \to \mathcal{O}_{\mathbb{P}^1}(2) \simeq T_{\mathbb{P}^1} \to T_{X|C} \to N_{C/X} \to 0,$$

which is very useful for calculating the dimensions appearing in Theorem 2.1.3 and in Theorem 2.1.7.

By Theorem 2.1.7 the exact sequence at the level of H^0 can be interpreted as saying that the tangent space to $\mathrm{Hilb}(X)$ at $[C]$ is given by factoring out the automorphisms of $C \simeq \mathbb{P}^1$ from the tangent space to $\mathrm{Hom}(\mathbb{P}^1, X)$ at $f : C \to X$. Indeed, $\mathrm{Hom}(\mathbb{P}^1, \mathbb{P}^1)$ contains the group subscheme of automorphism of $\mathbb{P}^1$, whose tangent space at the identity is exactly $H^0(T_{\mathbb{P}^1})$.

Definition 2.1.8 Let Y be a projective scheme over the fixed algebraically closed field K, let X be a scheme quasi-projective over K, let $B \subset Y$ be a closed subscheme and let $g : B \to X$ be a fixed morphism, which for every K-scheme S induces a morphism $g \times \mathbb{I}_S : B \times_K S \to X \times_K S$. Let $Hom(Y, X; g)(S)$ be the set of morphisms $Y \times_K S \to X \times_K S$ whose restriction to $B \times_K S \subset Y \times_K S$ coincides with $g \times \mathbb{I}_S$. This defines a functor

$$Hom(Y, X; g) : \{K - schemes\}^0 \to Sets,$$

which is a subfunctor of $Hom(Y, X)$. The natural restriction morphism

$$\rho(S) : Hom(Y, X)(S) \to Hom(B, X)$$

induces a morphism

$$\rho : Hom(Y, X) \to Hom(B, X)$$

such that $\rho^{-1}([g])$ is exactly $Hom(Y, X; g)$. From this follows the representability of $Hom(Y, X; g)$ by a projective scheme over K which will be indicated by $Hom(Y, X; g)$ and which is naturally a closed subscheme of $Hom(Y, X)$.

Fix an ample line bundle $H = \mathcal{O}(1)$ on X, a numerical polynomial $P(t) \in \mathbb{Q}[t]$ and let

$$Hom^P(Y, X; g) : \{K - Schemes\}^o \to Sets$$

be the functor defined for every K-scheme S by

$$Hom^P(Y, X; g)(S) = \{f \in Hom(Y, X; g)(S) \; : \; \chi(f^*(\mathcal{O}(m))) = P(m) \text{ for } m >> 0\}.$$

The functor $Hom^P(Y, X; g)$ is representable and we indicate by $\mathrm{Hom}^P(Y, X; g)$ the scheme representing it, which is called the *scheme of morphisms from Y to X with Hilbert polynomial $P = P(t)$ whose restriction to B is g.*

There is a *universal morphism*

$$\mathrm{ev} : Y \times \mathrm{Hom}(Y, X; g) \to X$$

such that for every scheme S and for every $\phi : Y \times_K S \to X \times_K S \in Hom(Y, X; g)(S)$ there exists a unique morphism $\tilde{\phi} : S \to \mathrm{Hom}(Y, X; g)$ such that

$$\phi = \mathrm{ev} \circ (\mathbb{I}_Y \times \tilde{\phi}).$$

Then, as always,

$$\mathrm{Hom}(Y, X; g) = \bigsqcup_P \mathrm{Hom}^P(Y, X; g).$$

Suppose for simplicity that X is smooth along $f(Y)$, where $f : Y \to X$ is a morphism which restricts to g on B. The restriction $\rho : \mathrm{Hom}(Y, X) \to \mathrm{Hom}(B, X)$ naturally induces a morphism of K-vector spaces

$$\rho : H^0(f^*(T_X)) \to H^0(g^*(T_X)),$$

so that

$$t_{[f]} \mathrm{Hom}(Y, X; g) = \ker(\rho : H^0(f^*(T_X)) \to H^0(g^*(T_X))) = H^0(f^*(T_X) \otimes \mathcal{I}_B),$$
(2.2)

where $\mathcal{I}_B$ is the ideal sheaf of B in Y.

Now we can finally state the following infinitesimal results for $\mathrm{Hom}(Y, X; g)$, see [40, Sect. 2.3].

Theorem 2.1.9 *Let Y and X be schemes projective over K, let B be a closed subscheme of Y and let $f : Y \to X$ be a morphism whose restriction to B is $g : B \to X$. Suppose moreover that X is smooth along $f(Y)$. Then:*

1.

$$t_{[f]} \mathrm{Hom}(Y, X; g) = H^0(f^*(T_X) \otimes \mathcal{I}_B),$$

yielding $\dim(\mathrm{Hom}(Y, X; g) \le h^0(f^*(T_X) \otimes \mathcal{I}_B).$

2.

$$\dim_{[f]}(\mathrm{Hom}(Y, X; g)) \geq h^0(f^*(T_X) \otimes \mathscr{I}_B) - h^1(f^*(T_X) \otimes \mathscr{I}_B).$$

In particular, if $h^1(f^*(T_X) \otimes \mathscr{I}_B) = 0$, *then*

$$\dim_{[f]}(\mathrm{Hom}(Y, X; g)) = h^0(f^*(T_X) \otimes \mathscr{I}_B),$$

the scheme $\mathrm{Hom}(Y, X; g)$ *is smooth at* $[f]$ *and there exists a unique irreducible component of* $\mathrm{Hom}(Y, X; g)$ *containing* $[f]$.

Definition 2.1.10 Let $0 \in \mathbb{P}^1$, let $x \in X$ and let $g : 0 \to X$ be a morphism such that $g(0) = x$. In this case $\mathrm{Hom}(\mathbb{P}^1, X; g)$ will be indicated by $\mathrm{Hom}(\mathbb{P}^1, X; 0 \to x)$.

Let $X \subset \mathbb{P}^N$ be a fixed embedding. Then

$$\mathrm{Hom}_d(\mathbb{P}^1, X) = \mathrm{Hom}^{dt+1}(\mathbb{P}^1, X)$$

parametrizes morphisms $f : \mathbb{P}^1 \to X$ such that the cycle $f_*(\mathbb{P}^1)$ has degree d. Let us recall that, if $C = f(\mathbb{P}^1)$, then $\deg(f_*(\mathbb{P}^1)) = \deg(f) \cdot \deg(C)$. In the same way one can define $\mathrm{Hom}_d(\mathbb{P}^1, X; g) = \mathrm{Hom}^{dt+1}(\mathbb{P}^1, X; g)$.

Before passing to the applications of our interest we need an explicit description of the differential of $\mathrm{ev} : Y \times \mathrm{Hom}(Y, X; g) \to X$.

Theorem 2.1.11 ([116, Proposition II.3.10]) *Let* $B \subset \mathbb{P}^1$ *be a scheme of length* $|B| \leq 2$ *and let* $g : B \to X$ *be a morphism to a smooth irreducible projective variety* X *of dimension* $n = \dim(X)$. *Let*

$$\mathrm{ev} : \mathbb{P}^1 \times \mathrm{Hom}(\mathbb{P}^1, X; g) \to X$$

be the evaluation morphism and let $f : \mathbb{P}^1 \to X$ *such that* $[f] \in \mathrm{Hom}(\mathbb{P}^1, X; g)$.
If

$$f^*(T_X) \otimes \mathscr{I}_B = \bigoplus_{i=1}^{n} \mathscr{O}_{\mathbb{P}^1}(\alpha_i),$$

then for every $p \in \mathbb{P}^1 \setminus B$ *we have*

$$\mathrm{rk}(d(\mathrm{ev})(p, [f])) = \#\{i \mid \alpha_i \geq 0\}.$$

We now prove two fundamental tools for the classification of *LQEL*-manifolds developed later in the text.

Corollary 2.1.12 ([116, Corollary II.3.11, Theorem II.3.12]) *Let the notation be as in Theorem 2.1.11, let $x \in X$ be a point, let $n = \dim(X)$ and suppose char(K)=0. Then*

i) *$(B = \emptyset)$ Suppose $V \subset \mathrm{Hom}(\mathbb{P}^1, X)$ is an irreducible component such that* ev : $\mathbb{P}^1 \times V \to X$ *is dominant. Then for $[f] \in V$ general $f^*(T_X)$ is generated by global sections.*

ii) *$(|B| = 1)$ Suppose $V \subset \mathrm{Hom}(\mathbb{P}^1, X; 0 \to x)$ is an irreducible component such that* ev : $\mathbb{P}^1 \times V \to X$ *is dominant. Then for $[f] \in V$ general $f^*(T_X)$ is ample.*

Proof Since ev : $\mathbb{P}^1 \times V \to X$ is dominant, there exists an open subset $U \subseteq X$ such that for every $(p, [f]) \in \mathrm{ev}^{-1}(U)$ we have

$$n = \mathrm{rk}(d(\mathrm{ev})(p, [f])) = \#\{i \mid \alpha_i \geq 0\},$$

where

$$f^*(T_X) = \bigoplus_{i=1}^{n} \mathcal{O}_{\mathbb{P}^1}(\alpha_i).$$

Thus if ev : $\mathbb{P}^1 \times V \to X$ is dominant, then $\alpha_i \geq 0$ for every $i = 1, \ldots, n$ and $f^*(T_X)$ is generated by global sections.

Reasoning as above and recalling that for $B = x$ we have $\mathcal{I}_B \simeq \mathcal{O}_{\mathbb{P}^1}(-1)$, letting

$$f^*(T_X) \otimes \mathcal{I}_B = \bigoplus_{i=1}^{n} \mathcal{O}_{\mathbb{P}^1}(\beta_i)$$

we deduce that $\beta_i \geq 0$ for every $i = 1, \ldots, n$ under the hypotheses of ii). Thus

$$f^*(T_X) = \bigoplus_{i=1}^{n} \mathcal{O}_{\mathbb{P}^1}(\alpha_i)$$

with $\alpha_i > 0$ for every $i = 1, \ldots, n$, as claimed.   □

The next result will allow us to count parameters without worrying about pathologies, which in this context are usually called *obstructions*.

Corollary 2.1.13 *Let the notation be as in Theorem 2.1.11 and assume char(K)=0. Then*

i) *if $X \subset \mathbb{P}^N$ and if through a general point of X there passes a rational curve of degree $d \geq 1$ in the given embedding, then there exists an open dense subset $U \subseteq X$ and an irreducible component $V \subseteq \mathrm{Hom}_d(\mathbb{P}^1, X)$ such that for every $[f] \in V$ with $f(\mathbb{P}^1) \cap U \neq \emptyset$ we have $f^*(T_X)$ generated by global sections.*

Let $C = f(\mathbb{P}^1) \subset X$ be such a curve and suppose that f is an isomorphism so that C is a smooth rational curve. Then

$$h^0(N_{C/X}) = -K_X \cdot C + n - 3,$$

the Hilbert scheme of rational curves of degree d is smooth at $[C]$ of dimension $-K_X \cdot C + n - 3$. There exists a unique irreducible component $\mathscr{C} \subseteq \mathrm{Hilb}^{dt+1}(X)$ containing $[C]$ and dominating X.

ii) Suppose $X \subset \mathbb{P}^N$ and let $x \in X$ be a point. If through x and a general point of X there passes a rational curve of degree $d \geq 1$ in the given embedding, then there exists an open dense subset $U \subseteq X$ and an irreducible component $V \subseteq \mathrm{Hom}_d(\mathbb{P}^1, X; 0 \to x)$ such that for every $[f] \in V$ with $f(\mathbb{P}^1) \cap U \neq \emptyset$ we have $f^*(T_X)$ ample.

Let $C = f(\mathbb{P}^1) \subset X$ be such a curve and suppose that f is an isomorphism so that C is a smooth rational curve. Then $h^0(N_{C/X}(-1)) = -K_X \cdot C - 2$, the Hilbert scheme of rational curves of degree d passing through x is smooth at $[C]$ of dimension $-K_X \cdot C - 2$ so that there exists a unique irreducible component $\mathscr{C}_x \subseteq \mathrm{Hilb}_x^{dt+1}(X)$ containing $[C]$ and dominating X.

Proof The hypotheses in i), respectively ii), imply that

$$\mathrm{ev} : \mathbb{P}^1 \times \mathrm{Hom}_d(\mathbb{P}^1, X) \to X$$

is dominant, respectively

$$\mathrm{ev} : \mathbb{P}^1 \times \mathrm{Hom}(\mathbb{P}^1, X; 0 \to x) \to X$$

is dominant. Thus there exists an irreducible component $V \subseteq \mathrm{Hom}(\mathbb{P}^1, X)$, respectively $V \subseteq \mathrm{Hom}(\mathbb{P}^1, X; 0 \to x)$, such that $\mathrm{ev} : \mathbb{P}^1 \times V \to X$ is dominant. Then we can take as $U \subseteq X$ the open non-empty subset constructed in the proof of the two parts of Corollary 2.1.12.

Let

$$N_{C/X} = \bigoplus_{j=1}^{n-1} \mathscr{O}_{\mathbb{P}^1}(\gamma_i).$$

From the exact sequence

$$0 \to \mathscr{O}_{\mathbb{P}^1}(2) \simeq T_{\mathbb{P}^1} \to T_{X|C} \to N_{C/X} \to 0,$$

we deduce that $N_{C/X}$ is generated by global sections, respectively ample, and that

$$\deg(N_{C/X}) = \deg(T_{X|C}) - 2 = -K_X \cdot C - 2.$$

By Riemann–Roch and by the fact that $\gamma_j \geq 0$ for every j, we deduce

$$h^0(N_{C/X}) = \sum_{j=1}^{n-1} h^0(\mathscr{O}_{\mathbb{P}^1}(\gamma_j)) = \sum_{j=1}^{n-1}(\gamma_j + 1) = -K_X \cdot C + n - 3,$$

as claimed in i).

In case ii), we have $\gamma_j > 0$ for every j so that

$$h^0(N_{C/X}(-1)) = \sum_{j=1}^{n-1} h^0(\mathscr{O}_{\mathbb{P}^1}(\gamma_j - 1)) = \sum_{j=1}^{n-1} \gamma_j = -K_X \cdot C - 2,$$

as claimed. The other conclusions follow from Theorem 2.1.3 since $h^1(N_{C/X}) = 0$, respectively $h^1(N_{C/X}(-1)) = 0$.   □

2.2 The Hilbert Scheme of Lines Contained in a Projective Variety and Passing Through a Point

We apply the tools presented in the previous section to the study of the Hilbert scheme of lines contained in a projective variety $X \subset \mathbb{P}^N$ and passing through a (general) point $x \in X$.

2.2.1 Notation, Definitions and Preliminary Results

Let $X \subset \mathbb{P}^N$ be a connected equidimensional projective variety of dimension $n \geq 1$, defined over a fixed algebraically closed field K of characteristic zero, which from now on will be simply called a *projective variety*. We shall always tacitly assume that $X \subset \mathbb{P}^N$ is also non-degenerate. If X is also smooth and irreducible, we shall call X a *manifold*.

Let $X_{\text{reg}} = X \setminus \text{Sing}(X)$ be the smooth locus of X. Recalling the definitions introduced in Sect. 1.1, $t_x X$ denotes the affine tangent space to X at x, $T_x X \subset \mathbb{P}^N$ is the projective tangent space to X at x and, for an arbitrary scheme Z and for a closed point $z \in Z$, $C_z Z$ is the affine tangent cone to Z at z.

Let $\mathscr{L}_{x,X} = \text{Hilb}_x^{t+1}(X)$ denote the Hilbert scheme of lines contained in X and passing through the point $x \in X$. For a line $L \subset X$ passing through x, we let $[L] \in \mathscr{L}_{x,X}$ be the corresponding point.

Let

$$\pi_x : \mathscr{H}_x \to \mathscr{L}_{x,X}$$

denote the universal family and let

$$\phi_x : \mathscr{H}_x \to X$$

be the tautological morphism. From now on we shall always suppose that $x \in X_{\text{reg}}$. Note that π_x admits a *tautological* section

$$s_x : \mathscr{L}_{x,X} \to \mathscr{E}_x \subset \mathscr{H}_x$$

sending $[L]$ to the point x on $\pi_x^{-1}([L])$. Thus $\mathscr{E}_x$ is contracted by ϕ_x to the point x, see also (2.3) below.

Consider the blowing-up

$$\sigma_x : \text{Bl}_x X \to X$$

of X at x. For every $[L] \in \mathscr{L}_{x,X}$ the line $L = \phi_x(\pi_x^{-1}([L]))$ is smooth at x so that [102, Lemma 4.3] and the universal property of the blowing-up ensure the existence of a morphism

$$\psi_x : \mathscr{H}_x \to \text{Bl}_x X$$

such that

$$\sigma_x \circ \psi_x = \phi_x.$$

Therefore we have the following commutative diagram

$$
\begin{array}{ccc}
\mathscr{H}_x & \xrightarrow{\psi_x} & \text{Bl}_x X \\
{\scriptstyle s_x}\Big\uparrow\Big\downarrow{\scriptstyle \pi_x} & {\scriptstyle \phi_x}\searrow & \Big\downarrow{\scriptstyle \sigma_x} \\
\mathscr{L}_{x,X} & & X.
\end{array}
\qquad (2.3)
$$

In particular, ψ_x maps the section $\mathscr{E}_x$ to E_x, the exceptional divisor of σ_x. Let

$$\tilde{\psi}_x : \mathscr{E}_x \to E_x$$

be the restriction of ψ_x to $\mathscr{E}_x$. We can define the morphism

$$\tau_x = \tau_{x,X} = \tilde{\psi}_x \circ s_x : \mathscr{L}_{x,X} \to \mathbb{P}((t_x X)^*) = E_x = \mathbb{P}^{n-1}, \qquad (2.4)$$

which associates to each line $[L] \in \mathscr{L}_{x,X}$ the corresponding tangent direction through x, that is

$$\tau_x([L]) = \mathbb{P}((t_xL)^*).$$

The morphism τ_x is clearly injective and we claim that τ_x is a closed immersion. Indeed, by taking in the previous construction $X = \mathbb{P}^N$ the corresponding morphism

$$\tau_{x,\mathbb{P}^N} : \mathscr{L}_{x,\mathbb{P}^N} \to \mathbb{P}((t_x\mathbb{P}^N)^*) = \mathbb{P}^{N-1}$$

is an isomorphism between $\mathscr{L}_{x,\mathbb{P}^N}$ and the exceptional divisor of $\mathrm{Bl}_x\,\mathbb{P}^N$. By definition the inclusion $X \subset \mathbb{P}^N$ induces a closed embedding

$$i_x : \mathscr{L}_{x,X} \to \mathscr{L}_{x,\mathbb{P}^N}.$$

If

$$j_x : \mathbb{P}((t_xX)^*) \to \mathbb{P}((t_x\mathbb{P}^N)^*)$$

is the natural closed embedding, then we have the following commutative diagram

$$
\begin{array}{ccc}
\mathscr{L}_{x,X} & \xrightarrow{\ \tau_{x,X}\ } & \mathbb{P}((t_xX)^*) \\
{\scriptstyle i_x}\downarrow & & \downarrow{\scriptstyle j_x} \\
\mathscr{L}_{x,\mathbb{P}^N} & \xrightarrow{\ \tau_{x,\mathbb{P}^N}\ } & \mathbb{P}((t_x\mathbb{P}^N)^*),
\end{array}
\tag{2.5}
$$

proving the claim.

For $x \in X_{\mathrm{reg}}$ such that $\mathscr{L}_{x,X} \neq \emptyset$, we shall always identify $\mathscr{L}_{x,X}$ with $\tau_x(\mathscr{L}_{x,X})$ and we shall naturally consider $\mathscr{L}_{x,X}$ as a subscheme of $\mathbb{P}^{n-1} = \mathbb{P}((t_xX)^*)$.

We shall denote by V_x the scheme-theoretic image of $\mathscr{H}_x$, that is

$$V_x := \phi_x(\mathscr{H}_x) \subset X.$$

Via (2.3) we deduce the following relation:

$$\mathbb{P}(C_x(V_x)) = \mathscr{L}_{x,X},
\tag{2.6}$$

as subschemes of $\mathbb{P}((t_xX)^*)$, where $\mathbb{P}(C_x(V_x))$ is the projectivized tangent cone to V_x at x, see Sect. 1.1.

2.2.2   Singularities of $\mathcal{L}_{x,X}$

We begin by studying the intrinsic geometry of $\mathcal{L}_{x,X} \subset \mathbb{P}^{n-1}$. When it is clear from the context which variety $X \subset \mathbb{P}^N$ we are considering we shall write $\mathcal{L}_x$ instead of $\mathcal{L}_{x,X}$.

The normal bundle $N_{L/X}$ is locally free, being a subsheaf of the locally free sheaf $N_{L/\mathbb{P}^N} \simeq \mathcal{O}_{\mathbb{P}^1}(1)^{N-1}$. If $L \cap X_{\mathrm{reg}} \neq \emptyset$, then $N_{L/X}$ is locally free of rank $n-1$ and more precisely

$$N_{L/X} \simeq \bigoplus_{i=1}^{n-1} \mathcal{O}_{\mathbb{P}^1}(a_i), \tag{2.7}$$

with $a_i \leq 1$ since $N_{L/X} \subset N_{L/\mathbb{P}^N}$.

If $N_{L/X}$ is also generated by global sections, then

$$N_{L/X} \simeq \mathcal{O}_{\mathbb{P}^1}(1)^{s(L,X)} \oplus \mathcal{O}_{\mathbb{P}^1}^{n-1-s(L,X)}. \tag{2.8}$$

Therefore if $L \subset X_{\mathrm{reg}}$ and if $N_{L/X}$ is generated by global sections, then Theorem 2.1.3 and/or Theorem 2.1.9 imply that $\mathcal{L}_x$ is unobstructed at $[L]$ since $h^1(N_{L/X}(-1)) = 0$. Thus $\mathcal{L}_x$ is smooth at $[L]$ and

$$\dim_{[L]}(\mathcal{L}_x) = h^0(N_{L/X}(-1)) = s(L, X),$$

where $s(L, X) \geq 0$ is the integer defined in (2.8).

For $x \in X_{\mathrm{reg}}$, let

$$S_x = S_{x,X} = \{[L] \in \mathcal{L}_x \text{ such that } L \cap \mathrm{Sing}(X) \neq \emptyset\} \subseteq \mathcal{L}_x.$$

Then $S_{x,X}$ has a natural scheme structure and the previous inclusion holds at the scheme-theoretic level. If X is smooth, then $S_{x,X} = \emptyset$. Moreover, if $L \subset X$ is a line passing through $x \in X_{\mathrm{reg}}$, clearly $[L] \notin S_{x,X}$ if and only if $L \subset X_{\mathrm{reg}}$.

We now prove that a singular point of $\mathcal{L}_x$ produces a line passing through x and through a singular point of X, a stronger condition than the mere existence of a singular point on X. The next results are probably well known to the experts in the field, at least for manifolds, see [93, Proposition 1.5] and also [160, Proposition 2.2]. In [47], the singularities of the Hilbert scheme of lines contained in a projective variety have been related to some geometrical properties of the variety. The next result is the first instance of the principle discussed in the Introduction showing that the property of smoothness passes from X to $\mathcal{L}_x$ for $x \in X$ general.

Proposition 2.2.1 ([161, Proposition 2.1]) *Let the notation be as above, let $X \subset \mathbb{P}^N$ be an irreducible projective variety of dimension $n \geq 2$ and let $x \in X_{\mathrm{reg}}$ be such* *that $\mathscr{L}_x \neq \emptyset$. If $x \in X_{reg}$ is a general point, then:*

1. *$\mathscr{L}_x \subset \mathbb{P}^{n-1}$ is smooth outside S_x, that is $\mathrm{Sing}(\mathscr{L}_x) \subseteq S_x$. In particular, if $X \subset \mathbb{P}^N$ is smooth and if $x \in X$ is general, then $\mathscr{L}_x \subset \mathbb{P}^{n-1}$ is a smooth variety.*
2. *If $\mathscr{L}_x^j$, $j = 1, \ldots, m$, are the irreducible components of $\mathscr{L}_x$ and if*

$$\dim(\mathscr{L}_x^l) + \dim(\mathscr{L}_x^p) \geq n - 1 \quad \text{for some } l \neq p,$$

then $\mathscr{L}_x$ is singular, X is singular and there exists a line $[L] \in \mathscr{L}_x$ such that $L \cap \mathrm{Sing}(X) \neq \emptyset$.

Proof There exists an open dense subset $U \subseteq X_{reg}$ such that for every line $L \subset X_{reg}$ such that $L \cap U \neq \emptyset$ the normal bundle $N_{L/X}$ is generated by global sections, see Corollary 2.1.13. Combining this result with the above discussion, we deduce that for every $x \in U$ the variety $\mathscr{L}_x \subset \mathbb{P}^{n-1}$ is smooth outside S_x, proving the first assertion.

The condition on the dimensions of two irreducible components of $\mathscr{L}_x$ in 2) implies that these components have to intersect in $\mathbb{P}^{n-1}$. A point of intersection is a singular point of $\mathscr{L}_x \subset \mathbb{P}^{n-1}$. This forces X to be singular by the first part and also the existence of a line $[L] \in S_x$, which by definition cuts $\mathrm{Sing}(X)$.   □

The next example shows that there exist cases in which $\mathrm{Sing}(\mathscr{L}_x) \subsetneq S_x$, a quite interesting phenomenon which, to the best of our knowledge, has not been explicitly pointed out before.

Example 2.2.2 (**$\mathrm{Sing}(\mathscr{L}_x) \subsetneq S_x$ for $x \in X$ general**) Let $X = V(f) \subset \mathbb{P}^4$ be a cubic hypersurface which is the projection of $Y = \mathbb{P}^1 \times \mathbb{P}^2 \subset \mathbb{P}^5$ from a point $p \in \mathbb{P}^5 \setminus Y$. By the homogeneity of the orbit $\mathbb{P}^5 \setminus Y$ the choice of the point p is irrelevant for the geometrical properties of X.

The locus of secant lines to Y passing through p, that is the entry locus of Y with respect to p, is a quadric $Q_p = \mathbb{P}^1 \times \mathbb{P}^1_p \subset \mathbb{P}^1 \times \mathbb{P}^2$ obtained in this way: Y has homogeneous ideal generated by three quadratic polynomials; through p there passes a pencil of quadric hypersurfaces vanishing on Y; the intersection of two distinct quadrics in this pencil is scheme-theoretically $Y \cup \mathbb{P}^3_p$ with $p \in \mathbb{P}^3_p$. Finally, $Q_p = \mathbb{P}^3_p \cap Y$, see also Exercise 1.5.14.

The projection from p induces a birational morphism $\pi_p : Y \to X$, which is an isomorphism outside $Y \setminus Q_p$ and which maps Q_p to a plane $\Pi = \pi_p(Q_p) \subset X$. Clearly Π is the singular set of X and it is also the locus of non-normal points of X.

Let us define the scheme $\mathrm{Sing}(X)$ as the base locus scheme of the Gauss map of X, that is the subscheme of $\mathbb{P}^4$ (and of X) defined by the partial derivatives of f. The support of $\mathrm{Sing}(X)$ is Π but the scheme $\mathrm{Sing}(X)$ has a conic $C_p \subset \Pi$ of *embedded points*, where $C_p \subset \Pi$ corresponds to the locus of tangent lines to Q_p (and/or to Y) passing through p. Outside C_p the scheme $\mathrm{Sing}(X)$ is reduced and its structure coincides with that of Π (see more details below).

Let $y \in Y$ be a point and let $\pi_1 : Y \to \mathbb{P}^1$ be the projection onto the first factor and let $\pi_2 : Y \to \mathbb{P}^2$ be the projection onto the second factor. Then $\mathscr{L}_{y,Y}$ is the natural projectivization of $(\pi_1(y) \times \mathbb{P}^2) \cup (\mathbb{P}^1 \times \pi_2(y))$. If $\mathscr{L}_1$ and $\mathscr{L}_2$ are the corresponding irreducible components, we have

$$\mathscr{L}_{y,Y} = \mathscr{L}_1 \sqcup \mathscr{L}_2.$$

If $[L] \in \mathscr{L}_1$, then $N_{L/Y} \simeq \mathscr{O}_L \oplus \mathscr{O}_L(1)$, while for $[L] = \mathscr{L}_2$, we have $N_{L/Y} \simeq \mathscr{O}_L \oplus \mathscr{O}_L$. These are the translations into the language of deformation theory of the properties $\dim(\mathscr{L}_1) = 1$ and $\dim(\mathscr{L}_2) = 0$, see also the proof of Proposition 2.2.1.

By the previous description of $Q_p = \mathbb{P}^1 \times \mathbb{P}^1_p$ we deduce that for $y \in Y \setminus Q_p$ we have that $\pi_p(L) \cap \mathrm{Sing}(X) \neq \emptyset$ if $[L] \in \mathscr{L}_1$ and $\pi_p(\mathbb{P}^1 \times \pi_2(y)) \cap \mathrm{Sing}(X) = \emptyset$.

Let $x \in X \setminus \Pi$ with $x = \pi_p(y)$. The variety Y is self-dual, see Exercise 1.5.21, thus, by Exercise 1.5.22, the dual of X is a general hyperplane section of Y corresponding to the hyperplane $p^\perp$, see Exercise 1.5.22. Therefore the Gauss image of X is a smooth surface $X^* \subset \mathbb{P}^4$ of degree 3, which is a rational normal scroll. The fiber of the Gauss map $\mathscr{G}_X$ of X passing through x is a line L_x. Since L_x is contracted by $\mathscr{G}_X$ the point of intersection $L_x \cap \Pi$ belongs to C_p. Otherwise $\mathrm{length}(L_x \cap \mathrm{Sing}(X)) = 1$ and $\mathscr{G}_X(L_x)$ would be a line since $\mathscr{G}_X$ is defined by quadratic equations.

We now calculate the normal bundle to each line L passing through a general point $x \in X$. For each line $L \subset X$ let us define the scheme $\Delta(L) = L \cap \mathrm{Sing}(X)$ (scheme-theoretic intersection). Thus $\Delta(L) = \emptyset$ if and only if $L = \pi_p(\mathbb{P}^1 \times \pi_2(y))$. Moreover, $\delta(L) = \mathrm{length}(\Delta(L)) \leq 2$ since $\mathrm{Sing}(X)$ is defined by quadratic equations and $\delta(L) = 2$ if and only if $L = L_x$.

We have the following exact sequence of locally free sheaves for each line $L \subset X$, not contained in $\mathrm{Sing}(X)$:

$$0 \to N_{L/X} \to N_{L/\mathbb{P}^4} \to N_{X/\mathbb{P}^4|L} \otimes \mathscr{O}_L(-\Delta(L)) \to 0, \qquad (2.9)$$

that is

$$0 \to \mathscr{O}_L(a_1) \oplus \mathscr{O}_L(a_2) \to \mathscr{O}_L(1) \oplus \mathscr{O}_L(1) \oplus \mathscr{O}_L(1) \to \mathscr{O}_L(3 - \delta(L)) \to 0.$$

Thus $a_1 + a_2 = \delta(L)$. Let us recall that $a_i \leq 1$ for $i = 1, 2$.

Let us consider the lines $L \subset X$ passing through (a general) $x \in X$, where as before $x = \pi_p(y)$. Then for $L = \pi_p(\mathbb{P}^1 \times \pi_2(y))$ we have $N_{L/X} \simeq \mathscr{O}_L \oplus \mathscr{O}_L$. Indeed, since $L \cap \mathrm{Sing}(X) = \emptyset$ by the previous description we deduce $a_i \geq 0$ and from $a_1 + a_2 = 0$ we get the previous expression. As expected, this line represents an isolated smooth point of $\mathscr{L}_{x,X}$.

Let us consider lines L coming from projections of lines in $\mathscr{L}_1$. If $L \neq L_x$, then $N_{L/X} \simeq \mathscr{O}_L \oplus \mathscr{O}_L(1)$ while $N_{L_x/X} \simeq \mathscr{O}_L(1) \oplus \mathscr{O}_L(1)$. Thus every line $L \neq L_x$ is a smooth point of the one-dimensional irreducible component $\mathscr{L}^1_{x,X}$ of $\mathscr{L}_{x,X}$ while $[L_x] \in \mathscr{L}^1_{x,X}$ is a singular point since $H^0(N_{L_x/X}(-1))$ has dimension 2.

In this example $S_{x,X} = \mathscr{L}_{x,X}^1$ and for every $[L] \in S_{x,X} \setminus [L_x]$ we have smooth points of $\mathscr{L}_{x,X}$ showing that Proposition 2.2.1 is not optimal since the inclusion $\mathrm{Sing}(\mathscr{L}_{x,X}) \subseteq S_{x,X}$ can be strict.

An explicit example of a cubic hypersurface as above is the following

$$X = V(x_0 x_3^2 + x_1 x_3 x_4 + x_2 x_4^2) \subset \mathbb{P}^4.$$

Then $\mathrm{Sing}(X)_{\mathrm{red}} = V(x_3, x_4)$ while $C_p = V(x_0 x_2 - x_1^2, x_3, x_4)$. Let us take $x = (0 : 0 : 0 : 1 : 0)$. The lines in $\mathbb{P}^4$ through x are naturally parametrized by $\mathbb{P}^3 = V(x_3) \subset \mathbb{P}^4$. The local expression of f (in the complement of $V(x_3)$) centered at x is $x_0 + x_1 x_4 + x_2 x_4^2$. The tangent space to X at x is $V(x_0)$ thus $\mathscr{L}_{x,X}$ has natural equations $V(x_3, x_0, x_1 x_4, x_2 x_4^2)$. The irreducible one-dimensional component $\mathscr{L}_{x,X}^1$ is the line $M = V(x_0, x_3, x_4)$ while the singular point of $\mathscr{L}_{x,X}$ is the point $(0 : 0 : 1 : 0 : 0)$, corresponding exactly to the fiber of the Gauss map through x. The other isolated component is the line corresponding to the point $(0 : 0 : 0 : 0 : 1)$.

Let us remark that the line M is tangent to C_p at the point $L_x \cap \Pi$. Thus the rulings of X, projection of the ruling by $\mathbb{P}^2$'s of Y, consists of the planes spanned by x and the tangent line to C_p at the point $L_x \cap \Pi$.

The behavior described above can be found in a quite large class of examples.

Proposition 2.2.3 Let $X = V(f) \subset \mathbb{P}^{n+1}$ be an *irreducible hypersurface of degree* $d \geq 2$ *with degenerate Gauss map, that is* $\dim(\mathscr{G}_x(X)) < \dim(X)$. *Let* $L \subset X$ *be a line passing through a general point* $x \in X$ *and contracted by the Gauss map of* X. *Then*

$$N_{L/X} \simeq \oplus \mathscr{O}_L(1)^{n-1}$$

so that $[L] \in \mathscr{L}_{x,X} \subset \mathbb{P}^{n-1}$ *is a singular point such that* $T_{[L]}\mathscr{L}_{x,X} = \mathbb{P}^{n-1}$.

Proof Since L is contracted by the Gauss map we have $\delta(L) = \mathrm{length}(\Delta(L)) = d - 1$. If $N_{L/X} \simeq \mathscr{O}_L(a_1) \oplus \ldots \oplus \mathscr{O}_L(a_{n-1})$, then from (2.9) we deduce

$$\sum_{i=1}^{n-1} a_i + 1 = n,$$

yielding $a_1 = \ldots = a_{n-1} = 1$ since $a_i \leq 1$ for every $i = 1, \ldots, n-1$. $\square$

Remark 2.2.4 Let us remark that the lines $L \subset X$ contracted by the Gauss map have the property $H^1(N_{L/X}) = 0$ and clearly $\dim_{[L]} \mathscr{L}_{x,X} < n - 1$ because $X \subset \mathbb{P}^{n+1}$ is not a hyperplane. Thus the line L is *obstructed* in X. This is not surprising since the embedding $L \subset X$ is not a regular embedding, see also part 3) of Theorem 2.1.3.

Corollary 2.2.5 Let $Y \subset \mathbb{P}^N$ be a manifold of dimension $n \geq 3$. If its general projection $X \subset \mathbb{P}^{n+1}$ has degenerate Gauss map, then the Hilbert scheme of lines *through a general point $x \in X$, $\mathscr{L}_{x,X} \subset \mathbb{P}^{n-1}$, is not empty and has singular points along a linear space of dimension* $\mathrm{def}(X) - 1 := n - \dim(X^*) - 1$.

Let $Y \subset \mathbb{P}^N$ be a dual defective manifold, let $\mathrm{def}(Y) = N - 1 - \dim(Y^*) > 0$, *let $y \in Y$ be a general point and let* $\pi : Y \to X \subset \mathbb{P}^{n+1}$ *be a general linear projection. Then $\mathscr{L}_{\pi(y),X}$ is singular along a linear space of dimension* $\mathrm{def}(Y) - 1$.

Proof By the general Duality Principle, see Exercise 1.5.22, the dual variety of X is a general linear section of Y^* with a $\mathbb{P}^{n+1}$. Thus $\mathrm{def}(X) = \mathrm{def}(Y)$ and the conclusions follow from the previous results. □

Example 2.2.6 Let $Y = \mathbb{G}(1,4) \subset \mathbb{P}^9$. Then $Y \simeq Y^*$ and $\mathrm{def}(Y) = 2$. Let $X \subset \mathbb{P}^7$ be the quintic hypersurface obtained by a general projection of Y. For a general point $x \in X$ the Hilbert scheme of lines $\mathscr{L}_{x,X}$ is singular along the line obtained by projectivizing the plane passing through x, which is the closure of the fiber through x of the Gauss map of X.

2.3 Equations for $\mathscr{L}_{x,X} \subset \mathbb{P}((t_x X)^*)$

We now expand and generalize the treatment outlined in Example 2.2.2 by looking at the equations defining $\mathscr{L}_x \subset \mathbb{P}^{n-1}$ for $x \in X_{\mathrm{reg}}$. We shall follow closely the paper [161].

Let

$$X = V(f_1, \ldots, f_m) \subset \mathbb{P}^N \qquad (*),$$

be a projective equidimensional connected variety, not necessarily irreducible, let $x \in X_{\mathrm{reg}}$, let $n = \dim(X)$ and let $c = \mathrm{codim}(X) = N - n$.

Thus $(*)$ means precisely that $X \subset \mathbb{P}^N$ is scheme-theoretically the intersection of $m \geq 1$ hypersurfaces of degrees $d_1 \geq d_2 \geq \ldots \geq d_m \geq 2$, where $d_i = \deg(f_i)$. Moreover, it is implicitly assumed that m is minimal, i.e. none of the hypersurfaces contains the intersection of the others. Define, following [105], the integer

$$d := \min\{\sum_{i=1}^{c}(d_i - 1) \text{ for expressions } (*) \text{ as above}\} \geq c.$$

With these definitions $X \subset \mathbb{P}^N$ (or more generally a scheme $Z \subset \mathbb{P}^N$) is called *quadratic* if it is scheme-theoretically an intersection of quadrics, which means that we can assume $d_1 = 2$ in $(*)$. Equivalently $X \subset \mathbb{P}^N$ is quadratic if and only if $d = c$. Let us recall that Mumford proved in [140] that every smooth projective variety in a suitable (re)embedding is isomorphic to a quadratic variety.

We can choose homogeneous coordinates $(x_0 : \ldots : x_N)$ on $\mathbb{P}^N$ such that

$$x = (1 : 0 : \ldots : 0),$$

and

$$T_x X = V(x_{n+1}, \ldots, x_N).$$

Let $\mathbb{A}^N = \mathbb{P}^N \setminus V(x_0)$ with affine coordinates $(y_1, \ldots, y_N)$, where $y_l = \frac{x_l}{x_0}$ for every $l = 1, \ldots, N$.

Let $\tilde{\mathbb{P}}^N = \mathrm{Bl}_x \mathbb{P}^N$ with exceptional divisor $E' \simeq \mathbb{P}((t_x\mathbb{P}^N)^*) = \mathbb{P}^{N-1}$ and let $\tilde{X} = \mathrm{Bl}_x X$ with exceptional divisor $E = \mathbb{P}((t_x X)^*) = \mathbb{P}^{n-1}$. Looking at the graph of the projection from x onto $V(x_0)$ we can naturally identify the projectivization of $\mathbb{A}^N \setminus \mathbf{0} = \mathbb{A}^N \setminus x$ with E' and with the projective hyperplane $V(x_0) = \mathbb{P}^{N-1}$.

Let

$$f_i = f_i^1 + f_i^2 + \cdots + f_i^{d_i},$$

with f_i^j homogeneous of degree j in the variables $(y_1, \ldots, y_N)$. So $f_1^1 = \ldots = f_m^1 = 0$ are the equations of $t_x X = T_x X \cap \mathbb{A}^N \subset \mathbb{A}^N$, which reduce to $y_{n+1} = \ldots = y_N = 0$ by the previous choice of coordinates, yielding

$$V(f_1^1, \cdots, f_m^1) = \mathbb{P}((t_x X)^*) \subset \mathbb{P}((t_x\mathbb{P}^N)^*) = \mathbb{P}^{N-1}.$$

With the previous identifications $\mathscr{L}_{x,\mathbb{P}^N} = E' = \mathbb{P}^{N-1} = \mathbb{P}((t_x\mathbb{P}^N)^*)$. We now write a set of equations defining $\mathscr{L}_x \subset E \subset E'$ as a subscheme of E' and of E. By definition $\mathbf{y} = (y_1 : \ldots : y_n)$ are homogeneous coordinates on $E \subset E'$. For every $j = 2, \ldots, m$ and for every $i = 1, \ldots, m$, let

$$\tilde{f}_i^j(\mathbf{y}) = f_i^j(y_1, \ldots, y_n, 0, 0, \ldots, 0, 0).$$

Then we have that $\mathscr{L}_x \subset E'$ is the scheme

$$V(f_1^1, f_1^2, \cdots, f_1^{d_1}, \cdots, f_m^1, f_m^2, \cdots, f_m^{d_m}) \subset E',$$

while $\mathscr{L}_x \subset E$ is the scheme

$$V(\tilde{f}_1^2, \cdots, \tilde{f}_1^{d_1}, \cdots, \tilde{f}_m^2, \cdots, \tilde{f}_m^{d_m}), \tag{2.10}$$

so that it is scheme-theoretically defined by at most $\sum_{i=1}^m (d_i - 1)$ equations.

The equations of

$$T_x X \cap X \cap \mathbb{A}^N = t_x X \cap X \cap \mathbb{A}^N,$$

as a subscheme of $\mathbb{A}^N$, are

$$V(f_1^1, \ldots, f_m^1, f_1^1 + f_1^2 + \cdots + f_1^{d_1}, \ldots, f_m^1 + f_m^2 + \cdots + f_m^{d_m}) =$$

$$V(f_1^1, \ldots, f_m^1, f_1^2 + \cdots + f_1^{d_1}, \ldots, f_m^2 + \cdots + f_m^{d_m}) \subset \mathbb{A}^N. \tag{2.11}$$

Thus the equations of $T_x X \cap X \cap \mathbb{A}^N = t_x X \cap X \cap \mathbb{A}^N$ as a subscheme of $t_x(X \cap \mathbb{A}^N) = t_x X$ are

$$V(\tilde{f}_1^2 + \cdots + \tilde{f}_1^{d_1}, \ldots, \tilde{f}_m^2 + \cdots + \tilde{f}_m^{d_m}) \subset t_x X = \mathbb{A}^n. \tag{2.12}$$

Let $I = \langle \tilde{f}_1^2 + \cdots + \tilde{f}_1^{d_1}, \ldots, \tilde{f}_m^2 + \cdots + \tilde{f}_m^{d_m} \rangle \subset \mathbb{C}[y_1, \ldots, y_n] = S$ and let I^{in} be the ideal generated by the *initial forms* of elements of I, see Definition 1.1.1. Note that if I is homogeneous and generated by forms of the same degree, then clearly $I = I^{\mathrm{in}}$. Then the affine tangent cone to $T_x X \cap X$ at x is

$$C_x(T_x X \cap X) = \mathrm{Spec}\left(\frac{S}{I^{\mathrm{in}}}\right)$$

so that

$$\mathbb{P}(C_x(T_x X \cap X)) = \mathrm{Proj}\left(\frac{S}{I^{\mathrm{in}}}\right), \tag{2.13}$$

see [142, III, Sect. 3] and Sect. 1.1.

Let $J \subset S$ be the homogeneous ideal generated by the polynomials in (2.10) defining $\mathscr{L}_{x,X}$ scheme-theoretically, that is $\mathscr{L}_{x,X} = \mathrm{Proj}(\frac{S}{J}) \subset \mathbb{P}((t_x X)^*)$. Clearly $I^{\mathrm{in}} \subseteq J$, yielding the closed embedding of schemes

$$\mathscr{L}_{x,X} \subseteq \mathbb{P}(C_x(T_x X \cap X)). \tag{2.14}$$

If $X \subset \mathbb{P}^N$ is quadratic, then $I = I^{\mathrm{in}} = J$. In conclusion we have proved the following results.

Proposition 2.3.1 ([161, Proposition 2.2]) *Let $X \subset \mathbb{P}^N$ be a (non-degenerate) projective variety, let $x \in X_{\mathrm{reg}}$ be a point and let the notation be as above. If $X \subset \mathbb{P}^N$ is quadratic, then*

$$T_x X \cap X \cap \mathbb{A}^N = C_x(T_x X \cap X) \subset t_x X \tag{2.15}$$

and

$$\mathbb{P}(C_x(V_x)) = \mathscr{L}_{x,X} = \mathbb{P}(C_x(T_x X \cap X)) \subset \mathbb{P}((t_x X)^*). \tag{2.16}$$

In particular, if $X \subset \mathbb{P}^N$ is quadratic, then the scheme $\mathscr{L}_{x,X} \subset \mathbb{P}((t_x X)^*)$ is quadratic.

2.3.1 V_x Versus $T_xX \cap X$ for a Quadratic Variety

The closed embedding (2.14) holds at the scheme-theoretic level. If $\mathscr{L}_{x,X}$ were reduced, or better smooth, it would be enough to prove that there exists an inclusion as sets. Since $x \in X_{\text{reg}}$ was arbitrary we cannot control a priori the structure of $\mathscr{L}_{x,X}$ even if $X \subset \mathbb{P}^N$ is a manifold. Proposition 2.2.1 implies that $\mathscr{L}_{x,X}$ is smooth as soon as X is a manifold and $x \in X$ is a general point.

Let us recall that $V_x = \phi_x(\mathscr{H}_x) \subset X$, see (2.3) for definitions/notation, and that $\mathbb{P}(C_x(V_x)) = \mathscr{L}_x$, see (2.6).

From now on in this subsection we shall assume $X \subset \mathbb{P}^N$ is quadratic. Then

1. $(V_x)_{\text{red}} = (T_xX \cap X)_{\text{red}}$;
2. if $X \subset \mathbb{P}^N$ is a manifold and if $x \in X$ is a general point, then

$$V_x = (T_xX \cap X)_{\text{red}};$$

3. the strict transforms of V_x and of $T_xX \cap X$ on $\text{Bl}_x X$ cut the exceptional divisor $E = \mathbb{P}((t_xX)^*)$ of $\text{Bl}_x X$ in the same scheme $\mathscr{L}_{x,X}$ [see (2.6) and (2.16)];
4. if $x \in X$ is a general point on a quadratic manifold $X \subset \mathbb{P}^N$ and if I^{in} is saturated, then $T_xX \cap X$ is reduced in a neighborhood of x so that it coincides with $\mathscr{C}_x$ in a neighborhood of x. Indeed, since $T_xX \cap X \cap \mathbb{A}^n = \text{Spec}(\frac{S}{I})$, with $I = I^{\text{in}} = J$ homogeneous and saturated, it follows that $T_xX \cap X$ is reduced at x. Thus in this case $T_xX \cap X$ is also reduced in a neighborhood of x.

Already for quadratic manifolds there exist many important differences between

$$\mathbb{P}(C_x(T_xX \cap X)) \subset \mathbb{P}((t_xX)^*)$$

and

$$C_x(T_xX \cap X) = T_xX \cap X \cap \mathbb{A}^N \subset t_xX$$

and also between $T_xX \cap X$ and the cone $V_x \subseteq T_xX \cap X$. We shall discuss some examples in order to analyze more closely these important schemes, which contain a lot of geometrical information.

Example 2.3.2 ($T_xX \cap X$ Non-reduced Only at x) Note that

$$t_x(T_xX \cap X) = t_xX$$

so that

$$\langle C_x(T_xX \cap X)\rangle = t_xX,$$

while in some cases $\mathbb{P}(C_x(T_xX \cap X))$ is degenerate in $\mathbb{P}((t_xX)^*)$.

Consider, for example, a rational normal scroll $X \subset \mathbb{P}^N$, different from the Segre varieties $\mathbb{P}^1 \times \mathbb{P}^{n-1}$, $n \geq 2$, and a general point $x \in X$. It is well known that $X \subset \mathbb{P}^N$ is quadratic so that $\mathscr{L}_{x,X} = \mathbb{P}(C_x(T_x X \cap X)) \subset \mathbb{P}((t_x X)^*)$ by (2.14). On the other hand, if $\mathbb{P}_x^{n-1}$ is the unique $\mathbb{P}^{n-1}$ of the ruling passing through $x \in X$, it is easy to see, using the same notation as above, that in this case

$$\mathscr{L}_{x,X} = \mathbb{P}(\mathbb{P}_x^{n-1} \cap \mathbb{A}^n) = \mathbb{P}^{n-2} \subset \mathbb{P}((t_x X)^*) = \mathbb{P}^{n-1}.$$

This is possible because in this example $T_x X \cap X$ and $C_x(T_x X \cap X)$ are not reduced at x. Indeed, the point $x \in C_x(T_x X \cap X)$ corresponds to the irrelevant ideal of S and I^{in} is not saturated since the equation defining the hyperplane $\mathscr{L}_{x,X}$ belongs to the saturation of I^{in} but not to I^{in} (I^{in} is generated by quadratic polynomials).

In the case of rational normal scrolls discussed in Example 2.3.2 we saw that $T_x X \cap X \setminus x = \mathscr{V}_x \setminus x$ as schemes, the affine tangent cones are different affine schemes but the projectivized tangent cones coincide.

By choosing suitable quadrics $Q_1, \ldots, Q_c$ we shall see in Sect. 2.3.3 that the complete intersection $Y = Q_1 \cap \ldots \cap Q_c$ coincides locally with X around x. Thus $T_x Y \cap Y$ and $T_x X \cap X$ coincide locally around x. In particular, the intersection of their strict transform on $\mathrm{Bl}_x X$ with the exceptional divisor is the same, so that $\mathscr{L}_{x,X} = \mathscr{L}_{x,Y}$ and the last scheme can be defined scheme-theoretically by $r \leq c$ linearly independent quadrics by (2.10).

In any case, the double nature of $T_x X \cap X$ as a subscheme of $T_x X$ and X plays a central role for its infinitesimal properties at x, measured precisely by

$$\mathbb{P}(C_x(T_x X \cap X)) \subset \mathbb{P}((t_x X)^*).$$

It is useful to consider $\mathbb{P}(C_x(T_x X \cap X)) \subset \mathbb{P}((t_x X)^*)$ as the base locus scheme of the restriction to the exceptional divisor over x of the projection of X from $T_x X$, as we shall see in the next section. In this way we shall find a further reason why for quadratic manifolds $\mathscr{L}_{x,X}$ can be defined scheme-theoretically by at most c quadratic equations for an arbitrary point $x \in X_{\mathrm{reg}}$.

2.3.2 Tangential Projection and Second Fundamental Form

There are several possible equivalent definitions of the projective second fundamental form $|II_{x,X}|$ of a connected equidimensional projective variety $X \subset \mathbb{P}^N$ at $x \in X_{\mathrm{reg}}$, see for example [107, 3.2 and the end of Sect. 3.5]. We use the one related to tangential projections, as in [107, Remark 3.2.11].

Suppose $X \subset \mathbb{P}^N$ is non-degenerate, as always, let $x \in X_{\mathrm{reg}}$ and consider the projection from $T_x X$ onto a disjoint $\mathbb{P}^{c-1}$

$$\pi_x : X \dashrightarrow W_x \subseteq \mathbb{P}^{c-1}. \tag{2.17}$$

The map π_x is not defined along the scheme $T_x X \cap X$, which contains x, and it is associated to the linear system of hyperplane sections cut out by hyperplanes containing $T_x X$, or equivalently by the hyperplane sections singular at x.

Let $\phi : \mathrm{Bl}_x X \to X$ be the blow-up of X at x, let

$$E = \mathbb{P}((t_x X)^*) = \mathbb{P}^{n-1} \subset \mathrm{Bl}_x X$$

be the exceptional divisor and let H be a hyperplane section of $X \subset \mathbb{P}^N$.

The induced rational map

$$\tilde{\pi}_x : \mathrm{Bl}_x X \dashrightarrow \mathbb{P}^{c-1}$$

is defined as a rational map along E since $X \subset \mathbb{P}^N$ is not a linear space. Indeed, the restriction of $\tilde{\pi}_x$ to E is given by a linear system in

$$|\phi^*(H) - 2E|_{|E} \subseteq |-2E_{|E}| = |\mathcal{O}_{\mathbb{P}((t_x X)^*)}(2)| = \mathbb{P}(S^2(t_x X)).$$

This means that to a hyperplane section H tangent to X at x we are associating the projectivization of the affine tangent cone to $H \cap X$ at x. Thus this linear system is empty if and only if the associated quadric hypersurface in $\mathbb{P}(t_x X)) = \mathbb{P}^{n-1}$ has rank 0. If $\mathrm{def}(X) = k$, then a local calculation due to Kleiman, see for example [51, 2.1 (a)], shows that the rank of the projectivized tangent cone to a general tangent hyperplane section at x is a quadric of rank equal to $n - k$. Thus $\tilde{\pi}_x$ is not defined along E if and only if $\mathrm{def}(X) = n$, i.e. if and only if $X \subset \mathbb{P}^N$ is a linear subspace. We have proved the following useful fact.

Proposition 2.3.3 *Let the notation be as above. The rational map* $\tilde{\pi}_x : \mathrm{Bl}_x X \dashrightarrow W_x \subset \mathbb{P}^{N-n-1}$ *is not defined along the exceptional divisor* $E \subset \mathrm{Bl}_x$ *if and only if* $X \subseteq \mathbb{P}^N$ *is a linear embedded* $\mathbb{P}^n$.

Consider the strict transform scheme of $T_x X \cap X$ on $\mathrm{Bl}_x X$, denoted from now on by $\tilde{T} = \mathrm{Bl}_x(T_x X \cap X)$. Then $\tilde{T}$ is the base locus scheme of $\tilde{\pi}_x$ and the restriction of $\tilde{\pi}_x$ to E has base locus scheme equal to

$$\tilde{T} \cap E = \mathbb{P}(C_x(T_x X \cap X)) = B_{x,X} \subset \mathbb{P}((t_x X)^*). \tag{2.18}$$

Definition 2.3.4 (Second Fundamental Form) The *second fundamental form*

$$|II_{x,X}| \subseteq \mathbb{P}(S^2(t_x X))$$

of a connected equidimensional non-degenerate projective variety $X \subset \mathbb{P}^N$ of dimension $n \geq 2$, $n < N$, at a point $x \in X_{\mathrm{reg}}$ is the non-empty linear system of quadric hypersurfaces in $\mathbb{P}((t_x X)^*)$ defining the restriction of $\tilde{\pi}_x$ to E.

The *base locus scheme of the second fundamental form of X at x* is indicated by

$$B_{x,X} \subset \mathbb{P}((t_x X)^*),$$

or simply by B_x.

Clearly $\dim(|II_{x,X}|) \leq N - n - 1$ and $\tilde{\pi}_x(E) \subseteq W_x \subseteq \mathbb{P}^{N-n-1}$.

From this point of view $B_x \subset E$ consists of *asymptotic directions*, i.e. of directions associated to lines having a contact to order two with X at x. Thus, when $X \subset \mathbb{P}^N$ is scheme-theoretically defined by equations of degree at most two, the scheme B_x consists of points determined by tangent lines to X at x and contained in X. Thus in this case we have $B_{x,X} = \mathscr{L}_{x,X}$.

The following result was classically well known and used repeatedly by Scorza in his papers on secant defective varieties, see [163] and [166].

Proposition 2.3.5 *Let $X \subset \mathbb{P}^N$ be a smooth irreducible non-degenerate variety of secant defect $\delta(X) = \delta \geq 1$. Then*

1. $\dim(|II_{X,x}|) = N - n - 1$ *for $x \in X$ a general point;*
2. $N \leq \frac{n(n+3)}{2}$ *and equality holds if and only if $|II_{X,x}|$ is the complete linear system of quadrics on $\mathbb{P}((\mathbf{T}_x X)^*) = \mathbb{P}^{n-1}$.*

Proof Let the notation be as above. To prove part 1) it is sufficient to show that $\dim(\tilde{\pi}_x(E)) = n - \delta$ because $\tilde{\pi}_x(E) \subseteq W_x$. Recall that $W_x \subset \mathbb{P}^{N-n-1}$ is a non-degenerate variety of dimension $n - \delta$ by Terracini's Lemma.

Let $TX = \cup_{x \in X} T_x X$ be the tangential variety to X. The following formula holds:

$$\dim(TX) = n + 1 + \dim(\tilde{\pi}_x(E)), \qquad (2.19)$$

see [187] (or [79, 5.6, 5.7] and [60, Theorem 3.3.1] for a modern reference).

The variety $X \subset \mathbb{P}^N$ is smooth and secant defective, so that $TX = SX$ by the Fulton–Hansen Theorem, see Corollary 3.2.2. Therefore $\dim(TX) = 2n + 1 - \delta$ and from (2.19) we get $\dim(\tilde{\pi}_x(E)) = n - \delta$, as claimed.

Since $|II_{X,x}| \subseteq |\mathscr{O}_{\mathbb{P}^{n-1}}(2)|$, we have $N - n - 1 \leq \dim(|\mathscr{O}_{\mathbb{P}^{n-1}}(2)|) = \binom{n+1}{2} - 1$ and the final statements of part 1) and part 2) follow. $\qquad\square$

Let $\tilde{I} \subset S$ be the homogeneous ideal generated by the $r \leq c$ linearly independent quadratic forms in the second fundamental form of X at x. Then via (2.18) we obtain

$$\mathrm{Proj}(\frac{S}{\tilde{I}}) = B_{x,X} = \mathbb{P}(C_x(T_x X \cap X)) = \mathrm{Proj}(\frac{S}{\tilde{I}^{\mathrm{in}}}) \subset \mathbb{P}((t_x X)^*). \qquad (2.20)$$

In conclusion we have proved the following results by combining (2.18) with (2.16) and (2.20).

Corollary 2.3.6 ([161, Corollary 2.5]) *Let $X \subset \mathbb{P}^N$ be a non-degenerate projective variety, let $x \in X_{\mathrm{reg}}$ be a point and let the notation be as above. Then:*

1. $\mathscr{L}_{x,X} \subseteq B_{x,X}$;
2. *if $X \subset \mathbb{P}^N$ is quadratic, then equality holds and $\mathscr{L}_{x,X} \subset \mathbb{P}((t_x X)^*)$ can be defined scheme-theoretically by $r \leq c$ linearly independent quadratic equations in the second fundamental form of X at x.*

Remark 2.3.7 The previous result has many important applications. We recall that, as proved in [105, Theorem, 2.4], if $X \subset \mathbb{P}^N$ is a quadratic manifold and if $c \leq \frac{n-1}{2}$, then, for $x \in X$ general, $\mathscr{L}_{x,X} \subset \mathbb{P}((t_x X)^*)$ is the complete intersection of the c linearly independent quadratic polynomials defining $|II_{x,X}|$. Then $\mathscr{L}_{x,X}$ has dimension $n - 1 - c$ from which it follows that $X \subset \mathbb{P}^N$ is a complete intersection, see [105, Theorem 3.8] and Sect. 5.2 for full details. This proves the Hartshorne Conjecture on Complete Intersections in the quadratic case and also leads to the classification of quadratic Hartshorne manifolds, see [105, Theorem 3.9] and Sect. 5.2.

2.3.3   Approach to $B_{x,X} = \mathscr{L}_{x,X}$ via [19]

For manifolds $X \subset \mathbb{P}^N$ there is another approach to the study of the relations between $B_{x,X}$ and $\mathscr{L}_{x,X}$, based on a construction of [19]. This construction generalizes an idea due to Severi, see *loc. cit.*, and it can be used to give a proof of a weaker form of Corollary 2.3.6 (in the sense that we shall prove it only for $x \in X$ general). This different treatment will also clearly illustrate the local nature of the second fundamental form in the projective setting. Let us remark that the general setting developed in the previous sections is unavoidable because a general point $x \in X$ is no longer necessarily general on the complete intersection $Y \supseteq X$ we now construct.

It was proved in [19] that given a manifold $X = V(f_1, \ldots, f_m) \subset \mathbb{P}^N$ as in (∗), we can choose $g_i \in H^0(\mathscr{I}_X(d_i))$, $i = 1, \ldots, c$ such that

$$Y = V(g_1, \ldots, g_c) = X \cup X', \tag{2.21}$$

where X' (if non-empty) meets X in a divisor D. Moreover, from (2.21) it follows that

$$\mathscr{O}_X(D) \simeq \det(\frac{\mathscr{I}_X}{\mathscr{I}_X^2}) \otimes \mathscr{O}_X(\sum_{i=1}^{c} d_i) \simeq \mathscr{O}_X(d - n - 1) \otimes \omega_X^*, \tag{2.22}$$

see also [19, pg. 597]. We now show the usefulness of this construction by proving some facts and results contained in [105, Theorem 2.4], see also Sect. 5.2.

Suppose that $X \subset \mathbb{P}^N$ is a quadratic manifold and take a point $x \in U = X \setminus \text{Supp}(D)$. By definition

$$Y \setminus \text{Supp}(D) = U \sqcup V,$$

where $V = X' \setminus \text{Supp}(D)$. Consider the two schemes $T_x X \cap X \cap U$ and $T_x Y \cap Y \cap U$. Since $t_x X = t_x Y$ and since $Y \cap U = X \cap U$, we obtain the following equality of schemes

$$C_x(T_x X \cap X) = C_x(T_x X \cap X \cap U) = C_x(T_x Y \cap Y \cap U) = C_x(T_x Y \cap Y).$$

Via (2.16) we deduce the following equality as subschemes of $\mathbb{P}((t_x X)^*)$:

$$\mathcal{L}_{x,Y} = \mathbb{P}(C_x(T_x Y \cap Y)) = \mathbb{P}(C_x(T_x X \cap X)) = \mathcal{L}_{x,X}. \tag{2.23}$$

Since $\mathcal{L}_{x,Y}$ can be defined scheme-theoretically by $r \leq c$ linearly independent quadratic equations, the same is true for $\mathcal{L}_{x,X}$. Now, no longer assuming that X is quadratic, since $x \in X$ is general, $\mathcal{L}_{x,X}$ is smooth and hence reduced. Clearly a line L passing through x is contained in X if and only if it is contained in Y, yielding $\mathcal{L}_{x,X} = (\mathcal{L}_{x,Y})_{\text{red}}$, see [105, Theorem 2.4]. Thus we have proved:

Proposition 2.3.8 *Let $X \subset \mathbb{P}^N$ be a manifold, let the notation be as above and let $x \in U$ be a general point. Then:*

1. *$\mathcal{L}_{x,X} = (\mathcal{L}_{x,Y})_{\text{red}}$ so that $\mathcal{L}_{x,X}$ can be defined set theoretically by the $r \leq d$ equations defining $\mathcal{L}_{x,Y}$ scheme-theoretically. In particular, if $d \leq n - 1$, then $\mathcal{L}_{x,X} \neq \emptyset$.*
2. *If $X \subset \mathbb{P}^N$ is quadratic, then $\mathcal{L}_{x,X} = \mathcal{L}_{x,Y}$ so that $\mathcal{L}_{x,X} \subset \mathbb{P}((t_x X)^*)$ is a quadratic manifold defined scheme-theoretically by at most c quadratic equations.*

2.3.4 Lines on Prime Fano Manifolds

Let $X \subset \mathbb{P}^N$ be a (non-degenerate) manifold of dimension $n \geq 2$. For a general point $x \in X$ we know that, if non-empty, $\mathcal{L}_x \subset \mathbb{P}^{n-1}$ is smooth by Proposition 2.2.1.

There are well-known examples where $\mathcal{L}_x \subset \mathbb{P}^{n-1}$ is not irreducible, such as $X = \mathbb{P}^a \times \mathbb{P}^b \subset \mathbb{P}^{ab+a+b}$ Segre embedded, and also examples where $\mathcal{L}_x \subset \mathbb{P}^{n-1}$ is degenerate, see Example 2.3.2 and also table (2.28) below.

A relevant class of manifolds where the properties of smoothness, irreducibility and non-degeneracy of $X \subset \mathbb{P}^N$ are transferred to $\mathcal{L}_x \subset \mathbb{P}^{n-1}$ consists of prime Fano manifolds of high index, which we now define.

A manifold $X \subset \mathbb{P}^N$ is called a *prime Fano manifold* if $-K_X$ is ample and if $\text{Pic}(X) \simeq \mathbb{Z}\langle \mathcal{O}(1) \rangle$. The *index of X* is the positive integer defined by $-K_X = i(X)H$, with H a hyperplane section of $X \subset \mathbb{P}^N$.

Let us recall some fundamental facts. Part 1) below is well known and follows from the previous discussion except for a fundamental Theorem of Mori which implies that for prime Fano manifolds of index greater than $\frac{n+1}{2}$, necessarily $\mathcal{L}_x \neq \emptyset$, see [136] and [116, Theorem V.1.6].

Proposition 2.3.9 *Let $X \subset \mathbb{P}^N$ be a projective manifold and let $x \in X$ be a general point. Then*

1. *If $\mathscr{L}_x \neq \emptyset$, then for every $[L] \in \mathscr{L}_x$ we have*

$$\dim_{[L]}(\mathscr{L}_x) = -K_X \cdot L - 2.$$

In particular, for prime Fano manifolds of index $i(X) \geq \frac{n+3}{2}$ the variety $\mathscr{L}_x \subset \mathbb{P}^{n-1}$ is irreducible (and in particular non-empty!).

2. *(Hwang [93]) If $X \subset \mathbb{P}^N$ is a prime Fano manifold of index $i(X) \geq \frac{n+3}{2}$, then $\mathscr{L}_x \subset \mathbb{P}^{n-1}$ is a non-degenerate manifold of dimension $i(X) - 2$.*

Proof We shall prove only the first part. The assertion on the dimension of $\mathscr{L}_x$ follows from Theorem 2.1.3 or from Theorem 2.1.9, see also part ii) of Corollary 2.1.13. Suppose now that $X \subset \mathbb{P}^N$ is a prime Fano manifold of index $i(X) \geq \frac{n+3}{2}$. By [136], see also [116, Theorem V.1.6], through a general point $x \in X$ there passes a rational curve $C_x \subset X$ such that $-K_X \cdot C_x \leq n + 1$. From $-K(X) = i(X)H, H \subset X$ a hyperplane section, we deduce

$$n + 1 \geq i(X)(H \cdot C_x) \geq \frac{n + 3}{2} \deg(C_x),$$

yielding $\deg(C_x) = 1$. Thus $\mathscr{L}_x \neq \emptyset$, $\mathscr{L}_x$ is equidimensional of dimension $i(X) - 2 \geq (n - 1)/2$ and $\mathscr{L}_x$ is irreducible by Proposition 2.2.1. $\square$

The condition $i(X) \geq (n + 3)/2$ implies that on a prime Fano manifold the dimension of the variety of lines $\mathscr{L}_x \subset \mathbb{P}^{n-1}$ is at least $(n - 1)/2$, yielding the irreducibility of $\mathscr{L}_x$. This last condition is also of interest for arbitrary manifolds covered by lines.

Indeed, if we fix some irreducible component, say $\mathscr{F}$, of the Hilbert scheme of lines on $X \subset \mathbb{P}^N$, such that X is covered by the lines in $\mathscr{F}$, then for a general point $x \in X$ we can define $\mathscr{F}_x = \{[\ell] \in \mathscr{F} \mid x \in \ell\}$ so that $\mathscr{F}_x \subseteq \mathscr{L}_x \subset \mathbb{P}^{n-1}$. When the dimension of $\mathscr{F}_x$ (and a fortiori of $\mathscr{L}_x$) is large, the study of manifolds covered by lines is greatly simplified by the following two facts:

First, we may reduce, via a Mori contraction, to the case where the Picard group is cyclic, that is of prime Fano manifolds via a result due to Beltrametti–Sommese–Wiśniewski, see [17]. Secondly, the variety $\mathscr{F}_x \subset \mathbb{P}^{n-1}$ inherits many of the good properties of $X \subset \mathbb{P}^N$ due to results of Hwang, proved in [93] and recalled in Proposition 2.3.9. One can also consult [16] for an application of these principles generalizing the result of [63] to the relative setting.

Theorem 2.3.10 *Let the notation be as above and let $\dim(\mathscr{F}_x) \geq \frac{n-1}{2}$. Then the following results hold:*

1. *(Beltrametti et al. [17]) There is a Mori contraction, say $\mathrm{cont}_{\mathscr{F}} : X \to W$, of the lines from $\mathscr{F}$; let F denote a general fiber of $\mathrm{cont}_{\mathscr{F}}$ and let f be its dimension;*

2. (Wiśniewski [195]) $\mathrm{Pic}(F) = \mathbb{Z}\langle H_F \rangle$, $i(F) = \dim(\mathscr{F}_x) + 2$ and F is covered by the lines from $\mathscr{F}$ contained in F;
3. (Hwang [93]) $\mathscr{F}_x \subseteq \mathbb{P}^{f-1}$ is smooth irreducible non-degenerate and it coincides with $\mathscr{L}_{x,F}$. In particular, F has only one maximal irreducible covering family of lines.

Let us finish this section by looking at another significant example in which meaningful geometrical properties of $X \subset \mathbb{P}^N$ are reflected in similar properties of $\mathscr{L}_x \subset \mathbb{P}^{n-1}$, when this is non-empty.

Example 2.3.11 ($\mathscr{L}_x$ of Smooth Fano Complete Intersections) Let $X \subset \mathbb{P}^N$ be a smooth complete intersection of type $(d_1, d_2, \ldots, d_c)$ with $d_c \geq 2$ and dimension $n \geq 1$. Recall that $-K_X = \mathscr{O}_X(n + 1 - d)$. Thus:

- if $n + 1 - d > 0$, then X is a Fano manifold and $i(X) = n + 1 - d$;
- if $n \geq 3$, then $\mathrm{Pic}(X) \simeq \mathbb{Z}\langle \mathscr{O}(1) \rangle$;
- if $i(X) \geq 2$, then $\mathscr{L}_x \neq \emptyset$ and for every $[L] \in \mathscr{L}_x$ we have

$$\dim_{[L]}(\mathscr{L}_x) = (-K_X \cdot L) - 2 = i(X) - 2 = n - 1 - d \geq 0.$$

Then $\mathscr{L}_x \subset \mathbb{P}^{n-1}$ is a smooth complete intersection of type

$$(2, \ldots, d_1; 2, \ldots, d_2; \ldots; 2, \ldots d_{c-1}; 2, \ldots, d_c)$$

which is scheme-theoretically defined by the d equations in (2.10), see also Theorem 2.4.3.

2.4 A Condition for Non-extendability

Definition 2.4.1 (Extension of a Projective Variety) Let us consider $H = \mathbb{P}^N$ as a hyperplane in $\mathbb{P}^{N+1}$. Let $Y \subset \mathbb{P}^N = H$ be a smooth (non-degenerate) irreducible variety of dimension $n \geq 1$. An irreducible variety $X \subset \mathbb{P}^{N+1}$ will be called *an extension of Y* if

1. $\dim(X) = \dim(Y) + 1$;
2. $Y = X \cap H$ as a scheme.

For every $p \in \mathbb{P}^{N+1} \setminus H$, the irreducible cone

$$X = S(p, Y) = \bigcup_{y \in Y} < p, y > \subset \mathbb{P}^{N+1}$$

is an extension of $Y \subset \mathbb{P}^N = H$, which will be called *trivial*.

For any extension $X \subset \mathbb{P}^{N+1}$ of $Y \subset \mathbb{P}^N$ necessarily #(Sing(X)) $< \infty$ since X is smooth along the hyperplane section $Y = X \cap H$. We also remark that in our definition Y is a fixed hyperplane section. In the classical approach it was usually required that H is a general hyperplane section of X, see for example [163]. Under these more restrictive hypotheses one can always assume that a general point on Y is also a general point on X.

2.4.1 Extensions of $\mathscr{L}_{x,Y} \subset \mathbb{P}^{n-1}$ via $\mathscr{L}_{x,X} \subset \mathbb{P}^n$

Let $y \in Y$ be a general point and let us consider an extension $X \subset \mathbb{P}^{N+1}$ of Y and an irreducible component $\mathscr{L}_{y,Y}^j$ of $\mathscr{L}_{y,Y} \subset \mathbb{P}^{n-1}$, which is a smooth irreducible variety by Proposition 2.2.1.

Proposition 2.4.2 ([161, Proposition 3.2]) *Let $X \subset \mathbb{P}^{N+1}$ be an irreducible projective variety which is an extension of the non-degenerate manifold $Y \subset \mathbb{P}^N$. Let $n = \dim(Y) \geq 1$ and let $y \in Y$ be an arbitrary point such that $\mathscr{L}_{y,Y} \neq \emptyset$. Then:*

1. $\mathscr{L}_{y,X} \cap \mathbb{P}((t_y Y)^*) = \mathscr{L}_{y,Y}$ *as schemes.*
2. *If $y \in Y$ is general and if $[L] \in \mathscr{L}_{y,Y}$, then $\dim_{[L]}(\mathscr{L}_{y,X}) = \dim_{[L]}(\mathscr{L}_{y,Y}) + 1$ and every $[L] \in \mathscr{L}_{y,Y}$ is a smooth point of $\mathscr{L}_{y,X}$.*
3. *If $y \in Y$ is general and if $\mathscr{L}_{y,Y}^j$ is an irreducible component of positive dimension, then there exists an irreducible component $\mathscr{L}_{y,X}^j$ such that $\mathscr{L}_{y,Y}^j = \mathscr{L}_{y,X}^j \cap \mathbb{P}((t_y Y)^*)$ as schemes.*
4. *If $y \in Y$ is general, then $\mathrm{Sing}(\mathscr{L}_{y,X}) \subseteq S_{y,X}$.*

Proof Let $Y = X \cap H$, with $H = \mathbb{P}^N \subset \mathbb{P}^{N+1}$ a hyperplane and let the notation be as in Sect. 2.3. The conclusion in 1) immediately follows from (2.10).

Let us pass to 2) and consider an arbitrary line $[L] \in \mathscr{L}_{y,Y}^j$, where $\mathscr{L}_{y,Y}^j$ is an irreducible component of the smooth not necessarily irreducible variety $\mathscr{L}_{y,Y}$. We have an exact sequence of normal bundles

$$0 \to N_{L/Y} \to N_{L/X} \to N_{Y/X|L} \simeq \mathscr{O}_{\mathbb{P}^1}(1) \to 0. \qquad (2.24)$$

Since $y \in Y$ is general, $N_{L/Y}$ is generated by global sections, see the proof of Proposition 2.2.1, so that (2.8) yields

$$N_{L/X} \simeq N_{L/Y} \oplus \mathscr{O}_{\mathbb{P}^1}(1) \simeq \mathscr{O}_{\mathbb{P}^1}(1)^{s(L,Y)+1} \oplus \mathscr{O}_{\mathbb{P}^1}^{n-s(L,Y)-1}. \qquad (2.25)$$

Thus $N_{L/X}$ is also generated by global sections and since $L \subset Y \subset X_{\mathrm{reg}}$ we deduce that $\mathscr{L}_{y,X}$ is smooth at $[L]$ and that $\dim_{[L]}(\mathscr{L}_{y,X}) = \dim_{[L]}(\mathscr{L}_{y,Y}) + 1$, proving 2).

Therefore if $y \in Y$ is general, there exists a unique irreducible component of $\mathscr{L}_{y,X} \subset \mathbb{P}((t_y X)^*)$, let us say $\mathscr{L}_{y,X}^j$, containing $[L]$. By the previous calculation

$\dim(\mathscr{L}^j_{y,X}) = s(L, Y) + 1 = \dim(\mathscr{L}^j_{y,Y}) + 1$. Recall that by part (1) we have

$$t_{[L]}\mathscr{L}_{y,Y} = t_{[L]}\mathscr{L}_{y,X} \cap \mathbb{P}((t_yY)^*)$$

so that

$$\mathscr{L}^j_{y,Y} \subseteq \mathscr{L}^j_{y,X} \cap \mathbb{P}((t_yY)^*) \subseteq \mathscr{L}_{y,Y} \subset \mathbb{P}^{n-1} = \mathbb{P}((t_yY)^*), \tag{2.26}$$

yielding that $\mathscr{L}^j_{y,Y}$ is an irreducible component of $\mathscr{L}^j_{y,X} \cap \mathbb{P}((t_yY)^*)$ as well as an irreducible component of the smooth variety $\mathscr{L}_{y,Y}$. Hence, if $\dim(\mathscr{L}^j_{y,Y}) \geq 1$, we have the equality $\mathscr{L}^j_{y,Y} = \mathscr{L}^j_{y,X} \cap \mathbb{P}((t_yY)^*)$ as schemes. Therefore, under this hypothesis, $\mathscr{L}^j_{y,X} \subset \mathbb{P}((t_yX)^*))$ (or better $(\mathscr{L}^j_{y,X})_{\mathrm{red}}$) is a projective extension of the smooth positive dimensional irreducible variety $\mathscr{L}^j_{y,Y} \subset \mathbb{P}((t_xY)^*)$. Indeed, $\dim(\mathscr{L}^j_{y,Y}) \geq 1$ forces $\dim(\mathscr{L}^j_{y,X}) \geq 2$ so that it is sufficient to recall that $\mathscr{L}_{y,X}$ is smooth along $\mathscr{L}_{y,Y}$ by the previous discussion, and also that an arbitrary hyperplane section of the irreducible variety $(\mathscr{L}^j_{y,X})_{\mathrm{red}}$ is connected by the Fulton–Hansen Theorem, see [69] and Theorem 3.1.7 here. In conclusion, if $\dim(\mathscr{L}^j_{y,Y}) \geq 1$, then equality as schemes holds in (2.26), proving part (3).

By Theorem 2.1.13 there exists a non-empty open subset $U \subseteq X$ such that $N_{\tilde{L}/X}$ is generated by global sections for every line $\tilde{L} \subset X_{\mathrm{reg}}$ intersecting U. If $U \cap Y \neq \emptyset$, then 4) clearly holds. Suppose $Y \cap U = \emptyset$. Let $[\tilde{L}] \in \mathscr{L}_{y,X} \setminus S_{y,X}$. If $\tilde{L} \cap U \neq \emptyset$, then $[\tilde{L}]$ is a smooth point of $\mathscr{L}_{y,X}$ by the previous analysis. If $\tilde{L} \cap U = \emptyset$, then $\tilde{L} \subset Y$ since now we are assuming $Y \cap U = \emptyset$. The generality of $y \in Y$ ensures that $N_{\tilde{L}/X}$ is generated by global sections by (2.25), concluding the proof of 4). $\qquad\square$

Now we are in position to prove the main result of this section and to give some applications.

Theorem 2.4.3 ([161, Theorem 3.3]) *Let the notation be as above and let $y \in Y$ be a general point. Then:*

1. *Suppose there exist two distinct irreducible components $\mathscr{L}^1_{y,X}$ and $\mathscr{L}^2_{y,X}$ of*

$$\mathscr{L}_{y,X} \subset \mathbb{P}((t_yX)^*),$$

   extending two irreducible components $\mathscr{L}^1_{y,Y}$, respectively $\mathscr{L}^2_{y,Y}$, of $\mathscr{L}_{y,Y}$ in the sense specified above. If $\mathscr{L}^1_{y,X} \cap \mathscr{L}^2_{y,X} \neq \emptyset$, then $X \subset \mathbb{P}^{N+1}$ is a cone over $Y \subset \mathbb{P}^N$ of vertex a point $p \in \mathbb{P}^{N+1} \setminus \mathbb{P}^N$.
2. *If $\mathscr{L}_{y,Y} \subset \mathbb{P}((t_yY)^*)$ is a manifold whose extensions are singular, then every extension of $Y \subset \mathbb{P}^N$ is trivial.*

Proof By the above discussion, we get that in both cases, for $y \in Y$ general, the variety $S_{y,X} \subseteq \mathscr{L}_{y,X}$ is not empty. Thus for every $y \in Y$ general there exists a line $L_y \subseteq X$ passing through y and through a singular point $p_y \in L_y \cap \mathrm{Sing}(X)$. Since Y

is irreducible and since Sing(X) consists of a finite number of points, there exists a $p \in$ Sing(X) such that $p \in L_y$ for $y \in Y$ general. This implies that $X = S(p, Y)$ is a cone over Y with vertex p. □

The first application is a quick proof of a result due to Scorza (see [166] and also [12, 197]), proved by him under the stronger assumption that $Y = X \cap H$ is a general hyperplane section of X.

Corollary 2.4.4 *Let* $1 \leq a \leq b$ *be integers, let* $n = a + b \geq 3$ *and let* $Y \subset \mathbb{P}^{ab+a+b}$ *be a manifold projectively equivalent to the Segre embedding* $\mathbb{P}^a \times \mathbb{P}^b \subset \mathbb{P}^{ab+a+b}$. *Then every extension of* Y *in* $\mathbb{P}^{ab+a+b+1}$ *is trivial.*

Proof For $y \in Y$ general, it is well known that $\mathscr{L}_{y,Y} = \mathscr{L}^1_{y,Y} \sqcup \mathscr{L}^2_{y,Y} \subset \mathbb{P}^{a+b-1} = \mathbb{P}^{n-1}$ with $\mathscr{L}^1_{y,Y} = \mathbb{P}^{a-1}$ and $\mathscr{L}^2_{y,Y} = \mathbb{P}^{b-1}$, both linearly embedded. Observe that $b - 1 \geq 1$ by the hypothesis $n \geq 3$. By (2.26) and the discussion following it, there exist two irreducible components $\mathscr{L}^j_{y,X}$, $j = 1, 2$, of $\mathscr{L}_{y,X} \subset \mathbb{P}^n = \mathbb{P}^{a+b}$ with $\dim(\mathscr{L}^1_{y,X}) = a$ and $\dim(\mathscr{L}^2_{y,X}) = b$. If $a \neq b$ then clearly $\mathscr{L}^1_{y,X} \neq \mathscr{L}^2_{y,X}$. If $a = b \geq 2$, then $\mathscr{L}^1_{y,X} \neq \mathscr{L}^2_{y,X}$ because an arbitrary hyperplane section of a variety of dimension at least 2 is connected, see [69] and Theorem 3.1.7. Since $a + b = n$, $\mathscr{L}^1_{y,X} \cap \mathscr{L}^2_{y,X} \neq \emptyset$ and the conclusion follows from the first part of Theorem 2.4.3. □

The previous result has some interesting consequences via iterated applications of the second part of Theorem 2.4.3. Indeed, let us consider the following homogeneous varieties (also known as irreducible hermitian symmetric spaces), in their homogeneous embedding, and the description of the Hilbert scheme of lines passing through a general point, see [93, Sect. 1.4.5] and also [184].

	Y	$\mathscr{L}_{y,Y}$	$\tau_y : \mathscr{L}_{y,Y} \to \mathbb{P}((t_y Y)^*)$
1	$\mathbb{G}(r, m)$	$\mathbb{P}^r \times \mathbb{P}^{m-r-1}$	Segre embedding
2	$SO(2r)/U(r)$	$\mathbb{G}(1, r-1)$	Plücker embedding
3	E_6	$SO(10)/U(5)$	miminal embedding
4	$E_7/E_6 \times U(1)$	E_6	Severi embedding
5	$Sp(r)/U(r)$	$\mathbb{P}^{r-1}$	quadratic Veronese embedding

$$(2.27)$$

There are also the following homogeneous contact manifolds with Picard number one associated to a complex simple Lie algebra $\mathbf{g}$, whose Hilbert scheme of lines passing through a general point is known. Let us observe that in these examples the variety $\mathscr{L}_{y,Y} \subset \mathbb{P}^{n-1} = \mathbb{P}((t_y Y)^*)$ is degenerate and its linear span is exactly $\mathbb{P}((D_y)^*) = \mathbb{P}^{n-2}$, there D_y is the tangent space at y of the distribution associated to the contact structure on Y. In this case there is the following factorization $\tau_y : \mathscr{L}_{y,Y} \to \mathbb{P}((D_y)^*) \subset \mathbb{P}((t_y Y)^*)$. For more details and the not universal notation one can consult [93, Sect. 1.4.6].

	g	$\mathscr{L}_{y,Y}$	$\tau_x : \mathscr{L}_{x,Y} \to \mathbb{P}((D_y)^*$
6	F_4	$Sp(3)/U(3)$	Segre embedding
7	E_6	$\mathbb{G}(2,5)$	Plücker embedding
8	E_7	$SO(12)/U(6)$	minimal embedding
9	E_8	$E_7/E_6 \times U(1)$	minimal embedding
10	$\mathfrak{so}_{m+4}$	$\mathbb{P}^1 \times Q^{m-2}$	Segre embedding

$$(2.28)$$

By case 1') we shall denote a variety as in 1) of (2.27) satisfying the following numerical conditions: $r < m - 1$; if $r = 1$, then $m \geq 4$. By 2') we shall denote a variety as in 2) with $r \geq 5$.

Corollary 2.4.5 *Let $Y \subset \mathbb{P}^N$ be a manifold as in Examples 1'), 2'), 3), 4), 7), 8), 9) above. Then every extension of Y is trivial.*

Proof In cases 2'), 3), 4) and 9) in the statement the variety $\mathscr{L}_{y,Y} \subset \mathbb{P}^{n-1}$ of one example is the variety $Y \subset \mathbb{P}^N$ occurring in the next one. Thus for these cases, by the second part of Theorem 2.4.3, it is sufficient to prove the result for case 1'). For this variety the conclusion follows from Corollary 2.4.4. For the remaining cases, the variety $\mathscr{L}_{y,Y} \subset \mathbb{P}^{n-1}$ is either as in case 1') with $(r,m) = (2,5)$ or as in case 2) with $r = 6$ and the conclusion follows once again by the second part of Theorem 2.4.3.

$\square$

The next result is also classical and well-known but we provide a direct geometric proof. Under the assumption that the hyperplane section $H \cap X = Y$ is general, it was proved by C. Segre for $n = 2$ in [169] and it was extended by Scorza in [163], see also [187], to arbitrary Veronese embeddings $v_d(\mathbb{P}^n) \subset \mathbb{P}^{N(d)}$, with $n \geq 2$ and $d \geq 2$. Modern proofs of the general cases are contained in [12] and also in [197]).

Proposition 2.4.6 *Let $n \geq 2$ and let $Y \subset \mathbb{P}^{\frac{n(n+3)}{2}}$ be a manifold projectively equivalent to the quadratic Veronese embedding $v_2(\mathbb{P}^n) \subset \mathbb{P}^{\frac{n(n+3)}{2}}$. Then every extension of Y is trivial.*

Proof Let $y \in Y$ be a general point and let $N = \frac{n(n+3)}{2}$. Since $\mathscr{L}_{y,Y} = \emptyset$, then $\mathscr{L}_{y,X} \subset \mathbb{P}^n$, if not empty, consists of at most a finite number of points and through $y \in X$ there passes at most a finite number of lines contained in X. Consider a conic $C \subset Y$ passing through y. Then $N_{C/Y} \simeq \mathscr{O}_{\mathbb{P}^1}(1)^{n-1}$. The exact sequence of normal bundles

$$0 \to N_{C/Y} \to N_{C/X} \to N_{Y/X|C} \simeq \mathscr{O}_{\mathbb{P}^1}(2) \to 0,$$

yields

$$N_{C/X} \simeq N_{C/Y} \oplus \mathscr{O}_{\mathbb{P}^1}(2) \simeq \mathscr{O}_{\mathbb{P}^1}(1)^{n-1} \oplus \mathscr{O}_{\mathbb{P}^1}(2).$$

Thus there exists a unique irreducible component $\mathscr{C}_{y,X}$ of the Hilbert scheme of conics contained in $X \subset \mathbb{P}^{N+1}$ passing through $y \in X$ to which $[C]$ belongs. Moreover, $\dim(\mathscr{C}_{y,X}) = n + 1$ and the conics parametrized by $\mathscr{C}_{y,X}$ cover X. Hence there exists a one-dimensional family of conics through y and a general point $x \in X$. By Bend and Break, see for example [40, Proposition 3.2], there is at least a singular conic through y and x. Since $X \subset \mathbb{P}^{N+1}$ is not a linear space, there exists no line joining y and a general x, i.e. the singular conics through x and y are reduced. Thus given a general point x in X, there exists a line $L_x \subset X$ through x, not passing through y, and a line $L_y \subset X$ through y such that $L_y \cap L_x \neq \emptyset$. Since there are a finite number of lines contained in X and passing through y, we can conclude that given a general point $x \in X$, there exists a fixed line passing through y, $\tilde{L}_y$, and a line L_x through x such that $L_x \cap \tilde{L}_y \neq \emptyset$.

Moreover, a general conic $[C_{x,y}] \in \mathscr{C}_{y,X}$ passing through a general point x is irreducible, does not pass through the finite set $\mathrm{Sing}(X)$ and has ample normal bundle verifying $h^0(N_{C_{x,y}/X}(-1)) = h^0(N_{C/X}(-1)) = n + 1$. This means that the deformations of $C_{x,y}$ keeping x fixed cover an open subset of X and also that through general points $x_1, x_2 \in X$ there passes a one-dimensional family of irreducible conics. The plane spanned by one of these conics contains x_1 and x_2 so that it has to vary with the conic. Otherwise the fixed plane would be contained in X and $X \subset \mathbb{P}^{N+1}$ would be a linearly embedded $\mathbb{P}^{N+1}$, which is contrary to our assumptions. In conclusion, through a general point $z \in < x_1, x_2 >$ there passes at least a one-dimensional family of secant lines to X so that

$$\dim(SX) \leq 2(n + 1) - 1 = 2n + 1 < N + 1 = \frac{n(n + 3)}{2} + 1, \qquad (2.29)$$

yielding $SX \subsetneq \mathbb{P}^{N+1}$.

Suppose the point $p_x = \tilde{L}_y \cap L_x$, for $y \in Y$ general, varies on $\tilde{L}_y$. Then the linear span of two general tangent spaces $T_{x_1}X$ and $T_{x_2}X$ would contain the line $\tilde{L}_y$. Since $T_z SX = < T_{x_1}X, T_{x_2}X >$ by the Terracini Lemma, we deduce that a general tangent space to SX contains $\tilde{L}_y$ and a fortiori y. Since $SX \subsetneq \mathbb{P}^{N+1}$, the variety $SX \subset \mathbb{P}^{N+1}$ would be a cone whose vertex, which is a linear space, contains $\tilde{L}_y$ and a fortiori $y \in Y$. By the generality of $y \in Y$ we deduce that $Y \subset \mathbb{P}^N$ is degenerate.

Thus $p_x = \tilde{L}_y \cap L_x$ does not vary with $x \in X$ general. Let us denote this point by p. Then clearly $X \subset \mathbb{P}^{N+1}$ is a cone with vertex p over Y. $\square$

Corollary 2.4.7 *Let $Y \subset \mathbb{P}^N$ be a manifold either as in 5) above with $r \geq 3$ or as in 6) above. Then every extension of Y is trivial.*

Proof By (2.27) we know that in case 5) with $r \geq 3$ we have $n - 1 = \frac{(r-1)(r+2)}{2}$ and the variety $\mathscr{L}_{y,Y} \subset \mathbb{P}^{n-1}$ is projectively equivalent to $v_2(\mathbb{P}^{r-1}) \subset \mathbb{P}^{\frac{(r-1)(r+2)}{2}}$. To conclude we apply Proposition 2.4.6 and the second part of Theorem 2.4.3. Case 6) follows from case 5) with $r = 3$ by the second part of Theorem 2.4.3. $\square$

Remark 2.4.8 Theorem 2.4.3 is another interesting incarnation of the Principle described in the Introduction while Corollaries 2.4.5 and 2.4.7, surely well known to everybody, were included only to show that they are immediate consequences of Scorza's result in [166], a fact which seems to have been overlooked until now. Last, but not least, our approach allows a direct construction of the cones as the unique possible projective extensions in a direct and geometric way, that is by firstly constructing explicitly the lines on the extension and then by showing that they all pass through a fixed point of the extension. As far as we know this application of the abstract tools introduced in Sect. 2.1 has not been noticed before.

Other conditions for non-extendability are presented in the survey paper [197].

Chapter 3
The Fulton–Hansen Connectedness Theorem, Scorza's Lemma and Their Applications to Projective Geometry

3.1 The Enriques–Zariski Connectedness Principle, the Fulton–Hansen Connectedness Theorem and the Generalizations of Some Classical Results in Algebraic Geometry

In the first chapter we introduced the main definitions and classical results concerning the geometry of embedded projective varieties. Many theorems in classical projective geometry deal with *general* objects.

For example, the classical Bertini Theorem on hyperplane sections, see Theorem 1.5.4, yields the smoothness of a general hyperplane section of a smooth projective variety. Since every hyperplane section of a smooth projective variety of dimension at least two is connected by the Enriques–Severi–Zariski Lemma, see [88, III.7.8 and III.7.9], we deduce that a general hyperplane section is irreducible for such varieties.

A more modern and refined version of Bertini's Theorem asserts that if $f : X \to \mathbb{P}^N$ is a morphism from a proper variety X such that $\dim(f(X)) \geq 2$, and if $H = \mathbb{P}^{N-1} \subset \mathbb{P}^N$ is a general hyperplane, then $f^{-1}(H)$ is irreducible, see [111, Theorem 6.10] or Theorem 3.1.3 below.

The **Enriques–Zariski Principle** says that *limits of connected varieties remain connected* and its incarnation in the previous setting is the following *generalization*: for an arbitrary $H = \mathbb{P}^{N-1} \subset \mathbb{P}^N$, $f^{-1}(H)$ is connected, see Theorem 3.1.4 below.

This result is particularly interesting because, as shown by Deligne and Jouanolou, a small generalization of it proved by Grothendieck, [80, XIII 2.3], yields a simplified proof of a deep Connectedness Theorem of Fulton and Hansen in [69], whose applications appear in different areas of algebraic geometry and

© Springer International Publishing Switzerland 2016
F. Russo, *On the Geometry of Some Special Projective Varieties*,
Lecture Notes of the Unione Matematica Italiana 18,
DOI 10.1007/978-3-319-26765-4_3

topology. Moreover, Deligne's proof generalizes to deeper statements involving higher homotopy groups when studying complex varieties, see [43, 44, 68, 70].

To illustrate this circle of ideas and the *connectedness principle*, we describe how the Theorem of Fulton–Hansen includes some classical theorems in Algebraic Geometry and how it generalizes them.

In our treatment we shall strictly follow the surveys [68] and [70]. The notes of Bădescu [11] and his book [12] are also interesting sources where the ideas of Grothendieck behind the previous results and their generalizations to d-connectedness and to weighted projective spaces are explained in great detail.

Now we recall some well-known important results emphasizing the connectedness properties with the aim of finding a common thread. When dealing with homotopy groups π_i, we are assuming $K = \mathbb{C}$ and referring to the classical topology. Let us list the following important and well-known results.

Four Classical Theorems:

1. (**Bézout**) Let X and Y be closed subvarieties of $\mathbb{P}^N$. If $\dim(X) + \dim(Y) \geq N$, then $X \cap Y \neq \emptyset$. If $\dim(X) + \dim(Y) > N$, then $X \cap Y$ is connected and more precisely $(\dim(X) + \dim(Y) - N)$-connected.

2. (**Bertini**) Let $f : X \to \mathbb{P}^N$ be a morphism, with X a proper variety, and let $L = \mathbb{P}^{N-l} \subset \mathbb{P}^N$ be a linear space. If $l \leq \dim(f(X))$, then $f^{-1}(L) \neq \emptyset$. If $l < \dim(f(X))$, then $f^{-1}(L)$ is connected.

3. (**Lefschetz**) If $X \subset \mathbb{P}^N$ is a closed irreducible subvariety of dimension n and if $L = \mathbb{P}^{N-l} \subset \mathbb{P}^N$ is a linear space containing $\mathrm{Sing}(X)$, then

$$\pi_i(X, X \cap L) = 0 \quad \text{for } i \leq n - l.$$

   Equivalently the morphism

$$\pi_i(X \cap L) \to \pi_i(X)$$

   is an isomorphism if $i < n - l$ and surjective if $i = n - l$.

4. (**Barth–Larsen**) If $X \subset \mathbb{P}^N$ is a closed irreducible non-singular subvariety of dimension n, then

$$\pi_i(\mathbb{P}^N, X) = 0 \quad \text{for } i \leq 2n - N + 1.$$

   (Recall that $\pi_i(\mathbb{P}^N) = \mathbb{Z}$ for $i = 0, 2$ and $\pi_i(\mathbb{P}^N) = 0$ for $i = 1, 3, 4, \ldots, 2N$.)

As recalled above, the classical theorems usually refer to properties of general linear sections, for which a better property can be expected. For example, in the case of Bertini's theorem one proves irreducibility of general linear sections; in the case of Bézout's theorem, when the intersection is transversal, one usually computes $\#(X \cap Y)$. In the classical Lefschetz theorem the variety is non-singular and L is general, etc.

Let us remark that the two parts of theorem 1) can be reformulated by means of homotopy groups. The first part is equivalent to

$$\pi_0(X \cap Y) \to \pi_0(X \times Y),$$

being surjective; the second one to the fact that the above morphism is an isomorphism. Similarly theorem 2) can be reformulated as

$$\pi_0(f^{-1}(L)) \to \pi_0(X)$$

being surjective, respectively an isomorphism.

A common look at the above theorems comes from the following observation of Hansen, [69, 70]. All the above results are statements about the non-emptiness, respectively connectedness, of the inverse image of $\Delta_{\mathbb{P}^N} \subset \mathbb{P}^N \times \mathbb{P}^N$ under a proper morphism

$$f : W \to \mathbb{P}^N \times \mathbb{P}^N$$

such that $\dim(f(W)) \geq N$, respectively $\dim(f(W)) > N$.

Suppose this is true and take $W = X \times Y$ for theorem 1) or $W = X \times L$ in theorem 2) and 3) at least to deduce the connectedness parts. Theorem 4) can be deduced by taking $W = X \times X$, see [70] and [68] for further details on this last case.

These properties can be explained from other points of view as consequences of the ampleness of the normal bundle of smooth subvarieties in $\mathbb{P}^N$. On the other hand the same positivity holds for $\Delta_{\mathbb{P}^N} \subset \mathbb{P}^N \times \mathbb{P}^N$ since

$$N_{\Delta_{\mathbb{P}^N}/\mathbb{P}^N \times \mathbb{P}^N} \simeq T_{\mathbb{P}^N}$$

and the tangent bundle to $\mathbb{P}^N$, $T_{\mathbb{P}^N}$, is ample, e.g. by considering the Euler sequence.

The above discussion and further generalizations by Faltings, Goldstein and Hansen revealed a connectedness principle, which we now state and which will later be justified.

Fulton Connectedness Principle, [68, pg. 18].

Let P be a smooth projective variety. Given a *suitable positive* embedding $Y \hookrightarrow P$ of codimension l and a proper morphism $f : W \to P$, $n = \dim(W)$,

$$
\begin{array}{ccc}
f^{-1}(Y) & \lhook\joinrel\longrightarrow & W \\
\downarrow & & \downarrow f \\
Y & \lhook\joinrel\longrightarrow & P,
\end{array}
$$

one *expects*

$$\pi_i(W, f^{-1}(Y)) \overset{\cong}{\to} \pi_i(P, Y) \quad \text{for } i \leq n - l - \text{"defect"}.$$

This defect should be measured by:

(a) lack of positivity of Y in P;
(b) singularities of W;
(c) dimensions of the fibers of f.

Usually $\pi_i(P, Y) = 0$ for small i, so the conclusion is that, as regards connectivity, $f^{-1}(Y)$ must look like W. If the defect is zero we deduce that:

(i) $f^{-1}(Y) \neq \emptyset$ if $n \geq l$;
(ii) $f^{-1}(Y)$ is connected and $\pi_1(f^{-1}(Y)) \to \pi_1(W)$ is surjective if $n > l$.

The most basic case is with $P = \mathbb{P}^N$ and $Y = \mathbb{P}^{N-l}$ a linear subspace. In this case the principle translates the theorems of Bertini and Lefschetz by taking $W = X$.

As we explained before, the case which allows us to include all the classical theorems is $P = \mathbb{P}^N \times \mathbb{P}^N$, and $Y = \Delta_{\mathbb{P}^N}$ diagonally embedded in P. Indeed, $W = X \times Y$ gives Bézout's Theorem, while theorems 2) and 3) are recovered by setting $W = X \times L$. Theorem 4) can be obtained with $W = X \times X$.

When $\mathbb{P}^N$ is replaced by other homogeneous spaces, one could measure the defect of positivity of its tangent bundle and one expects the principle to hold with this defect, see [13, 56, 75].

Why should one expect this connectedness principle to be valid? In some cases one can define a Morse function which measures distance from Y. Positivity should imply that all the Morse indices of this function are at least $n - l - 1$ (perhaps minus a defect). Then one constructs W from $f^{-1}(Y)$ by adding only cells of dimension at least $n - l - 1$, which yields the required vanishing of relative homotopy groups, see [68] for a proof giving theorems 3) and 4) above.

Before ending this long introduction to the connectedness theorem we recall for completeness the following statements for later reference. They are particular forms or consequences of results of Barth and Barth and Larsen. Chronologically part 2) has been stated before the Barth–Larsen Theorem involving higher homotopy groups and recalled above.

Theorem 3.1.1 *Let* $X \subset \mathbb{P}^N$ *be a smooth, irreducible projective variety and let* $H \subset X$ *be a hyperplane section.*

1. *(Barth–Larsen) If* $n \geq \frac{N+1}{2}$, *then* $\pi_1(X) = 1$.
2. *(Barth) If* $n \geq \frac{N+i}{2}$, *then the restriction map*

$$H^i(\mathbb{P}^N, \mathbb{Z}) \to H^i(X, \mathbb{Z})$$

   is an isomorphism.
3. *(Barth) If* $n \geq \frac{N+2}{2}$, *then*

$$\mathrm{Pic}(X) \simeq \mathbb{Z} < H > .$$

We can now finally state the following important Connectedness Theorem.

Theorem 3.1.2 (Fulton–Hansen Connectedness Theorem, [69]) *Let* X *be an irreducible variety, proper over an algebraically closed field* K. *Let* $f : X \to \mathbb{P}^N \times \mathbb{P}^N$ *be a morphism and let* $\Delta = \Delta_{\mathbb{P}^N} \subset \mathbb{P}^N \times \mathbb{P}^N$ *be the diagonal.*

1. *If* $\dim(f(X)) \geq N$, *then* $f^{-1}(\Delta) \neq \emptyset$.
2. *If* $\dim(f(X)) > N$, *then* $f^{-1}(\Delta)$ *is connected.*

To prove this we shall apply the next theorem, whose proof follows from [111, Theorem 6.10], where Jouanolou relaxes our hypothesis $K = \overline{K}$ and proves a more general result. Very recently a different proof valid in arbitrary characteristic and based on the generalized Hodge Index Theorem and on de Jong's Alterations has been obtained in [129].

Theorem 3.1.3 (Bertini's Theorem, See [111]) *Let* X *be an irreducible variety and let* $f : X \to \mathbb{P}^N$ *be a morphism. For a fixed integer* $l \geq 1$, *let* $\mathbb{G}(N - l, N)$ *be the Grassmann variety of linear subspaces of* $\mathbb{P}^N$ *of codimension* l. *Then*

1. *if* $l \leq \dim(\overline{f(X)})$, *then there is a non-empty open subset* $U \subseteq \mathbb{G}(N - l, N)$ *such that* $f^{-1}(L) \neq \emptyset$ *for every* $L \in U$;
2. *if* $l < \dim(\overline{f(X)})$, *then there is a non-empty open subset* $U \subseteq \mathbb{G}(N - l, N)$ *such that* $f^{-1}(L)$ *is irreducible for every* $L \in U$.

We now show that the Enriques–Zariski Principle recalled at the beginning of the section is an application of the next result, which is the key point towards Theorem 3.1.2. We pass from general linear sections to arbitrary ones and for simplicity we suppose $K = \overline{K}$, as always.

Theorem 3.1.4 ([69, 80], [111, Theorem 7.1]) *Let X be an irreducible variety and let $f : X \to \mathbb{P}^N$ be a morphism. Let $L = \mathbb{P}^{N-l} \subset \mathbb{P}^N$ be an arbitrary linear space of codimension l.*

1. If $l \leq \dim(\overline{f(X)})$ and if X is proper over K, then $f^{-1}(L) \neq \emptyset$.
2. If $l < \dim(\overline{f(X)})$ and if X is proper over K, then $f^{-1}(L)$ is connected.

More generally for an arbitrary irreducible variety X, if $f : X \to \mathbb{P}^N$ is proper over some open subset $V \subseteq \mathbb{P}^N$, and if $L \subseteq V$, then, when the hypotheses on the dimensions are satisfied, the same conclusions hold for $f^{-1}(L)$.

Proof (According to [111]) We prove the second part of the theorem from which the statements in 1) and 2) follow.

Let $W \subseteq \mathbb{G}(N - l, N)$ be the open subset consisting of linear spaces contained in V and let

$$Z = \{(x, L') \in X \times W : f(x) \in L'\} \subset \{(x, L') \in X \times \mathbb{G}(N-l, N) : f(x) \in L'\} = \mathscr{I}.$$

The scheme Z is irreducible since it is an open subset of the Grassmann bundle $p_1 : \mathscr{I} \to X$. Since f is proper over V, the second projection $p_2 : Z \to W$ is a proper morphism. Consider its Stein factorization:

so that the morphism q is proper with connected fibers and surjective, while r is finite.

By Theorem 3.1.3 r is dominant and hence surjective if $l \leq \dim(\overline{f(X)})$, respectively generically one-to-one and surjective if $l < \dim(\overline{f(X)})$. In the first case $p_2 : Z \to W$ is surjective so that $f^{-1}(L) \neq \emptyset$ for every $L \in W$. In the second case, since W is smooth, it follows that r is one-to-one everywhere, so that $f^{-1}(L) = q^{-1}(r^{-1}(L))$ is connected for every $L \in W$. □

Remark 3.1.5 The original proof of Grothendieck used an analogous local theorem proved via local cohomology. His method has been used and extended by Hartshorne, Ogus, Speiser and Faltings. Faltings proved with similar techniques a connectedness theorem for other homogeneous spaces, see [56], at least in characteristic zero. It seems that a different proof of a special case of the above result was also given by Barth in 1969.

We are finally in a position to prove the Fulton–Hansen Theorem.

Proof (of Theorem 3.1.2, According to Deligne, [43]) The idea is to pass from the diagonal embedding $\Delta \subset \mathbb{P}^N \times \mathbb{P}^N$ to a linear embedding $L = \mathbb{P}^N \subset \mathbb{P}^{2N+1}$, a well known classical trick.

In $\mathbb{P}^{2N+1}$ separate the $2N + 2$ coordinates into $[X_0 : \ldots : X_N]$ and $[Y_0 : \ldots : Y_N]$ and regard these two sets as coordinates on each factor of $\mathbb{P}^N \times \mathbb{P}^N$. The two N dimensional linear subspaces $H_1 : X_0 = \ldots = X_N = 0$ and $H_2 : Y_0 = \ldots = Y_N = 0$ of $\mathbb{P}^{2N+1}$ are disjoint.

Let $V = \mathbb{P}^{2N+1} \setminus (H_1 \cup H_2)$. Since there is a unique secant line to $H_1 \cup H_2$ passing through each $p \in V$, there is a morphism

$$\phi : V \to H_1 \times H_2 = \mathbb{P}^N \times \mathbb{P}^N,$$

which associates to p the points

$$\phi(p) = (p_1, p_2) = (\langle H_2, p \rangle \cap H_1, \langle H_1, p \rangle \cap H_2).$$

In coordinates,

$$\phi([X_0 : \ldots : X_N : Y_0 : \ldots : Y_N] = ([X_0 : \ldots : X_N], [Y_0 : \ldots : Y_N]).$$

Then $\phi^{-1}(\phi(p)) = \langle p_1, p_2 \rangle \setminus \{p_1, p_2\} \simeq \mathbb{A}_K^1 \setminus 0$.

Let $L = \mathbb{P}^N \subset V$ be the linear subspace of $\mathbb{P}^{2N+1}$ defined by $X_i = Y_i$, $i = 0, \ldots, N$. Then

$$\phi_{|L} : L \xrightarrow{\simeq} \Delta$$

is an isomorphism. Given $f : X \to \mathbb{P}^N \times \mathbb{P}^N$ we construct the following Cartesian diagram

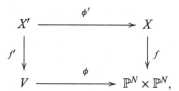

where

$$X' = V \times_{\mathbb{P}^N \times \mathbb{P}^N} X.$$

Clearly ϕ' induces an isomorphism between $f'^{-1}(L)$ and $f^{-1}(\Delta)$. To prove the theorem it is sufficient to verify the corresponding assertion for $f'^{-1}(L)$. To this end we apply Theorem 3.1.4. Let us verify the hypotheses.

Since $\phi'^{-1}(x) \simeq \phi^{-1}(f(x)) = \mathbb{A}_K^1 \setminus 0$ for every $x \in X$, the scheme X' is irreducible and of dimension $\dim(X) + 1$. The morphism f is proper, so that also $f' : X' \to V$ is proper and moreover $\dim(f'(X')) = \dim(f(X)) + 1$. If

$\dim(f(X)) \geq N$, then $\dim(f'(X')) \geq N+1 = \mathrm{codim}(L, \mathbb{P}^{2N+1})$. If $\dim(f(X)) > N$, then $\dim(f'(X')) > N+1 = \mathrm{codim}(L, \mathbb{P}^{2N+1})$. $\square$

Let us list some immediate consequences.

Corollary 3.1.6 (Generalized Bézout Theorem) *Let X and Y be closed subvarieties of* $\mathbb{P}^N$*. Then:*

1. $X \cap Y \neq \emptyset$ *if* $\dim(X) + \dim(Y) \geq N$.
2. $X \cap Y$ *is connected (and more precisely* $(\dim(X) + \dim(Y) - N)$*-connected) if* $\dim(X) + \dim(Y) > N$.

Proof Let $Z = X \times Y$ and let $f = i_X \times i_Y : Z \to \mathbb{P}^N \times \mathbb{P}^N$, where i_X and i_Y are the inclusions in $\mathbb{P}^N$. Then $X \cap Y \simeq f^{-1}(\Delta_{\mathbb{P}^N})$ and the conclusions follow from the Fulton–Hansen Theorem. $\square$

Corollary 3.1.7 (Generalized Bertini Theorem) *Let* $X \subset \mathbb{P}^N$ *be an irreducible non-degenerate variety.*

If $\dim(X) \geq 2$, *then every hyperplane section is connected. In particular, if X is also smooth, a general hyperplane section is smooth and irreducible.*

Proof Let $Z = X \times H$ and let $f = i_X \times i_H : Z \to \mathbb{P}^N \times \mathbb{P}^N$, where i_X and i_Y are the inclusions in $\mathbb{P}^N$ and $H = \mathbb{P}^{N-1}$ is a hyperplane. Then $X \cap H \simeq f^{-1}(\Delta_{\mathbb{P}^N})$ and the conclusions follow from the Fulton–Hansen Theorem. $\square$

3.2   Zak's Applications to Projective Geometry

In this section we come back to projective geometry and apply the Fulton–Hansen theorem to prove some interesting and non-classical results in projective geometry. Most of the applications are due to Fyodor L. Zak, see [70, 125, 198], and they will be significant improvements of the classical material presented in the first chapter. Other applications can be found in [68–70].

We begin with the following key result, which refines a result of Johnson, [109].

Theorem 3.2.1 ([69, 198]) *Let* $Y \subseteq X \subset \mathbb{P}^N$ *be a closed subvariety of dimension* $r = \dim(Y) \leq \dim(X) = n$, *with X irreducible and projective. Then, either*

1. $\dim(T^*(Y, X)) = r + n$ *and* $\dim(S(Y, X)) = r + n + 1$, *or*
2. $T^*(Y, X) = S(Y, X)$.

Proof We can suppose first that Y is irreducible and then apply the same argument to each irreducible component of Y. We know that $T^*(Y, X) \subseteq S(Y, X)$ and that, by construction, $\dim(T^*(Y, X)) \leq r + n$. Suppose $\dim(T^*(Y, X)) = r + n$. Since $S(Y, X)$ is irreducible and since $\dim(S(Y, X)) \leq r + n + 1$, the conclusion holds.

Suppose now $\dim(T^*(Y, X)) = t < r + n$. There exists an $L = \mathbb{P}^{N-t-1}$ such that $L \cap T^*(Y, X) = \emptyset = L \cap X$. The projection $\pi_L : \mathbb{P}^N \setminus L \to \mathbb{P}^t$ restricts to a finite morphism on X and on Y, since $L \cap X = \emptyset$. Then $(\pi_L \times \pi_L)(X \times Y) \subset \mathbb{P}^t \times \mathbb{P}^t$ has dimension $r + n > t$ by hypothesis. By Theorem 3.1.2, the closed set

$$\tilde{\Delta} = (\pi_L \times \pi_L)^{-1}(\Delta_{\mathbb{P}^t}) \subset Y \times X$$

is connected and contains the closed set $\Delta_Y \subset Y \times X$ so that Δ_Y is closed in $\tilde{\Delta}$.
We claim that

$$\Delta_Y = \tilde{\Delta}.$$

This yields $L \cap S(Y, X) = \emptyset$ and hence $\dim(S(Y, X)) \leq N - 1 - \dim(L) = t$.
Suppose $\tilde{\Delta} \setminus \Delta_Y \neq \emptyset$. We shall find $y' \in Y$ such that

$$\emptyset \neq T_{y'}^*(Y, X) \cap L \subseteq T^*(Y, X) \cap L,$$

contrary to our assumption on L.

Indeed, if $\tilde{\Delta} \setminus \Delta_Y \neq \emptyset$, the connectedness of $\tilde{\Delta}$ implies the existence of $(y', y') \in \overline{\tilde{\Delta} \setminus \Delta_Y} \cap \Delta_Y$. Let the notation be as in Definition 1.2.1, i.e. $p_2(p_1^{-1}(y, x)) = < x, y >$ if $x \neq y$ and $p_2(p_1^{-1}(y, x)) = T_y^*(Y, X)$ if $x = y \in Y$.
For every $(y, x) \in \tilde{\Delta} \setminus \Delta_Y$ we have $< y, x > \cap L \neq \emptyset$ by definition of π_L since, for $y \neq x$, we have $\pi_L(y) = \pi_L(x)$ if and only if $< y, x > \cap L \neq \emptyset$. Thus the same conclusion holds for (y', y') so that $p_2(p_1^{-1}(y, x)) \cap L \neq \emptyset$ forces $p_2(p_1^{-1}(y', y')) \cap L \neq \emptyset$. This contradiction proves the claim, concluding the proof. $\qquad\square$

Corollary 3.2.2 *Let $X \subset \mathbb{P}^N$ be an irreducible projective variety of dimension n. Then either*

1. $\dim(T^*X) = 2n$ and $\dim(SX) = 2n + 1$, or
2. $T^*X = SX$.

The following result well illustrates the passage from general to arbitrary linear spaces.

Theorem 3.2.3 (Zak's Theorem on Tangency) *Let $X \subset \mathbb{P}^N$ be an irreducible projective non-degenerate variety of dimension n. Let $L = \mathbb{P}^m \subset \mathbb{P}^N$ be a linear subspace, $n \leq m \leq N - 1$, which is J-tangent along the closed set $Y \subseteq X$. Then $\dim(Y) \leq m - n$.*

Proof Without loss of generality we can firstly suppose that Y is irreducible and then apply the conclusion to each irreducible component. By hypothesis and by definition we get $T^*(Y, X) \subseteq L$. Since $X \subseteq S(Y, X)$ and since X is non-degenerate, $S(Y, X)$ is not contained in L so that $T^*(Y, X) \neq S(Y, X)$. By Theorem 3.2.1 we have

$$\dim(Y) + n = \dim(T^*(Y, X)) \leq \dim(L) = m.$$

$\qquad\square$

We now come back to the problem of tangency and to contact loci of smooth varieties, providing two notable applications of the Theorem on Tangency. We begin with the finiteness of the Gauss map of a smooth variety.

Corollary 3.2.4 (Gauss Map Is Finite for Smooth Varieties, Zak) *Let $X \subsetneq \mathbb{P}^N$ be a smooth irreducible non-degenerate projective variety of dimension n. Then the Gauss map $\mathscr{G}_X : X \to \mathbb{G}(n, N)$ is finite. If, moreover, char(K)=0, then $\mathscr{G}_X$ is finite and birational onto the image, i.e. X is a normalization of $\mathscr{G}_X(X)$.*

Proof It is sufficient to prove that $\mathscr{G}_X$ has finite fibers. For every $x \in X$, $\mathscr{G}_X^{-1}(\mathscr{G}_X(x))$ is the locus of points at which the tangent space $T_x X$ is tangent. By Theorem 3.2.3 it has dimension less than or equal to $\dim(T_x X) - n = 0$.

If char(K)=0, then every fiber $\mathscr{G}_X^{-1}(\mathscr{G}_X(x))$ is linear by Theorem 1.5.10 and of dimension zero by the first part, so that it reduces to a point as a scheme. ▢

The next result reveals a special feature of non-singular varieties, since the result is clearly false for cones, see Exercise 1.5.16.

Corollary 3.2.5 (Lower Bound for the Dimension of Dual Varieties) *Let $X \subset \mathbb{P}^N$ be a smooth projective non-degenerate variety and let $X^* \subset \mathbb{P}^{N*}$ be its dual variety.*
Then $\dim(X^*) \geq \dim(X)$. In particular, if X^* is also smooth, then $\dim(X^*) = \dim(X)$.

Proof By the theorem on the dimension of the fibers of a morphism, letting the notation be as in Definition 1.5.1, $\dim(X^*) = N - 1 - \dim(p_2^{-1}([H]))$, $[H] \in X^*$ a general point. By Theorem 3.2.3, $\dim(p_2^{-1}([H])) \leq N - 1 - \dim(X)$ and the conclusion follows. ▢

Remark 3.2.6 In Exercise 1.5.16, we saw that $(\mathbb{P}^1 \times \mathbb{P}^n)^* \simeq \mathbb{P}^1 \times \mathbb{P}^n$ for every $n \geq 1$. In [51], L. Ein shows that if $N \geq 2/3 \dim(X)$, if X is smooth, if char(K)=0 and if $\dim(X) = \dim(X^*)$, then $X \subset \mathbb{P}^N$ is either a hypersurface, or $\mathbb{P}^1 \times \mathbb{P}^n \subset \mathbb{P}^{2n+1}$ Segre embedded, or $\mathbb{G}(1, 4) \subset \mathbb{P}^9$ Plücker embedded, or the 10-dimensional spinor variety $S^{10} \subset \mathbb{P}^{15}$. In the last three cases $X \simeq X^*$. For a different proof relating dual and secant defectiveness, see Corollary 4.4.9.

We apply the Theorem on Tangency to deduce some strong properties of the hyperplane sections of varieties of small codimension. By the generalized Bertini Theorem proved in the previous section an arbitrary hyperplane section of a variety of dimension at least 2 is connected. When the codimension of the variety is small with respect to the dimension, some further restrictions for the scheme structure appear.

If $X \subset \mathbb{P}^N$ is a non-singular irreducible non-degenerate variety, we recall that for every $H \in X^*$

$$\text{Sing}(H \cap X) = \{x \in X : T_x X \subset H\},$$

i.e. it is the locus of points at which H is tangent to X. By Theorem 3.2.3 we get

$$\dim(\text{Sing}(H \cap X)) \leq N - 1 - \dim(X),$$

i.e.

$$\text{codim}(\text{Sing}(H \cap X), H \cap X) \geq 2\dim(X) - N.$$

Clearly $H \cap X$ is a Cohen–Macaulay scheme of dimension $\dim(X) - 1$. In particular, every local ring of $H \cap X$ satisfies Serre's Condition S_k.

Let us recall that a Noetherian ring A satisfies Serre's Condition S_k if $\text{depth}(A_p) \geq \min\{k, \text{ht}(p)\}$ for every prime ideal $p \subset A$. Therefore a noetherian ring A satisfies condition S_k for every k if and only if it is Cohen–Macaulay.

A noetherian ring A satisfies condition R_k if the ring R_p is a regular local ring for every prime ideal $p \subset A$ with $\text{ht}(p) \leq k$.

These conditions are useful when verifying that a Noetherian ring A is reduced or normal. Indeed, we have that A is reduced if and only if it satisfies conditions R_0 and S_1, see [130, pg. 125]. In particular, $H \cap X$ is reduced if and only if it is generically smooth.

Moreover, A is normal if and only if it satisfies conditions R_1 and S_2 (Serre's Criterion of normality), see for example [130, Theorem 39]. Thus $H \cap X$ is normal if and only if its singular locus has codimension at least two.

Thus if $N \leq 2\dim(X) - 1$, then $H \cap X$ is a reduced scheme, being non-singular in codimension zero. The condition forces $\dim(X) \geq 2$ so that $H \cap X$ is also connected by Bertini's Theorem.

If $N \leq 2\dim(X) - 2$, which forces $\dim(X) \geq 3$, then $H \cap X$ is also non-singular in codimension 1 so that it is normal by Serre's Criterion of normality stated above. Since it is connected and integral, it is also irreducible. The case of the Segre 3-fold $\mathbb{P}^1 \times \mathbb{P}^2 \subset \mathbb{P}^5$ shows that this last result cannot be improved, since a hyperplane containing a $\mathbb{P}^2$ of the ruling yields a reducible, reduced, hyperplane section.

Clearly in the same way, if $N \leq 2\dim(X) - k - 1, k \geq 0$, then $X \cap H$ is connected, Cohen–Macaulay and non-singular in codimension k. We summarize these results in the following Corollary to the Theorem on Tangency.

Corollary 3.2.7 (Zak) *Let $X \subset \mathbb{P}^N$ be a smooth non-degenerate projective variety of dimension n. The following holds:*

1. if $N \leq 2n - 1$, then every hyperplane section is connected and reduced;
2. if $N \leq 2n - 2$, then every hyperplane section is irreducible and normal;
3. let $k \geq 2$. If $N \leq 2n - k - 1$, then every hyperplane section is irreducible, normal and non-singular in codimension k.

3.3   Tangential Invariants of Algebraic Varieties and Scorza's Lemma

We introduce some projective invariants of an irreducible non-degenerate variety $X \subset \mathbb{P}^N$ such that $SX \subsetneq \mathbb{P}^N$. These invariants measure the tangential behavior of $X \subset \mathbb{P}^N$ and the relative position of two general tangent spaces.

Consider a general point $p \in SX$, $p \in\, <x,y>$, $x,y \in X$ general points, and take

$$T_p SX = \langle T_x X, T_y X \rangle.$$

Definition 3.3.1   The *contact locus of* $T_p SX$ *on* X is

$$\Gamma_p = \Gamma_p(X) = \overline{\{x \in X_{\text{reg}} \ : \ T_x X \subseteq T_p SX\}} \subset X.$$

For a general hyperplane $H \subset \mathbb{P}^N$ containing $T_p SX$, we define the *contact locus of* H *on* $X \subset \mathbb{P}^N$:

$$\Xi_p(H) = \Xi_p(X,H) = \overline{\{x \in X_{\text{reg}} \ : \ T_x X \subseteq H\}} \subset X.$$

The contact locus of $T_p SX$ on X is called the *tangential contact locus of* X in [28]. By Terracini's Lemma and by definition we get

$$\Sigma_p \subseteq \Gamma_p \subseteq \Xi_p(H) \tag{3.1}$$

for every H containing $T_p SX$, where Σ_p is the entry locus of X with respect to p introduced in Definition 1.4.5. A monodromy argument shows that the irreducible components of Γ_p, respectively of $\Xi_p(H)$, through x and y are uniquely determined and have the same dimension (and in the second case that this dimension does not depend on the choice of a general $H \supseteq T_p SX$). We define this dimension as $\gamma(X)$, respectively $\xi(X)$. In particular we deduce

$$\delta(X) \leq \gamma(X) \leq \xi(X).$$

Let

$$\pi_x : X \dashrightarrow W_x \subset \mathbb{P}^{N-n-1}$$

be a general tangential projection and let $\tilde{\gamma}(X)$ be the dimension of the general fiber of the Gauss map

$$\mathcal{G}_{W_x} : W_x \dashrightarrow \mathbb{G}(n - \delta(X), N - n - 1)$$

of the irreducible non-degenerate variety $W_x = \overline{\pi_x(X)} \subset \mathbb{P}^{N-n-1}$ of dimension $n - \delta(X)$. Set $\tilde{\xi}(X) = \mathrm{def}(W_x)$, the dual defect of $W_x \subset \mathbb{P}^{N-n-1}$. Thus

$$\tilde{\gamma}(X) \leq \tilde{\xi}(X). \tag{3.2}$$

Since $\dim(W_x) = n - \delta(X)$, a general fiber of π_x has pure dimension $\delta(X)$.

The following result generalizes the ideas behind the proof of Scorza's Lemma, which we shall describe below.

Lemma 3.3.2 *Let* $X \subset \mathbb{P}^N$ *be an irreducible non-degenerate variety such that* $SX \subsetneq \mathbb{P}^N$ *and let* $\pi_x : X \dashrightarrow W_x \subset \mathbb{P}^{N-n-1}$ *be a general tangential projection. Then:*

1. $\xi(X) = \delta(X) + \tilde{\xi}(X)$;
2. $\gamma(X) = \delta(X) + \tilde{\gamma}(X)$;
3. $0 \leq \gamma(X) - \delta(X) \leq \xi(X) - \delta(X) \leq n - 1 - \delta(X)$.

Proof Let us prove 1) and 3), the proof of 2) being similar. Let the notation be as above. Consider π_x also as a map from $\mathbb{P}^N \setminus T_x X$ to $\mathbb{P}^{N-n-1}$. Define $\tilde{H} = \pi_x(H) \subset \mathbb{P}^{N-n-1}$ and let $\hat{\Xi} = \pi_x(\Xi_p(X))$. For every point $z \in \Xi_p(X) \setminus (T_x X \cap X)$ we get $\pi_x(T_z X) \subseteq T_{\pi_x(z)} W_x$. Thus a smooth point $\pi_x(z) \in W_x$, with $z \in \Xi_p(X) \cap X_{\mathrm{reg}}$, is contained in the contact locus of $\tilde{H}$ on W_x. Note that by generic smoothness we can assume that $\pi_x(T_y X) = \pi_x(T_p S X) = T_{\pi_x(y)} W_x$ and that $\pi_x(y)$ is a smooth point of W_x. Therefore $\hat{\Xi}$ is contained in the contact locus, let us say $\tilde{\Xi}$, of $\tilde{H}$ on W_x, yielding $\Xi_p(X) \subseteq \pi_x^{-1}(\tilde{\Xi})$. On the other hand, by reversing the argument, we immediately see that the irreducible component of $\pi_x^{-1}(\tilde{\Xi})$ passing through y coincides with the irreducible component of $\Xi_p(X)$ passing through y. By the generality assumptions every irreducible component of $\pi_x^{-1}(\tilde{\Xi})$ has dimension $\dim(\tilde{\Xi}) + \delta(X) = \tilde{\xi}(X) + \delta(X)$, proving part 1). The last inequality in part 3) follows from the fact that $X \subset \mathbb{P}^N$ is non-degenerate. $\qquad\square$

The following results of Scorza reveals that (smooth) varieties with *good* tangential behaviour and positive secant defect have as entry loci quadric hypersurfaces, provided their secant varieties do not fill the ambient space. It also easily implies [145, Proposition 2.1], where another condition assuring the quadratic entry locus property is introduced.

Theorem 3.3.3 (Scorza's Lemma, [163, footnote pg. 170 Opere Scelte Vol. I] and [166]) *Let* $X \subset \mathbb{P}^N$ *be an irreducible non-degenerate variety of secant defect* $\delta(X) = \delta \geq 1$ *such that* $SX \subsetneq \mathbb{P}^N$. *Suppose that a general tangential projection* $\pi_x(X) = W_x \subset \mathbb{P}^{N-n-1}$ *is an irreducible variety having birational Gauss map, i.e.* $\tilde{\gamma}(X) = 0$; *equivalently suppose that* $\gamma(X) = \delta$. *Let* $y \in X$ *be a general point. Then*

1. *the irreducible component of the closure of the fiber of the rational map*

$$\pi_x : X \dashrightarrow W_x \subset \mathbb{P}^{N-n-1}$$

passing through y is either an irreducible quadric hypersurface of dimension δ or a linear space of dimension δ, the last case occurring only for singular varieties.

2. *There exists on $X \subset \mathbb{P}^N$ a $2(n - \delta)$-dimensional family $\mathcal{Q}$ of quadric hypersurfaces of dimension δ such that through two general points of $x, y \in X$ there passes a unique quadric $Q_{x,y}$ of the family $\mathcal{Q}$. Furthermore, the quadric $Q_{x,y}$ is smooth at the points x and y and it consists of the irreducible components of Σ_p passing through x and y, $p \in < x, y >$ general.*

3. *If X is smooth, then a general member of $\mathcal{Q}$ is smooth.*

Proof Let $p \in < x, y >$ be a general point. Then $\pi_x(y) = y' \in W_x$ and $\pi_y(x) = x' \in W_y$. By definition of π_x, respectively π_y,

$$\langle T_x X, T_y X \rangle = \langle T_x X, T_{y'} W_x \rangle = \langle T_{x'} W_y, T_y X \rangle \tag{3.3}$$

and these linear spaces have dimension $2n + 1 - \delta = \dim(T_p SX)$ by Terracini's Lemma.

The cones $S(T_x X, W_x)$ and $S(T_y X, W_y)$ contain X. Thus, by (3.3), we deduce that Γ_p is contained in the contact locus on $S(T_x X, W_x)$ of $\langle T_x X, T_{y'} W_x \rangle$, which is $\langle T_x X, y' \rangle = \langle T_x X, y \rangle$, and also in the contact locus on $S(T_y X, W_y)$ of $\langle T_{x'} W_y, T_y X \rangle$, which is $\langle T_y X, x' \rangle = \langle T_y X, x \rangle$. This follows from the hypothesis on W_x, respectively W_y.

In particular, the irreducible component through y of the contact locus on X of

$$\langle T_x X, T_{y'} W_x \rangle,$$

respectively the irreducible component through x of the contact locus of $\langle T_{x'} W_y, T_y X \rangle$, is contained in $\langle T_x X, y \rangle$, respectively $\langle x, T_y X \rangle$. Thus the irreducible components of Γ_p through x and through y have dimension $\gamma(X) = \delta$, coincide with Σ_p^x and Σ_p^y and are contained in $\langle y, T_x X \rangle \cap \langle x, T_y X \rangle = \mathbb{P}^{\delta+1}$.

The line $< x, y >$ is a general secant line to X, contained in $\mathbb{P}^{\delta+1}$, so that

$$\{x, y\} \subseteq < x, y > \cap (\Sigma_p^x \cup \Sigma_p^y) \subseteq < x, y > \cap X = \{x, y\},$$

where the last equality is scheme-theoretical by the Trisecant Lemma. Thus the hypersurface $\Sigma_p^x \cup \Sigma_p^y \subset \mathbb{P}^{\delta+1}$ has degree two and the points x, y are smooth points of the quadric hypersurface $\Sigma_p^x \cup \Sigma_p^y$. Therefore either $\Sigma_p^x = \Sigma_p^y$ is an irreducible quadric hypersurface, or $\Sigma_p^x \cup \Sigma_p^y$ is a rank 2 quadric hypersurface in $\mathbb{P}^{\delta+1}$. Since Σ_p is equidimensional, Terracini's Lemma implies that $\Sigma_p \setminus \mathrm{Sing}(X) = (\Sigma_p^x \cup \Sigma_p^y) \setminus \mathrm{Sing}(X)$.

Let $\mathcal{Q}$ be the family of quadric hypersurfaces generated in this way. A count of parameters shows that the family has dimension $2(n - \delta)$, while the smoothness of the entry loci at x and y assures us that there is a unique quadric of the family through x and y.

If X is smooth, the arguments of [66, pg. 964–966] yield the smoothness of the general entry locus of X. All the other assertions now easily follow. □

Scorza repeatedly used the previous result in [163] and [166] when $\xi(X) = \delta(X) = 1$, even though his argument actually proves Theorem 3.3.3. The fact that varieties with $\xi(X) = \delta(X) \geq 1$ have quadratic entry loci can also be obtained via a strengthening of Terracini's Lemma, a result obtained by Terracini in [186], reproved recently by Chiantini and Ciliberto and which now we recall. We define $\xi_k(X) = \xi_k(X, H)$ as the dimension of the contact locus on X of a hyperplane $H \supseteq S^k X$, if any. Reasoning as for $k = 1$ we see that $\xi_k(X)$ is well defined.

Theorem 3.3.4 ([29, Theorem 2.4]) *Let $X \subset \mathbb{P}^N$ be a non-degenerate irreducible variety such that $S^k X \subsetneq \mathbb{P}^N$. Then for general $p \in S^k X$ we have*

$$\dim(\langle \Xi_p(H) \rangle) = (k + 1)(\xi_k(X) - n) + s_k(X)$$

and moreover $S^k \Xi_p(H) = \langle \Xi_p(H) \rangle$.

3.4 Severi's Characterization of the Veronese Surface Versus Mori's Characterization of Projective Spaces

One of the most well-known results in the classical theory of projective surfaces is Severi's characterization of the Veronese surface in $\mathbb{P}^5$ as the unique irreducible surface, not a cone, whose secant variety does not fill the whole space. See also the discussion at the beginning of Sect. 1.4 for the relation of this result to earlier work of del Pezzo and Scorza.

The proof presented below, based on tangential projections and its connections with the second fundamental form, has the advantage of revealing a very interesting parallel with the proof of the well-known *abstract* characterization of $\mathbb{P}^n$, due to Mori [137], as the unique smooth variety with ample tangent bundle.

On one hand secant defectiveness (or some maximality condition as in Theorem 3.4.4) assures the existence of a complete family of irreducible conics such that through two general points there passes a unique conic of the family, see the proof of Theorem 3.4.1 or of Theorem 3.4.4. On the other hand the ampleness of the tangent bundle implies the existence of a family of (*abstract*) *lines* joining two arbitrary points of the manifold, see [136].

Our proof also points out the first instance of the importance of studying rational curves naturally appearing on secant defective varieties and their relations with the second fundamental form and with tangential projections.

All the ideas behind the approach we shall present are essentially due to Scorza, see [166], who first tried to classify all irreducible, not necessarily smooth, secant defective varieties and who first realized the importance of the *conic connectedness condition* for embedded varieties. In our opinion the theory developed in [166] can be considered as the first incarnation, *ante litteram*, of the notion of rational connectedness:

> Invece per le V_4 di prima e terza specie arrivo a caratterizzarle tutte valendomi della teoria dei sistemi lineari sopra una varietà algebrica e, per le ultime, della circostanza che esse contengono un sistema ∞^6 di coniche così che per ogni loro coppia di punti passa una e una sola conica. Inoltre la natura dei ragionamenti è tale da mostrare come i risultati ottenuti possano estendersi, almeno per la maggior parte, alle varietà (di *prima* e *ultima* specie) a un numero qualunque di dimensioni, …
>
> Gaetano Scorza, [166, *Opere Scelte*, vol. 1, pg. 253].

Some of the above cited pioneering ideas of Scorza have been rewritten in modern language and further developed in the work in progress [30].

Theorem 3.4.1 (Severi's Characterization of the Veronese Surface in $\mathbb{P}^5$, [176])
Let $X \subset \mathbb{P}^N$ $N \geq 5$ be an irreducible surface such that $\dim(SX) = 4$. Then either X is a cone over an irreducible curve or $N = 5$ and X is projectively equivalent to the Veronese surface $v_2(\mathbb{P}^2) \subset \mathbb{P}^5$.

Proof For general $x \in X$, let $\pi_x : X \dashrightarrow W_x \subseteq \mathbb{P}^{N-3}$ be the tangential projection. Since $\dim(SX) = 4$, W_x is an irreducible curve so that $\tilde{\gamma}(X) = 0$. By Scorza's Lemma, Theorem 3.3.3, we deduce the existence of a two-dimensional family $\mathscr{C}$ of conics contained in X such that through two general points $x_1, x_2 \in X$ there passes a unique conic C_{x_1,x_2} of the family $\mathscr{C}$.

If the general conic in $\mathscr{C}$ is reducible, then X is a cone over a curve by Corollary 1.3.4. On the contrary if the general conic in $\mathscr{C}$ is irreducible, then $N = 5$ and W_x is a conic since a general conic in $\mathscr{C}$ dominates W_x via π_x.

In particular, the tangential projection of a conic passing through a general point z is an isomorphism onto W_x. The conic through x and z is contracted by π_x so that the intersection of two general conics of the family is transversal and consists of a unique point, see also the proof of Corollary 5.4.2.

Consider $\psi : \mathrm{Bl}_x X \to X$ and let $\tilde{\pi}_x$ be the map induced by π_x on X, see Sect. 2.3.2. Let E be the exceptional divisor of ψ. Since $\tilde{\pi}_x(E) = W_x$ by the previous analysis, $\tilde{\pi}_x$ is given by the complete linear system $|\mathscr{O}_{\mathbb{P}^1}(2)|$. Therefore there is no line passing through x and every conic of $\mathscr{C}$ passing through x is irreducible.

Let $\mathscr{C}_x \subset \mathrm{Hilb}_x^{2t+1}(X)$ be an irreducible (rational) curve parametrizing conics of the family passing through a general point $x \in X$. After normalizing and letting $\tilde{\mathscr{C}}_x \to \mathscr{C}$ be the normalization morphism, we can suppose that we have the following diagram:

where $\mathscr{F}$ is a smooth surface such that $\pi : \mathscr{F} \to \tilde{\mathscr{C}}_x$ is a $\mathbb{P}^1$–bundle, $\tilde{\mathscr{C}}_x \simeq \mathbb{P}^1$ and ϕ is birational.

The fibers of π are sent into conics through x, while the strict transform on $\mathscr{F}$ of general conics through a general point z are sent into sections of π disjoint from the tautological section $E = \phi^{-1}(x)$, which is contracted to the smooth point x. Thus $E^2 = -e, e > 0$ and $\mathscr{F} \simeq \mathbb{F}_e$ as a $\mathbb{P}^1$–bundle. Let $i : X \to \mathbb{P}^5$ be the embedding of X and let f be the class of a general fiber of π. Then we can suppose that $i \circ \phi : \mathbb{F}_e \to \mathbb{P}^5$ is given by a linear system of the form $|\alpha E + \beta f|$.

If $H \subset \mathbb{P}^5$ is a hyperplane, then, by definition, $\phi^*(H) = \alpha E + \beta f$. Thus $2 = f \cdot \phi^*(H) = \alpha, 0 = \phi^*(H) \cdot E = -\alpha e + \beta$, yielding $\phi^*(H) = 2(E + ef)$. Since the inverse image by ϕ of a general conic passing through a general point of X produces a section E' of π disjoint from E, we get $2 = E' \cdot \phi^*(H) = 2e$, yielding $e = 1$ and $\phi^*(H) = 2(E + f)$. In particular, we have $\deg(X) = [2(E + f)]^2 = 4$.

In conclusion, $\mathscr{F} \simeq \mathbb{F}_1$, the morphism ϕ contracts E to a smooth point and it is an isomorphism outside E. Thus $X \simeq \mathbb{P}^2$ and $\mathscr{O}_X(1) \simeq \mathscr{O}_{\mathbb{P}^2}(2)$, concluding the proof. $\qquad\square$

The following is an interesting generalization to higher dimension proved by Edwards in [50].

Corollary 3.4.2 *Let $X \subset \mathbb{P}^N$ be an irreducible non-degenerate variety of dimension $n \geq 3$ such that $\dim(S^k X) = n + 2k$ for some $k \geq 1$ and such that $N \geq n + 2k + 1$ (equivalently $S^k X \subsetneq \mathbb{P}^N$). Let $b = \dim(\mathrm{Vert}(X))$. Then either $b = n - 2$ and $X \subset \mathbb{P}^N$ is a cone over a curve; or $k = 1$, $N = n + 3$, $b = n - 3$ and $X \subset \mathbb{P}^{n+3}$ is a cone over a Veronese surface in $\mathbb{P}^5$.*

Proof From Corollary 1.2.3 we deduce $\dim(SX) = n + 2$ and $\delta(X) = n - 1$. Thus the general tangential projection of X is an irreducible curve $W_x \subseteq \mathbb{P}^{N-n-1}$ and by Scorza's Lemma there is a quadric of dimension $\delta(X) = n - 1 \geq 2$ passing through two general points of X. In particular, there is a line passing through a general point of X so that X is a cone by Corollary 1.3.4.

Let $L = \mathbb{P}^{N-n+2} \subset \mathbb{P}^N$ be a general linear space and let

$$Y = L \cap X \subset L = \mathbb{P}^{N-n+2}.$$

Then Y is an irreducible surface such that $\dim(SY) = 4$. Thus by Theorem 3.4.1 either Y is a cone over a curve or $N - n + 2 = 5$ and Y is projectively equivalent to the Veronese surface. The conclusions now easily follow. $\qquad\square$

We introduce another definition used frequently throughout the paper.

Definition 3.4.3 (Linear Normality) A non-degenerate irreducible variety $X \subset \mathbb{P}^N$ is said to be *linearly normal* if the linear system of hyperplane sections is complete, i.e. if the injective, restriction morphism

$$H^0(\mathscr{O}_{\mathbb{P}^N}(1)) \xrightarrow{r} H^0(\mathscr{O}_X(1))$$

is surjective.

If a variety $X \subset \mathbb{P}^N$ is not linearly normal, then the complete linear system $|\mathcal{O}_X(1)|$ is of projective dimension greater than N and it embeds X as a variety $X' \subset \mathbb{P}^M$, $M > N$. Moreover, there exists a linear space $L = \mathbb{P}^{M-N-1}$ such that $L \cap X' = \emptyset$ and such that $\pi_L : X' \to X \subset \mathbb{P}^N$ is an isomorphism.

Indeed, if $V = r(H^0(\mathcal{O}_{\mathbb{P}^N}(1))) \subsetneq H^0(\mathcal{O}_X(1))$ and if $U \subset H^0(\mathcal{O}_X(1))$ is a complementary subspace of V in $H^0(\mathcal{O}_X(1))$, then one can take $\mathbb{P}^M = \mathbb{P}(H^0(\mathcal{O}_X(1)))$, $L = \mathbb{P}(U)$ and the claim follows from the fact that $\pi_L : X' \simeq X \to X \subset \mathbb{P}^N = \mathbb{P}(V)$ is given by the very ample linear system $|V|$. On the contrary, if X is an isomorphic linear projection of a variety $X' \subset \mathbb{P}^M$, $M > N$, then X is not linearly normal.

Now we are in a position to provide a suitable generalization of the characterization of the Veronese surface, [176], we have just proved under the smoothness assumption. This result was first obtained by Gallarati in [71] and later reproved by Zak, see [198, Chap. V], in a different way. As we shall see, our proof is parallel to the celebrated Mori Characterization of Projective Spaces yielding projective incarnations to the steps of Mori's proof in arbitrary dimension $n \geq 2$.

Theorem 3.4.4 *Let $X \subset \mathbb{P}^N$, $N \geq \frac{n(n+3)}{2}$, be a smooth non-degenerate variety of secant defect $\delta(X) \geq 1$. Then $N = \frac{n(n+3)}{2}$ and $X \subset \mathbb{P}^{\frac{n(n+3)}{2}}$ is projectively equivalent to $v_2(\mathbb{P}^n) \subset \mathbb{P}^{\frac{n(n+3)}{2}}$.*

Proof Proposition 2.3.5 yields $\tilde{\pi}_x(E) = W_x = \pi_x(X) \subset \mathbb{P}^{N-n-1}$, $N = \frac{n(n+3)}{2}$, that $X \subset \mathbb{P}^N$ is linearly normal and that through x there passes no line contained in X (a line contained in X and passing through x would produce a base point in the linear system $|II_{x,X}|$). Therefore $\pi_x(X) = W_x \subset \mathbb{P}^{\frac{(n-1)(n+2)}{2}}$, being equal to $\tilde{\pi}_x(E)$, is projectively equivalent to $v_2(\mathbb{P}^{n-1}) \subset \mathbb{P}^{\frac{(n-1)(n+2)}{2}}$. Indeed, $\tilde{\pi}_x(E)$ is the image of $\mathbb{P}^{n-1}$ by the complete linear system $|\mathcal{O}_{\mathbb{P}^{n-1}}(2)|$. In particular, $\delta(X) = \dim(X) - \dim(W_x) = 1$.

Scorza's Lemma implies that for a general $x \in X \subset \mathbb{P}^{\frac{n(n+3)}{2}}$, from now on fixed, there exists an irreducible family $\mathscr{C}$ of irreducible conics of dimension $2n - 2$ such that through x and a general point of X there passes a unique conic of the family.

Let $V_x \subseteq X$ be the locus described by the conics in $\mathscr{C}$ passing through x, indicated by $\mathscr{C}_x$. By construction $V_x = X$. We claim that through every point $y \in X \setminus x$ there passes at most a finite number of conics in $\mathscr{C}$.

Indeed, suppose on the contrary that there exist infinitely many conics through x and such a $y \in X \setminus x$. Since there are no lines passing through x, every such conic is irreducible and hence smooth. Thus we could construct the following diagram:

where $\mathscr{F}$ is a smooth surface, $\tilde{\mathscr{C}}_{x,y}$ is a smooth curve and $\pi : \mathscr{F} \to \tilde{\mathscr{C}}_{x,y}$ is a $\mathbb{P}^1$-bundle having two sections: E_x, corresponding to *conics through x*, and E_y, corresponding to *conics through y*.

The Hodge Index Theorem applied to the lattice generated by E_x, E_y and the class f of a fiber of π shows that this is impossible because $E_x^2 < 0$ and $E_y^2 < 0$, being curves contracted by ϕ. This is the basic step for the so-called *Bend and Break* of Mori Theory, reproduced here in its simplest possible incarnation, see [116, Theorem II.5.4] for the original and strongest version.

Consider the diagram

where $\mathscr{F}$ is the universal family and ϕ is the tautological morphism and where $\mathscr{C}_x$ is (an irreducible component of) the family of conics in $\mathscr{C}$ passing through x. Let $\tilde{E} = \phi^{-1}(x)$, scheme-theoretically.

Every conic in $\mathscr{C}_x$ is irreducible since there are no lines through x. Therefore through every point of $X \setminus x$ there passes a unique conic of $\mathscr{C}_x$ by Zariski's Main Theorem and by the property of $V_x = X$ proved above (a finite birational morphism onto a normal variety is an isomorphism). Thus ϕ is an isomorphism between $\mathscr{F} \setminus \tilde{E}$ and $X \setminus x$, contracting $\tilde{E}$ to a point.

For an arbitrary point $y \in X \setminus x$, there exists a unique smooth conic through x and y, yielding $(T_x X \cap X)_{\mathrm{red}} = x$. In particular, $\tilde{\pi}_x : \mathrm{Bl}_x X \to W_x$ is a morphism so that given a conic through x and a tangential direction through x there exists a unique smooth conic through x tangent to this direction.

This implies that $\phi : \mathscr{F} \to X$ factors through $\mathrm{Bl}_x X \to X$, inducing an isomorphism between $\tilde{E}$ and E, see the discussion leading to the construction of diagram (2.3). Therefore $\mathscr{F} \simeq \mathrm{Bl}_x X$ and $\tilde{E} \simeq \mathbb{P}^{n-1}$.

From [116, Lemma V.3.7.8] we deduce that $\mathrm{Bl}_x(X) \to \mathbb{P}^{n-1}$ is isomorphic over $\mathbb{P}^{n-1}$ to $\mathbb{P}(\mathscr{O}_{\mathbb{P}^{n-1}} \oplus \mathscr{O}_{\mathbb{P}^{n-1}}(-1))$ and that X is isomorphic to $\mathbb{P}^n$. Now it is clear that the embedding of X in $\mathbb{P}^N$ is given by the complete linear system $|\mathscr{O}_{\mathbb{P}^n}(2)|$, concluding the proof. $\square$

Chapter 4
Local Quadratic Entry Locus Manifolds and Conic Connected Manifolds

4.1 Definitions and First Geometrical Properties

In the following definition, we consider varieties having the *simplest possible* entry locus. These manifolds were systematically studied for the first time in [160] and subsequently in [103, 104].

Definition 4.1.1 (Quadratic Entry Locus Manifolds = *QEL*-Manifolds) A smooth irreducible non-degenerate projective variety $X \subset \mathbb{P}^N$ is said to be a *quadratic entry locus manifold of type $\delta \geq 0$*, briefly a *QEL-manifold of type δ*, if for general $p \in SX$ the entry locus $\Sigma_p(X)$ is a quadric hypersurface of dimension $\delta = \delta(X)$, where $\delta(X) = 2\dim(X) + 1 - \dim(SX)$ is the secant defect of X.

For $p \in SX \setminus X$ recall that $C_p(X)$ is the cone described by the secant (and tangent) lines to X passing through p. Let us remark that the Trisecant Lemma, see Proposition 1.4.3, ensures that, as soon as $C_p(X)$ is a linear space and $\mathrm{codim}(X) \geq 2$, the general entry locus $\Sigma_p(X)$ is necessarily a quadric hypersurface in $C_p(X)$. Moreover, the smoothness of the general entry locus of a QEL-manifold is a consequence of the smoothness of X (see, for example, [66, pp. 964–966]).

The next result, which is essentially contained in the proof of [188, Proposition 2.8], shows that the class of QEL-manifolds is sufficiently large and interesting.

Proposition 4.1.2 ([188, Proposition 2.8]) *A smooth non-degenerate variety $X \subset \mathbb{P}^N$, scheme-theoretically defined by quadratic equations whose Koszul syzygies are generated by the linear ones, is a QEL-manifold.*

The class of QEL-manifolds is not stable under isomorphic projection. So, we extend this notion, following [103, 104, 160], see also Proposition 4.1.5 part (ii) and the discussion after the next definition.

© Springer International Publishing Switzerland 2016
F. Russo, *On the Geometry of Some Special Projective Varieties*,
Lecture Notes of the Unione Matematica Italiana 18,
DOI 10.1007/978-3-319-26765-4_4

Definition 4.1.3 (Local Quadratic Entry Locus Manifolds = *LQEL*-Manifolds)
A smooth irreducible non-degenerate projective variety $X \subset \mathbb{P}^N$ is said to be a *local quadratic entry locus manifold of type* $\delta \geq 0$, briefly an *LQEL-manifold of type* δ, if, for general $x, y \in X$ distinct points, there is a quadric hypersurface of dimension $\delta = \delta(X)$ contained in X and passing through x, y.

Note that, for $\delta = 0$, being an LQEL-manifold imposes no restriction on X. Moreover, any *QEL*-manifold is *LQEL* but the converse is not true. Indeed, the isomorphic projection of a *QEL*-manifold $X \subset \mathbb{P}^N$ of type $\delta > 0$ with $N > \dim(SX)$ to $\mathbb{P}^{\dim(SX)}$ produces an example of a *LQEL*-manifold of type δ such that the entry locus through a general point of $\mathbb{P}^{\dim(SX)}$ has $\deg(SX) \geq 3$ irreducible components, which are smooth quadrics of dimension δ.

A further generalization is given by the following class of manifolds.

Definition 4.1.4 (Conic-Connected Manifold = *CC*-Manifold, [103, 104, 106, 112, 160]) A smooth irreducible non-degenerate projective variety $X \subset \mathbb{P}^N$ is said to be a *conic-connected manifold*, briefly a *CC-manifold*, if through two general points of X there passes an irreducible conic contained in X.

Smooth cubic hypersurfaces in $\mathbb{P}^4$ are *CC*-manifolds but not *LQEL*-manifolds since their secant defect is equal to their dimension. Thus these two classes are also distinct.

Let us collect some consequences of the above definitions in the following proposition, whose easy proof is left to the reader.

Proposition 4.1.5 *Let $X \subset \mathbb{P}^N$ be an irreducible non-degenerate smooth projective variety.*

(i) *If X is a QEL-manifold and $SX = \mathbb{P}^N$, then X is linearly normal.*
(ii) *If $X' \subset \mathbb{P}^M$, $M \leq N - 1$, is an isomorphic projection of X, then X' is an LQEL-manifold if and only if X is an LQEL-manifold.*
(iii) *If X is an (L)QEL-manifold of type $\delta \geq 1$, then a general hyperplane section is an (L)QEL-manifold of type $\delta - 1$.*

All examples of LQEL-manifolds we are aware of are obtained by isomorphic projections of QEL-manifolds. Therefore, one could ask if any linearly normal LQEL-manifold is a QEL-manifold. We do not know the answer.

Lemma 4.1.6 *Let $X \subset \mathbb{P}^N$ be an LQEL-manifold with $\delta(X) > 0$ and let $x, y \in X$ be general points. There is a unique quadric hypersurface of dimension δ, say $Q_{x,y}$, passing through x, y and contained in X. Moreover, $Q_{x,y}$ is irreducible.*

Proof Uniqueness follows from the fact that the general entry locus passing through two general points is smooth at these points, which is an easy consequence of Terracini's Lemma, see e.g. Exercise 1.5.24 or [92, Proposition 3.3]. To see that $Q_{x,y}$ is irreducible, we may assume $\delta = 1$ by passing to general hyperplane sections by Proposition 4.1.5. Clearly we can assume that X is not covered by lines passing through a fixed general point $x \in X$ because otherwise X would be a linear space, which it is not an *LQEL*-manifold. Let $y \in X$ be a general point and suppose that

$Q_{x,y}$ were reducible. Then $Q_{x,y} = L_x \cup L_y$ with $L_x \subset X$ a line passing through x and $L_y \subset X$ a line passing through y. Since $x, y \in X$ are general the normal bundles $N_{L_x/X}$ and $N_{L_y/X}$ are generated by global sections by Corollary 2.1.13. Therefore the conic $L_x \cup L_y$ can be smoothed inside X keeping x fixed by [116, II.7.6.1]. This means that the general conic passing through x is irreducible and hence that through x and a general $y' \in X$ there passes an irreducible conic, contrary to the previous claim. □

A monodromy argument shows that if X is an *LQEL*-manifold, the general entry locus is a union of finitely many quadric hypersurfaces of dimension δ. Moreover, *LQEL*-manifolds of type $\delta \geq 1$ are *CC*-manifolds.

4.2 Qualitative Properties of *CC*-Manifolds and of *LQEL*-Manifolds

We describe the family of conics naturally appearing on *CC*-manifolds and on *LQEL*-manifolds of type $\delta > 0$, relating them to intrinsic invariants via the tools and the theories introduced in Chap. 2.

Proposition 4.2.1 ([103, Proposition 3.2]) *Let $X \subset \mathbb{P}^N$ be a CC-manifold of secant defect δ. Let $C = C_{x,y}$ be a general conic through the general points $x, y \in X$ and let $c = [C]$ be the point representing C in the Hilbert scheme of X. Let $\mathscr{C}_x$ be the unique irreducible component of the Hilbert scheme of conics passing through x and which contains the point c.*

1. *X is a simply connected manifold such that $H^0(\Omega_X^{\otimes m}) = 0$ for every $m \geq 1$ and $H^i(\mathscr{O}_X) = 0$ for every $i > 0$.*
2. *We have $n + \delta \geq -K_X \cdot C = \dim(\mathscr{C}_x) + 2 \geq n + 1$ and the locus of conics through x and y is contained in the linear space*

$$\mathbb{P}^{\delta+1} = \langle T_x X, y \rangle \cap \langle x, T_y X \rangle.$$

3. *The equality $-K_X \cdot C = n + \delta$ holds if and only if $X \subset \mathbb{P}^N$ is an LQEL-manifold.*
4. *If $\delta \geq 3$, then $X \subset \mathbb{P}^N$ is a Fano manifold with $\mathrm{Pic}(X) \simeq \mathbb{Z}\langle \mathscr{O}_X(1) \rangle$ and index $i(X) = \frac{\dim(\mathscr{C}_x)}{2} + 1$.*

Proof The variety X is clearly rationally connected. The conclusions of part (1) are contained in [117, Proposition 2.5] and also in [40, Corollary 4.18].

We have the universal family $g : \mathscr{F}_x \to \mathscr{C}_x$ and the tautological morphism $f : \mathscr{F}_x \to X$, which is surjective. Since $C \in \mathscr{C}_x$ is a general conic and since $x \in X$ and y are general points, we get $\dim(\mathscr{C}_x) = -K_X \cdot C - 2$.

Indeed, by Corollary 2.1.13,

$$T_{X|C} \simeq \bigoplus_{i=1}^{n} \mathscr{O}_{\mathbb{P}^1}(a_i)$$

is ample. Hence $a_i > 0$ for every $i = 1, \ldots, n$, and $N_{C/X}$ is ample, being a quotient of $TX_{|C}$. Thus $\mathscr{C}_x$ is smooth at $C \in \mathscr{C}_x$ and of dimension

$$H^0(N_{C/X}(-1)) = H^0(T_{X|C}(-1)) - 2 = -2 + \sum_{i=1}^{n} a_i = -K_X \cdot C - 2.$$

Take a general point $y \in X$ and a general $p \in \langle x, y \rangle$. The conics passing through x and y are parameterized by $g(f^{-1}(y))$, which has pure dimension

$$\dim(\mathscr{F}_x) - n = \dim(\mathscr{C}_x) + 1 - n = -K_X \cdot C - 1 - n.$$

We claim that the locus of conics through x and y, denoted by $\mathscr{C}_{x,y}$, has dimension $-K_X \cdot C - n$ and is clearly contained in the irreducible component of the entry locus (with respect to p) through x and y. Indeed, conics through x, y and another general point $z \in \mathscr{C}_{x,y}$ have to be finitely many. Otherwise, their locus would fill up the plane $\langle x, y, z \rangle$ and this would imply that the line $\langle x, y \rangle$ is contained in X. But we have excluded linear spaces from the definition of CC and LQEL-manifolds. Therefore $\delta \geq -K_X \cdot C - n$, that is $-K_X \cdot C \leq n + \delta$. The locus of conics is contained in $\langle T_x X, y \rangle \cap \langle x, T_y X \rangle$, which is a linear space of dimension $\delta + 1$ by Terracini's Lemma. This proves (2).

If $-K_X \cdot C = n + \delta$, then, for $p \in \langle x, y \rangle$ general, the irreducible component $\Sigma_{x,y}^p$ of the entry locus passing through x and y coincides with the locus of conics through x and y, so that it is contained in $\langle T_x X, y \rangle \cap \langle x, T_y X \rangle = \mathbb{P}^{\delta+1}$. Thus $\Sigma_{x,y}^p$ is a quadric hypersurface by the Trisecant Lemma, see Proposition 1.4.3, and by the generality of x and y (if $\delta = n$, $X \subset \mathbb{P}^{n+1}$ is a quadric hypersurface). So, (3) is proved.

Finally, (4) follows from the Barth–Larsen Theorem, Theorem 3.1.1 here, from the fact that X contains moving conics and from (2).   □

Corollary 4.2.2 ([160, Theorem 2.1]) *Let $X \subset \mathbb{P}^N$ be an LQEL-manifold of type $\delta \geq 1$. Then:*

1. *There exists on X an irreducible family of conics $\mathscr{C}$ of dimension $2n + \delta - 3$, whose general member is smooth. This family describes an open subset of an irreducible component of the Hilbert scheme of conics on X.*
2. *Given a general point $x \in X$, let $\mathscr{C}_x$ be the family of conics in $\mathscr{C}$ passing through x. Then $\mathscr{C}_x$ has dimension $n + \delta - 2$, equal to the dimension of the irreducible components of $\mathscr{C}_x$ describing dense subsets of X.*
3. *Given two general points $x, y \in X$, the locus $Q_{x,y}$ of the family $\mathscr{C}_{x,y}$ of smooth conics in $\mathscr{C}$ passing through x and y is a smooth quadric hypersurface of dimension δ. The family $\mathscr{C}_{x,y}$ is irreducible and of dimension $\delta - 1$.*
4. *A general conic $C \in \mathscr{C}_x$ intersects $T_x X$ only at x. Moreover, the tangent lines to smooth conics contained in X and passing through $x \in X$ describe an open subset of $\mathbb{P}((t_x X)^*)$.*

Proof Part (4) is the definition of an *LQEL*-variety. Indeed, the plane spanned by every conic through x and y contains the line $\langle x, y \rangle$, which is a general secant line to X. Thus a conic through x and y is contained in the entry locus of every $p \in \langle x, y \rangle$ not on X, so that for $p \in \langle x, y \rangle$ general it is contained in the smooth quadric hypersurface $Q_{x,y}$, the unique irreducible component of Σ_p passing through x and y.

Let us prove parts (3) and (5). Fixing two general points $x, y \in X$ there exists a smooth quadric hypersurface $Q_{x,y} \subset X \subset \mathbb{P}^N$ of dimension $\delta \geq 1$ through x and y. Thus there exists a smooth conic passing through x and a general point $y \in X$. In particular, there exists an irreducible family of smooth conics passing through x, let us say $\mathscr{C}_x^1$, whose members describe a dense subset of X.

Let $\widetilde{\mathscr{C}_x}$ be the universal family over $\mathscr{C}_x$ and let $\pi : \widetilde{\mathscr{C}_x} \to X$ be the tautological morphism. By part (4) and the Theorem on the dimension of the fibers we get that $\dim(\widetilde{\mathscr{C}_x^1}) = n + \delta - 1$. Thus $n + \delta - 2 = \dim(\mathscr{C}_x^1)$ and parts (3) and (5) are proved. Part (2) now easily follows in the same way, counting dimensions.

Let $C \subset X$ be a general smooth conic passing through $x \in X$. If we fix a direction through x, there exists a conic in $\mathscr{C}_x^1$ tangent to the fixed direction. If the direction does not correspond to a line contained in X, then the conic is irreducible, and hence smooth at x.

Consider the map $\tau_x : \mathscr{C}_x \dashrightarrow \mathbb{P}((t_x X)^*)$, which associates to a conic in $\mathscr{C}_x$ its tangent line at x. The closure of the image of τ_x in $\mathbb{P}((t_x X)^*)$ has dimension $n - 1$ by the above analysis, containing the open set parametrizing the tangent directions not corresponding to lines through x and contained in X.   □

On an *LQEL*-manifold of type $\delta \geq 2$ there are also lines coming from the entry loci and we proceed to investigate them. Indeed, if $\delta \geq 2$, then $\mathscr{L}_{x,X}$ is non-empty so that it is a smooth variety by Proposition 2.2.1.

We are now in position to prove a fundamental result on the geometry of lines on an *LQEL*-manifold, which, via the study of the projective geometry of $\mathscr{L}_x \subset \mathbb{P}^{n-1}$ and of its dimension, will yield significant obstructions for the existence of *LQEL*-manifolds of type $\delta \geq 3$. The most relevant part for future applications is part (4). Part (1) holds more generally for every smooth secant defective variety, as we have seen in Proposition 2.3.5.

Theorem 4.2.3 ([160, Theorem 2.3]) *Suppose that $X \subset \mathbb{P}^N$ is an LQEL-manifold of type δ.*

1. ([166, p. 282, Opere Complete, vol. I]) If $\delta \geq 1$, then

$$\widetilde{\pi}_x : \mathbb{P}((t_x X)^*) \dashrightarrow W_x \subseteq \mathbb{P}^{N-n-1}$$

is dominant, so that $\dim(|II_{x,X}|) = N - n - 1$ and $N \leq \frac{n(n+3)}{2}$.

2. If $\delta \geq 2$, the smooth, not necessarily irreducible, variety $\mathscr{L}_x \subset \mathbb{P}^{n-1}$ is non-degenerate and it consists of irreducible components of the base locus scheme of $|II_{x,X}|$. Moreover, the closure of the irreducible component of a general fiber of $\widetilde{\pi}_x$ passing through a general point $p \in \mathbb{P}((t_x X)^*)$ is a linear space $\mathbb{P}_p^{\delta-1}$, cutting scheme-theoretically $\mathscr{L}_x$ in a quadric hypersurface of dimension $\delta - 2$.

3. If $\mathscr{L}_x \subset \mathbb{P}((t_x X)^*)$ is irreducible and if $\delta \geq 2$, then $S\mathscr{L}_x = \mathbb{P}((t_x X)^*)$ and $\mathscr{L}_x \subset \mathbb{P}((t_x X)^*)$ is a QEL-manifold of type $\delta(X) - 2$.

4. *If* $\delta \geq 3$, *then*

a) $\mathrm{Pic}(X) \simeq \mathbb{Z}\langle \mathscr{O}_X(1) \rangle$.

b) *For any line* $L \subset X$, $-K_X \cdot L = \frac{n+\delta}{2}$, *so that* $i(X) = \frac{n+\delta}{2}$, *where* $i(X)$ *is the index of the Fano manifold* X. *In particular*, $n + \delta \equiv 0 \pmod 2$, *that is* $n \equiv \delta \pmod 2$.

c) *There exists on* X *an irreducible family of lines of dimension* $\frac{3n+\delta}{2} - 3$ *such that for a general* L *in this family*

$$TX_{|L} = \mathscr{O}_{\mathbb{P}^1}(2) \oplus \mathscr{O}_{\mathbb{P}^1}(1)^{\frac{n+\delta}{2}-2} \oplus \mathscr{O}_{\mathbb{P}^1}^{\frac{n-\delta}{2}+1}.$$

d) *If* $x \in X$ *is general, then* $\mathscr{L}_x \subset \mathbb{P}((t_x X)^*)$ *is a QEL-manifold of dimension* $\frac{n+\delta}{2} - 2$, *of type* $\delta(X) - 2$ *and such that* $S\mathscr{L}_x = \mathbb{P}((t_x X)^*)$.

Proof Part (1) is classical and as we said above holds for every smooth secant defective variety, see Proposition 2.3.5. Since its proof is self-contained and elementary for *LQEL*-varieties, we include it for the reader's convenience. It suffices to show that, via the restriction of $\widetilde{\pi}_x$, the exceptional divisor $E = \mathbb{P}((t_x X)^*)$ dominates $W_x \subseteq \mathbb{P}^{N-n-1}$. Take a general point $y \in X$. By part (6) of Corollary 4.2.2, there exists a conic $C_{x,y}$ through x and y, cutting $T_x X$ only at x. Thus $\pi_x(C_{x,y}) = \pi_x(y) \in W_x$ is a general point and clearly $\widetilde{\pi}_x(\mathbb{P}(t_x C_{x,y})) = \pi_x(C_{x,y}) = \pi_x(y)$. Therefore the restriction of $\widetilde{\pi}_x$ to E is dominant as a map to $W_x \subseteq \mathbb{P}^{N-n-1}$, yielding $\dim(|II_{x,x}|) = N-n-1$. In particular, $N-n-1 = \dim(|II_{x,x}|) \leq \dim(|\mathscr{O}_{\mathbb{P}^{n-1}}(2)|) = \frac{n(n+1)}{2} - 1$ and $N \leq \frac{n(n+3)}{2}$.

Suppose from now on $\delta \geq 2$. If $y \in X$ is a general point and if $C_{x,y}$ is a smooth conic through x and y the point $\mathbb{P}(t_x C_{x,y})$ is a general point of $\mathbb{P}((t_x X)^*)$, by Corollary 4.2.2 part (6). Consider the unique quadric hypersurface $Q_{x,y}$ of dimension $\delta \geq 2$ through x and y, the irreducible component through x and y of the entry locus of a general $p \in \langle x, y \rangle$. Then $C_{x,y} \subset Q_{x,y}$ and $T_x C_{x,y} \subset T_x Q_{x,y}$. Take a line L_x through x and contained in $Q_{x,y}$, which can be thought of as a point of $\mathscr{L}_x \subset \mathbb{P}((t_x X)^*)$. The plane $\langle L_x, T_x C_{x,y} \rangle$ is contained in $T_x Q_{x,y}$ so that it cuts $Q_{x,y}$ at least in another line L'_x, clearly different from $T_x C_{x,y}$. Thus $T_x C_{x,y}$ belongs to the pencil generated by L_x and L'_x, which projectivized in $\mathbb{P}((t_x X)^*)$ simply means that through the general point $\mathbb{P}(t_x C_{x,y}) \in \mathbb{P}((t_x X)^*)$ there passes the secant line $\langle \mathbb{P}(t_x L_x), \mathbb{P}(t_x L'_x) \rangle$ to $\mathscr{L}_x$. Therefore $\mathscr{L}_x \subset \mathbb{P}((t_x X)^*)$ is non-degenerate and the join of $\mathscr{L}_x$ with itself equals $\mathbb{P}((t_x X)^*)$.

For an irreducible $\mathscr{L}_x \subset \mathbb{P}((t_x X)^*)$ this means precisely $S\mathscr{L}_x = \mathbb{P}((t_x X)^*)$. The scheme $T_x Q_{x,y} \cap Q_{x,y}$ is a quadric cone with vertex x and base a smooth quadric hypersurface of dimension $\delta - 2$. The lines in $T_x Q_{x,y} \cap Q_{x,y}$ describe a smooth quadric hypersurface of dimension $\delta - 2$, $\widetilde{Q}_{x,y} \subset \mathscr{L}_x \subset \mathbb{P}((t_x X)^*)$, whose linear span $\langle \widetilde{Q}_{x,y} \rangle = \mathbb{P}^{\delta-1}$ passes through $r = \mathbb{P}(t_x C_{x,y})$. Since $\widetilde{\pi}_x : \mathbb{P}((t_x X)^*) \dashrightarrow W_x \subseteq \mathbb{P}^{N-n-1}$ is given by a linear system of quadrics vanishing on $\mathscr{L}_x$, the whole $\mathbb{P}^{\delta-1}$ is contracted by $\widetilde{\pi}_x$ to $\widetilde{\pi}_x(r)$.

The closure of the irreducible component of $\widetilde{\pi}_x^{-1}(\widetilde{\pi}_x(r))$ passing through r has dimension $n-1-\dim(W_x) = \delta-1$ so that it coincides with $\langle \widetilde{Q}_{x,y} \rangle = \mathbb{P}^{\delta-1}$. This also shows that $\mathscr{L}_x$ is an irreducible component of the support of the base locus scheme of $|II_{x,X}|$ and also that, when irreducible, $\mathscr{L}_x \subset \mathbb{P}((t_xX)^*)$ is a *QEL*-manifold of type $\delta - 2$. Indeed, in this case $r \in S\mathscr{L}_x$ is a general point and every secant or tangent line to the smooth irreducible variety $\mathscr{L}_x$ passing through r is contracted by $\widetilde{\pi}_x$, since the quadrics in $|II_{x,X}|$ vanish on $\mathscr{L}_x$. Thus every secant line through r is contained in $\langle \widetilde{Q}_{x,y} \rangle$ and the entry locus with respect to r is precisely $\widetilde{Q}_{x,y}$. This concludes the proof of parts (2) and (3).

Suppose $\delta \geq 3$ and let us concentrate on part (4). Item a) follows directly from the Barth–Larsen Theorem, see [15] and Theorem 3.1.1 here, but we provide a direct proof using the geometry of *LQEL*-varieties. There are lines through a general point $x \in X$, for example the ones constructed from the family of entry loci. Reasoning as in Proposition 2.2.1, we get

$$TX_{|L} = \mathcal{O}_{\mathbb{P}^1}(2) \oplus \mathcal{O}_{\mathbb{P}^1}(1)^{m(L)} \oplus \mathcal{O}_{\mathbb{P}^1}^{n-m(L)-1}$$

for every line L through x. Thus if such a line comes from a general entry locus, we get

$$2 + m(L) = -K_X \cdot L = \frac{-K_X \cdot C}{2} = \frac{n+\delta}{2},$$

yielding $m(L) = \frac{n+\delta}{2} - 2$.

We define R_x to be the locus of points on X which can be joined to x by a connected chain of lines whose numerical class is $\frac{1}{2}[C]$, $C \in \mathscr{C}$ a general conic. By construction we get $R_x = X$, so that the Picard number of X is one by [116, IV.3.13.3]. Since the variety X is simply connected, being rationally connected, see Theorem 4.2.2, we deduce $\mathrm{Num}(X) = \mathrm{NS}(X) = \mathrm{Pic}(X) \simeq \mathbb{Z}\langle \mathcal{O}(1) \rangle$. Thus $X \subset \mathbb{P}^N$ is a Fano variety, $\mathscr{L}_x \subset \mathbb{P}^{n-1}$ is equidimensional of dimension $\frac{n+\delta}{2} - 2$ and $m(L) = \frac{n+\delta}{2} - 2$ for every line L through x. Moreover, $\mathscr{L}_x$ is irreducible by Proposition 2.2.1.

The fact that $\mathscr{L}_x \subset \mathbb{P}((t_xX)^*)$ is a *QEL*-manifold of type $\delta(X) - 2$ such that $S\mathscr{L}_x = \mathbb{P}((t_xX)^*)$ follows from part (3) above. Therefore all the assertions are now proved.   □

Example 4.2.4 (Segre Varieties $X = \mathbb{P}^l \times \mathbb{P}^m \subset \mathbb{P}^{lm+l+m}$, $l \geq 1$, $m \geq 1$) By Proposition 4.1.2, we know that $X = \mathbb{P}^l \times \mathbb{P}^m \subset \mathbb{P}^{lm+l+m}$ is a *QEL*-manifold, see also Exercise 1.5.15. We calculate its type, that is we determine $\delta(X)$.

The locus of lines through a point $x \in X$ is easily described, being the union of the two linear spaces of the rulings through x, so that $\mathscr{L}_x = \mathbb{P}^{l-1} \sqcup \mathbb{P}^{m-1} \subset \mathbb{P}^{l+m-1}$. Letting the notation be as in Theorem 4.2.2, we have $C \equiv L_1 + L_2$, where the lines L_1 and L_2 belong to different rulings. Then

$$n + \delta = -K_X \cdot C = (-K_X \cdot L_1) + (-K_X \cdot L_2) = (l-1) + 2 + (m-1) + 2 = n + 2,$$

so that $\delta(\mathbb{P}^l \times \mathbb{P}^m) = 2$ for every $l, m \geq 1$.

Example 4.2.5 (Grassmann Varieties of Lines $\mathbb{G}(1, r) \subset \mathbb{P}^{\binom{r+1}{2}-1}$) It is well known that $\mathscr{L}_x \subset \mathbb{P}((t_x\mathbb{G}(1, r))^*) \simeq \mathbb{P}^{2r-3}$ is projectively equivalent to the Segre variety $\mathbb{P}^1 \times \mathbb{P}^{r-2} \subset \mathbb{P}^{2r-3}$. Moreover, $\mathbb{G}(1, r) \subset \mathbb{P}^{\binom{r+1}{m+1}-1}$ is a *QEL*-manifold, for example by Proposition 4.1.2, and we determine its type δ. Take $x, y \in \mathbb{G}(1, r)$ general. They represent two lines $l_x, l_y \subset \mathbb{P}^r$, $r \geq 3$, which are skew so that $\langle l_x, l_y \rangle = \mathbb{P}^3_{x,y} \subseteq \mathbb{P}^r$. The Plücker embedding of the lines in $\mathbb{P}^3_{x,y}$ is a $\mathbb{G}(1, 3)_{x,y} \subseteq \mathbb{G}(1, r)$ passing through x and y. Therefore $\delta(\mathbb{G}(1, r)) \geq 4$. Thus $r - 1 = \dim(\mathscr{L}_x) = -K_X \cdot L - 2$, where $L \subset \mathbb{G}(m, r)$ is an arbitrary line, yielding $-K_X = (r + 1)H$, H a hyperplane section. Finally, $r + 1 = \frac{2(r-1)+\delta}{2}$ by Theorem 4.2.3, that is $\delta = 4$.

Example 4.2.6 (Spinor Variety $S^{10} \subset \mathbb{P}^{15}$ and E_6-Variety $X \subset \mathbb{P}^{26}$) Let us analyze the ten-dimensional spinor variety $S^{10} \subset \mathbb{P}^{15}$. It is scheme-theoretically defined by ten quadratic forms defining a map $\phi : \mathbb{P}^{15} \dashrightarrow \phi(\mathbb{P}^{15}) \subset \mathbb{P}^9$. The closure of the image $Q = \overline{\phi(\mathbb{P}^{15})} \subset \mathbb{P}^9$ is a smooth eight-dimensional quadric hypersurface and the closure of every fiber is a $\mathbb{P}^7$ cutting X along a smooth quadric hypersurface, see for example [53]. In particular, $\delta(X) = 6$ and $X \subset \mathbb{P}^{15}$ is a Fano manifold of index $i(X) = 8 = n - 2$ such that $\mathrm{Pic}(X) = \mathbb{Z}\langle \mathcal{O}_X(1)\rangle$. It is a so-called Mukai variety with $b_2(X) = 1$ and by the above description it is a *QEL*-manifold of type $\delta = 6$.

For every $x \in X$ the variety $X^1 := \mathscr{L}_x(X) \subset \mathbb{P}^9$ is a variety of dimension $\frac{n+\delta(X)}{2} - 2 = 6$, defined by $\mathrm{codim}(X) = 5$ quadratic equations yielding a dominant map $\widetilde{\pi}_x : \mathbb{P}^9 \dashrightarrow \mathbb{P}^4$. The general fiber of $\widetilde{\pi}_x$ is a linear $\mathbb{P}^5$ cutting $\mathscr{L}_x \subset \mathbb{P}^9$ along a smooth quadric hypersurface of dimension 4; see Theorem 4.2.3. It is now easy to deduce (and well known) that $\mathscr{L}_x \simeq \mathbb{G}(1, 4) \subset \mathbb{P}^9$ Plücker embedded. From $\mathscr{L}_x \subset \mathbb{P}^9$ we can construct the locus of tangent lines and obtain $X^2 := \mathscr{L}_x(X^1) \subset \mathbb{P}^5$, the Segre threefold $\mathbb{P}^1 \times \mathbb{P}^2 \subset \mathbb{P}^5$; see Example 4.2.5.

We can begin the process with the 16-dimensional variety $X = E_6 \subset \mathbb{P}^{26}$, a Fano manifold of index $i(X) = 12$ with $b_2(X) = 1$ and with $\delta(X) = 8$. This is a *QEL*-manifold of type $\delta = 8$, being the center of a $(2, 2)$ special Cremona transformation, see [53]. By applying the above constructions one obtains $X^1 = \mathscr{L}_x(X) = S^{10} \subset \mathbb{P}^{15}$, see [198, IV]. One could also apply [138], since $X^1 \subset \mathbb{P}^{15}$ has dimension 10 and type $\delta = 6$ so that it is a Fano manifold of index $i(X) = (n+\delta)/2 = 8 = n - 2$. Hence $X^2 = \mathscr{L}_x(X^1) = \mathbb{G}(1, 4) \subset \mathbb{P}^9$ and finally $X^3 = \mathscr{L}_x(X^2) = \mathbb{P}^1 \times \mathbb{P}^2 \subset \mathbb{P}^5$.

The examples discussed above and the results of part (4) in Theorem 4.2.3 suggest to iterate the process, whenever possible, of attaching to an *LQEL*-manifold of type $\delta \geq 3$ a non-degenerate *QEL*-manifold $\mathscr{L}_x \subset \mathbb{P}^{n-1}$ of type $\delta - 2$ such that $S\mathscr{L}_x = \mathbb{P}^{n-1}$. If $r \geq 1$ is the largest integer such that $\delta - 2r \geq 1$, and if $X \subset \mathbb{P}^N$ is an *LQEL*-manifold of type δ, then the process can be iterated r times, obtaining *QEL*-manifolds of type $\delta - 2k \geq 3$ for every $k = 1, \ldots, r - 1$.

Definition 4.2.7 (Manifolds of Lines X^k) Let $X \subset \mathbb{P}^N$ be an *LQEL*-manifold of type $\delta \geq 3$. Let

$$r_X = \sup\{r \in \mathbb{N} : \delta \geq 2r + 1\} = \left[\frac{\delta - 1}{2}\right],$$

where $[\alpha]$ is the integer part of $\alpha \in \mathbb{R}$.

For every $k = 1, \ldots, r_X - 1$, we define inductively

$$X^k = X^k(z_0, \ldots, z_{k-1}) = \mathscr{L}_{z_{k-1}}(X^{k-1}(z_0, \ldots, z_{k-2})),$$

where $z_i \in X^i$, $i = 0, \ldots, k-1$, is a general point and where $X^0 = X$.

The process is well defined by Theorem 4.2.3 since for every $k = 1, \ldots, r_X - 1$, the variety X^k is a *QEL*-manifold of type $\delta(X^k) = \delta - 2k \geq 3$.

The *QEL*-manifold X^k depends on the choices of the general points $z_0, \ldots, z_{k-1}$ used to define it. The type and dimensions of the X^k's are well defined and we are interested in the determination of these invariants.

The following result is crucial for the rest of the paper. Its proof is a direct consequence of iterating part (4), d) of Theorem 4.2.3.

Theorem 4.2.8 (Divisibility Theorem for the Defect of an *LQEL*-Manifold, [160, Theorem 2.8]) *Let $X \subset \mathbb{P}^N$ be an LQEL-manifold of type $\delta \geq 3$. Then:*

1. *For every $k = 1, \ldots, r_X$, the variety $X^k \subset \mathbb{P}^{\frac{n+(2^{k-1}-1)\delta}{2^{k-1}} - 2k+1}$ is a QEL-manifold of type $\delta(X^k) = \delta - 2k$, of dimension $\dim(X^k) = \frac{n+(2^k-1)\delta}{2^k} - 2k$, such that $SX^k = \mathbb{P}^{\frac{n+(2^{k-1}-1)\delta}{2^{k-1}} - 2k+1}$; in particular, $\operatorname{codim}(X^k) = \frac{n-\delta}{2^k} + 1$.*

2.

$$2^{r_X} \text{ divides } n - \delta, \tag{4.1}$$

that is

$$n \equiv \delta \pmod{2^{r_X}}.$$

Remark 4.2.9 Much weaker forms of the Divisibility Theorem were proposed in [145, Theorem 0.2] after long computations with Chern classes.

A new proof of the Divisibility Theorem via K-theory has been recently obtained by Nash, see [143] for details.

The hypothesis $\delta \geq 3$ is clearly sharp for the congruence established in part (2) of Theorem 4.2.8, or for its weaker form proved in part (4) of Theorem 4.2.3. Indeed, for the Segre varieties $X_{l,m} = \mathbb{P}^l \times \mathbb{P}^m \subset \mathbb{P}^{lm+l+m}$, $1 \leq l \leq m$, of odd dimension $n = l + m$ we have $\delta(X_{l,m}) = 2$, see Example 4.2.4, so that $n \not\equiv 2 = \delta \pmod{2}$.

It is worthwhile to remark that the weak form of the parity result $n \equiv \delta(X)$ (mod 2) is not true for arbitrary smooth secant defective varieties with $\delta(X) \geq 3$, showing that the *LQEL* hypothesis is crucial for the previous results.

Indeed, if $X \subset \mathbb{P}^N$ is a smooth non-degenerate complete intersection with $N \leq 2n - 2$ and such that $n \not\equiv N - 1 \pmod{2}$, then $SX = \mathbb{P}^N$ by Corollary 1.5.6. Thus for $N \leq 2n - 2$, we get $\delta(X) = 2n + 1 - N \geq 3$ and $\delta \equiv N - 1 \pmod{2}$.

Infinite series of secant defective smooth varieties $X \subset \mathbb{P}^N$ of dimension n with $SX \subsetneq \mathbb{P}^N$, $\delta(X) \geq 3$ and such that $n \not\equiv \delta(X) \pmod{2^{r_X}}$ can be constructed in the following way. Take $Z \subset \mathbb{P}^N$ a smooth *QEL*-manifold of type $\delta \geq 4$ and dimension

n such that $SZ \subsetneq \mathbb{P}^N$. Consider a $\mathbb{P}^{N+1}$ containing the previous $\mathbb{P}^N$ as a hyperplane, take $p \in \mathbb{P}^{N+1} \setminus \mathbb{P}^N$ and let $Y = S(p, Z) \subset \mathbb{P}^{N+1}$ be the cone over Z of vertex p. If $W \subset \mathbb{P}^{N+1}$ is a general hypersurface of degree $d > 1$, not passing through p, then $X = W \cap Y \subset \mathbb{P}^{N+1}$ is a smooth non-degenerate variety of dimension n such that $SX = S(p, SZ) \subsetneq \mathbb{P}^{N+1}$, see Exercise 1.5.13. Thus $\delta(X) = \delta(Z) - 1 = \delta - 1 \geq 3$ and $n \not\equiv \delta(X) \pmod{2^{r_X}}$ since $n \equiv \delta \pmod{2^{r_X}}$. Clearly also $n \not\equiv \delta(X) \pmod 2$.

One can take, for example, $Z_n = \mathbb{G}(1, \frac{n}{2} + 1) \subset \mathbb{P}^{\frac{n(n+6)}{8}}$, $n \geq 8$, which are *QEL*-manifolds of dimension $n \geq 8$ and type $\delta = 4$ such that $SZ \subsetneq \mathbb{P}^{\frac{n(n+6)}{8}}$.

4.3   Classification of *LQEL*-Manifolds with $\delta \geq \dim(X)/2$

In this section we classify several classes of *LQEL*-manifolds following [160]. Ionescu and Russo [104] contains the complete classification of *CC*-manifolds with $\delta(X) \leq 2$ and hence that of *LQEL*-manifolds of type $\delta = 1, 2$, see Theorem 4.4.1 below.

Proposition 4.3.1 *Let $X^n \subset \mathbb{P}^N$ be an LQEL-manifold of type $\delta = n - 1 \geq 1$. Then $n = 2$ or $n = 3$, $N \leq 5$ and $X \subset \mathbb{P}^N$ is projectively equivalent to one of the following:*

1. *$\mathbb{P}^1 \times \mathbb{P}^2 \subset \mathbb{P}^5$ Segre embedded, or one of its hyperplane sections;*
2. *the Veronese surface $v_2(\mathbb{P}^2) \subset \mathbb{P}^5$ or one of its isomorphic projections in $\mathbb{P}^4$.*

Proof Theorem 4.2.3 part (4) yields $n - 1 = \delta \leq 2$. Thus either $n = 2$ and $\delta = 1$, or $n = 3$ and $\delta = 2$.

Suppose first $n = 2$. Let $C \subset X$ be a general entry locus. Since $-K_X \cdot C = 3$ by Theorem 4.2.2, we get $C^2 = 1$ via the Adjunction Formula. Moreover, $h^1(\mathcal{O}_X) = 0$ by Theorem 4.2.2, so that $h^0(\mathcal{O}_X(C)) = h^0(\mathcal{O}_{\mathbb{P}^1}(1)) + 1 = 3$ and $|C|$ is base point free.

The birational morphism $\phi = \phi_{|C|} : X \to \mathbb{P}^2$ sends a conic C into a line. Thus $\phi^{-1} : \mathbb{P}^2 \dashrightarrow X \subset \mathbb{P}^N$ is given by a sublinear system of $|\mathcal{O}_{\mathbb{P}^2}(2)|$ of dimension at least four and the conclusion for $n = 2$ is now immediate. If $n = 3$, we get the conclusion by passing to a hyperplane section, taking into account Proposition 4.1.5.   □

The first relevant application of the Divisibility Property is the following classification of *LQEL*-manifolds of type $\delta > \frac{n}{2}$, which was left as an open problem posed in [112, 0.12.6].

Corollary 4.3.2 (Classification of *LQEL*-Manifolds with $\frac{n}{2} < \delta < n$, [160, Corollary 3.1]) *Let $X^n \subset \mathbb{P}^N$ be an LQEL-manifold of type δ with $\frac{n}{2} < \delta < n$. Then $X^n \subset \mathbb{P}^N$ is projectively equivalent to one of the following:*

  i) *the Segre threefold $\mathbb{P}^1 \times \mathbb{P}^2 \subset \mathbb{P}^5$;*
 ii) *the Plücker embedding $\mathbb{G}(1, 4) \subset \mathbb{P}^9$;*
iii) *the ten-dimensional spinor variety $S^{10} \subset \mathbb{P}^{15}$;*

iv) *a general hyperplane section of* $\mathbb{G}(1, 4) \subset \mathbb{P}^9$;
v) *a general hyperplane section of* $S^{10} \subset \mathbb{P}^{15}$.

Proof By assumption $\delta > 0$. If $\delta \leq 2$, then $n = 3$ and $\delta = 2 = n - 1$. Therefore $N = 5$ and X is projectively equivalent to the Segre threefold $\mathbb{P}^1 \times \mathbb{P}^2 \subset \mathbb{P}^5$ by Proposition 4.3.1.

From now on we can assume $\delta \geq 3$ and that $X \subset \mathbb{P}^N$ is a Fano manifold with $\mathrm{Pic}(X) = \mathbb{Z}\langle \mathcal{O}(1) \rangle$. By Theorem 4.2.8 there exists an integer $m \geq 1$ such that $2\delta > n = \delta + m2^{r_X}$, so that

$$\delta > m2^{r_X}. \tag{4.2}$$

Suppose $\delta = 2r_X + 2$. From $2r_X + 2 > m2^{r_X}$ it follows that $m = 1$ and $r_X \leq 2$. Hence either $\delta = 4$ and $n = 6$ and $X \subset \mathbb{P}^N$ is a Fano manifold of index $i(X) = (n+\delta)/2 = 5 = n-1$ or $\delta = 6$ and $n = 10$ and $X \subset \mathbb{P}^N$ is a Fano manifold as above and of index $i(X) = (n + \delta)/2 = 8 = n - 2$. In the first case by [65, Theorem 8.11] we get case ii). In the second case we apply [138], obtaining case iii).

Suppose $\delta = 2r_X + 1$. From (4.2) we get $2r_X + 1 > m2^{r_X}$ forcing $m = 1$ and $r_X = 1, 2$. Therefore either $\delta = 3$ and $n = \delta + m2^{r_X} = 5$; or $\delta = 5$ and $n = \delta + m2^{r_X} = 9$. Reasoning as above, we get cases iv) and v). $\square$

Another interesting application of Theorem 4.2.8 concerns the classification of *LQEL* manifolds of type $\delta = \frac{n}{2}$. For such varieties we get immediately that $n = 2, 4, 8$ or 16 and among them we find the so-called Severi varieties, which we shall introduce in the next chapter. Indeed, by [198, IV.2.1, IV.3.1, IV.2.2], see Corollary 5.4.2, Severi varieties are *LQEL*-manifolds of type $\delta = \frac{n}{2}$.

Once we know that $n = 2, 4, 8$ or 16, it is rather simple to classify Severi varieties, as we shall show in Proposition 5.4.3, see also [198, IV.4] and [119]. For $n = 2, 4$ the result is classical and well known while in our approach the $n = 8$ case follows from the classification of Mukai manifolds, [138]. The less obvious case is $n = 16$.

What is notable, in our opinion, is not the fact that this proof is short, easy, natural, immediate and almost self-contained but the perfect parallel between our argument based on the Divisibility Theorem and some proofs of Hurwitz's Theorem on the dimension of composition algebras over a field such as the one contained in [118, V.5.10], see also [39, Chap. 10. Sect. 36]. Surely this connection is well known today, see [198, pp. 89–91], but the other proofs of the classification of Severi varieties did not make this parallel so transparent.

Moreover, our approach also *explains* why a Severi Variety, introduced in the next chapter (or more generally an *LQEL*-manifold of dimension n and type $n/2$), has dimension at most 16. Indeed, a variety of the previous kind with $n > 16$ would produce an example of a smooth non-degenerate manifold $\mathscr{L}_x \subset \mathbb{P}^{n-1}$ violating the Hartshorne Conjecture on Complete Intersections, see Sect. 5.3 below.

Concerning the next result and the word *generalization*, we would like to quote Hermann Weyl: "*Before you can generalize, formalize and axiomatize, there must be a mathematical substance*", [194]. There is no doubt that the mathematical substance in this classification problem is entirely due to Fyodor Zak, who firstly brilliantly solved it in [196].

Corollary 4.3.3 (Classification of *LQEL*-Manifolds with $\delta = \frac{n}{2}$, [160, Corollary 3.2]) *Let $X^n \subset \mathbb{P}^N$ be an LQEL-manifold of type $\delta = \frac{n}{2}$. Then $n = 2, 4, 8$ or 16 and $X^n \subset \mathbb{P}^N$ is projectively equivalent to one of the following:*

   i) the cubic scroll $S(1, 2) \subset \mathbb{P}^4$;
  ii) the Veronese surface $v_2(\mathbb{P}^2) \subset \mathbb{P}^5$ or one of its isomorphic projections in $\mathbb{P}^4$;
 iii) the Segre fourfold $\mathbb{P}^1 \times \mathbb{P}^3 \subset \mathbb{P}^7$;
 iv) a general four-dimensional linear section $X \subset \mathbb{P}^7$ of $\mathbb{G}(1, 4) \subset \mathbb{P}^9$;
  v) the Segre fourfold $\mathbb{P}^2 \times \mathbb{P}^2 \subset \mathbb{P}^8$ or one of its isomorphic projections in $\mathbb{P}^7$;
 vi) a general eight-dimensional linear section $X \subset \mathbb{P}^{13}$ of $S^{10} \subset \mathbb{P}^{15}$;
vii) the Plücker embedding $\mathbb{G}(1, 5) \subset \mathbb{P}^{14}$ or one of its isomorphic projections in $\mathbb{P}^{13}$;
viii) the E_6-variety $X \subset \mathbb{P}^{26}$ or one of its isomorphic projections in $\mathbb{P}^{25}$.

Proof By assumption n is even. If $n < 6$, then $n = 2$ or $n = 4$. If $n = 2$, the conclusion is well known, see [176] or Proposition 4.3.1.

If $n = 4$, then $\delta = 2 = n - 2$. If H is a hyperplane section and if $C \in \mathscr{C}$ is a general conic, then $(K_X + 3H)\cdot C = -n - \delta + 2n - 2 = 0$ by part (5) of Theorem 4.2.2. Suppose $X \subset \mathbb{P}^N$ is a scroll over a curve, which is rational by Theorem 4.2.2. Since for a rational normal scroll either $SX = \mathbb{P}^N$ or $\dim(SX) = 2n + 1$, we get $N = \dim(SX) = 2n + 1 - \delta = 7$ so that $X \subset \mathbb{P}^7$ is a rational normal scroll of degree 4, which is the case described in iii).

If $X \subset \mathbb{P}^N$ is not a scroll over a curve, $|K_X + 3H|$ is generated by global sections, see [99, Theorem 1.4], and since through two general points of X there passes such a conic, we deduce $-K_X = 3H$. Thus $X \subset \mathbb{P}^N$ is a del Pezzo manifold, obtaining cases iii), iv) or v) by [65, Theorem 8.11].

Suppose from now on $n \geq 6$, $\delta = n/2 \geq 3$ and hence that $X \subset \mathbb{P}^N$ is a Fano manifold with $\text{Pic}(X) = \mathbb{Z}\langle \mathscr{O}(1) \rangle$. By Theorem 4.2.8, 2^{r_X} divides $n - \delta = \frac{n}{2} = \delta$ so that $2^{r_X + 1}$ divides n and $\delta = \frac{n}{2}$ is even. By definition of r_X, $\frac{n}{2} = 2r_X + 2$, so that, for some integer $m \geq 1$,

$$m2^{r_X + 1} = n = 4(r_X + 1).$$

Therefore either $r_X = 1$ and $n = 8$, or $r_X = 3$ and $n = 16$.

In the first case we get that $X \subset \mathbb{P}^N$ is a Fano manifold as above and of index $i(X) = (n + \delta)/2 = 6 = n - 2$ and we are in cases vi) and vii) by [138].

In the remaining cases $\mathscr{L}_x \subset \mathbb{P}^{15}$ is a ten-dimensional *QEL*-manifold of type $\delta = 6$ so that $\mathscr{L}_x \subset \mathbb{P}^{15}$ is projectively equivalent to $S^{10} \subset \mathbb{P}^{15}$ by Corollary 4.3.2. By the Main Theorem of [135] we are in case viii) since $X \simeq E_6 \subset \mathbb{P}^{26}$. □

4.4 Classification of Conic-Connected Manifolds and of Manifolds with Small Dual

The following Classification Theorem is one of the main results of [104]. As CC-manifolds are stable under isomorphic projection, one may always assume $X \subset \mathbb{P}^N$ to be linearly normal.

Theorem 4.4.1 (Classification of *CC*-Manifolds [104, Theorem 2.1]) *Let $X^n \subset \mathbb{P}^N$ be a smooth irreducible linearly normal non-degenerate CC-manifold of dimension n. Then either $X \subset \mathbb{P}^N$ is a Fano manifold with $\mathrm{Pic}(X) \simeq \mathbb{Z}\langle \mathscr{O}_X(1)\rangle$ and of index $i(X) \geq \frac{n+1}{2}$, or it is projectively equivalent to one of the following:*

(i) $v_2(\mathbb{P}^n) \subset \mathbb{P}^{\frac{n(n+3)}{2}}$.

(ii) The projection of $v_2(\mathbb{P}^n)$ from the linear space $\langle v_2(\mathbb{P}^s)\rangle$, where $\mathbb{P}^s \subset \mathbb{P}^n$ is a linear subspace; equivalently $X \simeq \mathrm{Bl}_{\mathbb{P}^s}(\mathbb{P}^n)$ embedded in $\mathbb{P}^N$ by the linear system of quadric hypersurfaces of $\mathbb{P}^n$ passing through $\mathbb{P}^s$; alternatively $X \simeq \mathbb{P}_{\mathbb{P}^r}(\mathscr{E})$ with $\mathscr{E} \simeq \mathscr{O}_{\mathbb{P}^r}(1)^{\oplus n-r} \oplus \mathscr{O}_{\mathbb{P}^r}(2)$, $r = 1, 2, \ldots, n-1$, embedded by $|\mathscr{O}_{\mathbb{P}(\mathscr{E})}(1)|$. Here $N = \frac{n(n+3)}{2} - \binom{s+2}{2}$ and s is an integer such that $0 \leq s \leq n-2$.

(iii) A hyperplane section of the Segre embedding $\mathbb{P}^a \times \mathbb{P}^b \subset \mathbb{P}^{N+1}$. Here $n \geq 3$ and $N = ab + a + b - 1$, where $a \geq 2$ and $b \geq 2$ are such that $a + b = n + 1$.

(iv) $\mathbb{P}^a \times \mathbb{P}^b \subset \mathbb{P}^{ab+a+b}$ Segre embedded, where a, b are positive integers such that $a + b = n$.

Corollary 4.4.2 *A CC-manifold is a Fano manifold with second Betti number $b_2 \leq 2$; for $b_2 = 2$ it is also rational.*

The previous Theorem reduces the classification of CC-manifolds to the study of Fano manifolds having large index and Picard group $\mathbb{Z}$. The next result, essentially due to Hwang–Kebekus [94, Theorem 3.14], shows that, conversely, such Fano manifolds are conic-connected. Note that we slightly improve the bound on the index given in [94]. The first part of the Proposition is well known.

Proposition 4.4.3 ([103, Proposition 2.4]; Cf. Also [94]) *Let $X \subset \mathbb{P}^N$ be a Fano manifold with $\mathrm{Pic}(X) \simeq \mathbb{Z}\langle H \rangle$ and $-K_X = i(X)H$, H being the hyperplane section and $i(X)$ the index of X. Let as always $\mathscr{L}_x \subset \mathbb{P}((t_xX)^*) = \mathbb{P}^{n-1}$ be the Hilbert scheme of lines through a general point $x \in X$.*

(i) If $i(X) \geq \frac{n+3}{2}$ and $S\mathscr{L}_x = \mathbb{P}^{n-1}$, then $X \subset \mathbb{P}^N$ is a CC-manifold.

(ii) If $i(X) > \frac{2n}{3}$, then $X \subset \mathbb{P}^N$ is a CC-manifold.

Proof Part (i) follows from [94, Theorem 3.14].

To prove part (ii), first observe that $i(X) \geq \frac{n+3}{2}$, unless $n \leq 6$. If $n \leq 6$, $i(X) > \frac{2n}{3}$ gives $i(X) \geq n - 1$, so the conclusion follows by the classification of del Pezzo manifolds (see [65]). To conclude, using part (i), it is enough to see that $S\mathscr{L}_x = \mathbb{P}^{n-1}$.

By [93, Theorem 2.5], the variety $Y_x \subset \mathbb{P}^{n-1}$ is non-degenerate and by hypothesis

$$n - 1 < \frac{3i(X) - 6}{2} + 2 = \frac{3\dim(\mathscr{L}_x)}{2} + 2,$$

so that $S\mathscr{L}_x = \mathbb{P}^{n-1}$ by Zak's Linear Normality Theorem, see Theorem 5.1.6 here. $\square$

Corollary 4.4.4 *If $X \subset \mathbb{P}^{n+r}$ is a smooth non-degenerate complete intersection of multi-degree $(d_1, d_2, \ldots, d_r)$ with $n > 3\left(\sum_1^r d_i - r - 1\right)$, then X is a CC-manifold.*

The classification of *CC* and of *LQEL*-manifolds has been recently applied in different contexts, see, for example, [64]. We shall now quote some interesting results due to K. Han, who first gave a notable application of Scorza's Lemma and then used it and Theorem 4.4.1 to classify secant defective manifolds $X \subset \mathbb{P}^N$ with $N \geq n(n + 1)/2 + 2$, improving significantly Theorem 3.4.4. Here are Han's contributions:

Theorem 4.4.5 ([82, Theorem 2.1]) *Let $X \subset \mathbb{P}^N$ be a non-degenerate secant defective manifold of dimension $n \geq 2$ with $SX \subsetneq \mathbb{P}^N$. If $N \geq \frac{n(n+3)}{2} - \epsilon$ with $0 \leq \epsilon \leq n - 2$, then $X \subset \mathbb{P}^N$ is an LQEL-manifold of type $\delta = 1$.*

Corollary 4.4.6 ([82, Corollary 2.3]) *Let $X \subset \mathbb{P}^N$ be a non-degenerate secant defective manifold of dimension $n \geq 2$ with $SX \subsetneq \mathbb{P}^N$. If $N \geq \frac{n(n+1)}{2} + 2$, then X is projectively equivalent to one of the following:*

1. *the second Veronese embedding $v_2(\mathbb{P}^n) \subset \mathbb{P}^{\frac{n(n+3)}{2}}$;*
2. *an isomorphic projection of $v_2(\mathbb{P}^n)$ into $\mathbb{P}^{\frac{n(n+3)}{2} - \epsilon}$ with $1 \leq \epsilon \leq n - 2$;*
3. *the projection of $v_2(\mathbb{P}^n)$ from the linear space $\langle v_2(\mathbb{P}^s) \rangle$, where $\mathbb{P}^s \subset \mathbb{P}^n$ is a linear subspace with $s \geq 0$ and with $s + 2 \leq n - 2$;*
4. *an isomorphic projection of a manifold as in (3) into $\mathbb{P}^N$ with $N \geq \frac{n(n+1)}{2} + 2$.*

4.4.1 Classification of Varieties with Small Dual

For an irreducible variety $Z \subset \mathbb{P}^N$, we defined $\operatorname{def}(Z) = N - 1 - \dim(Z^*)$ as the dual defect of $Z \subset \mathbb{P}^N$, where $Z^* \subset \mathbb{P}^{N*}$ is the dual variety of $Z \subset \mathbb{P}^N$.

In [51, Theorem 2.4] it is proved that if $\operatorname{def}(X) > 0$, then $\operatorname{def}(X) \equiv n \pmod{2}$, a result usually attributed to Landman. Here we shall provide an easy and direct proof of this fundamental fact following directly [106]. This will show once more, if needed, the unity and simplicity of the approach to these problems via the study of $\mathscr{L}_x$ and of its infinitesimal properties.

Proposition 4.4.7 *Let $X \subset \mathbb{P}^N$ be a smooth irreducible non-degenerate variety and assume that $\mathrm{def}(X) > 0$. Then*

(i) (**Landman Parity Theorem**, [51, Theorem 2.4]) *through a general point $x \in X$ there passes a line $L_x \subset X$ such that*

$$N_{L_x/X} \simeq \mathscr{O}_{\mathbb{P}^1}^{\frac{n-\mathrm{def}(X)}{2}} \oplus \mathscr{O}_{\mathbb{P}^1}(1)^{\frac{n+\mathrm{def}(X)-2}{2}}.$$

In particular, $-K_X \cdot L_x = \frac{n+\mathrm{def}(X)+2}{2}$ and $\mathrm{def}(X) \equiv n \pmod 2$;
(ii) ([51, Theorem 3.2]) $\mathrm{def}(X) = n - 2$ *if and only if $X \subset \mathbb{P}^N$ is a scroll over a smooth curve, i.e. it is a $\mathbb{P}^{n-1}$-bundle over a smooth curve, whose fibers are linearly embedded.*

Proof (According to [106, Lemma 1.1 and Proposition 3.1]) Recall that by definition of dual variety there exists a morphism

$$p_2 : \mathbb{P}(N_{X/\mathbb{P}^N}(-1)) \to X^*$$

defined by a sublinear system of $\mathscr{O}_{\mathbb{P}(N_{X/\mathbb{P}^N}(-1))}(1)$, see Exercise 1.5.17.
From the existence of the surjection

$$\mathscr{O}_X^{N+1} \to N_{X/\mathbb{P}^N}(-1) \to 0,$$

induced by Euler's sequence twisted by $\mathscr{O}(-1)$ and by the sequence of the normal/tangent bundles twisted by $\mathscr{O}(-1)$, we deduce that the morphism p_2 factors through $p_2 : X \times \mathbb{P}^{N*} \to \mathbb{P}^{N*}$. In particular, the fibers of $p_2 : \mathbb{P}(N_{X/\mathbb{P}^N}(-1)) \to X^*$ map isomorphically onto X via the structural morphism

$$\pi : \mathbb{P}(N_{X/\mathbb{P}^N}(-1)) \to X.$$

The variety X is ruled by the $\mathbb{P}^{\mathrm{def}(X)}$'s which are images via π of the general fibers of p_2. Therefore, through a general point $x \in X$ there pass lines L_x which are isomorphic projections onto X of lines $L_x \subset \mathbb{P}^{\mathrm{def}(X)} = p_2^{-1}(y)$, $y \in X^*$ general. For simplicity let $X' = \mathbb{P}(N_{X/\mathbb{P}^N}(-1))$ and let $\mathscr{E} = N_{X/\mathbb{P}^N}(-1)$.
Thus $\deg(N_{L_x/X'}) = \deg(N_{L_x/\mathbb{P}^{\mathrm{def}(X)}}) = \mathrm{def}(X) - 1$. Since $L_x \cdot \mathscr{O}_{\mathbb{P}(\mathscr{E})}(1) = 0$ and since

$$K_{X'} = -\mathrm{rk}(\mathscr{E})\mathscr{O}_{\mathbb{P}(\mathscr{E})}(1) + \pi^*(\det(\mathscr{E}) + K_X),$$

we deduce, via the Adjunction Formula:

$$\mathrm{def}(X) - 1 = \deg(N_{L_x/X'}) = -K_{X'} \cdot L_x + 2 = -K_X \cdot L_x + 2 - \deg(N_{X/\mathbb{P}^N}(-1)_{|L_x}) =$$
$$= \deg(N_{L_x/X}) - \deg(N_{X/\mathbb{P}^N}(-1)_{|L_x}). \tag{4.3}$$

From

$$0 \to N_{L_x/X}(-1) \to N_{L_x/\mathbb{P}^N}(-1) \simeq \mathscr{O}_{\mathbb{P}^1}^{N-1} \to N_{X/\mathbb{P}^N}(-1)_{|L_x} \to 0$$

we deduce

$$-\deg(N_{X/\mathbb{P}^N}(-1)_{|L_x}) = \deg(N_{L_x/X}(-1)).$$

Letting

$$N_{L_x/X} \simeq \mathscr{O}_{\mathbb{P}^1}^{n-a-1} \oplus \mathscr{O}_{\mathbb{P}^1}(1)^a,$$

we get from (4.3)

$$\det(X) - 1 = a - (n - a - 1),$$

yielding

$$a = \frac{n + \det(X) - 2}{2}$$

and proving part (i).

To prove part (ii) we remark that $\det(X) = n - 2$ implies $K_X \cdot L_x = -n$ and hence $(K_X + (n-1)H) \cdot L_x = -1 < 0$. Thus $\mathscr{O}(K_X + (n-1)H)$ is not generated by global sections and X is a scroll in $\mathbb{P}^{n-1}$ over a smooth curve by [99, Theorem 1]. □

Let us recall that Zak's Theorem on Tangency implies that $\dim(X^*) \geq \dim(X)$ for a smooth non-degenerate variety $X \subset \mathbb{P}^N$, see Corollary 3.2.5.

We combine the geometry of CC and LQEL-manifolds to give a new proof of [51, Theorem 4.5]. Our approach avoids the use of Beilinson's spectral sequence and more sophisticated computations as in [51, 4.2, 4.3, 4.4].

Proposition 4.4.8 ([103, Proposition 4.3]) *Let $X \subset \mathbb{P}^N$ be a smooth irreducible non-degenerate variety. Assume that X is a Fano manifold with $\mathrm{Pic}(X) \cong \mathbb{Z}\langle \mathscr{O}_X(1)\rangle$ and let $x \in X$ be a general point.*

(i) If $\det(X) > 0$ and $\det(X) > \frac{n-6}{3}$, then X is a CC-manifold with $\delta \geq \det(X) + 2$.
    Moreover, if $\delta = \det(X)+2$, then X is an LQEL-manifold of type $\delta = \det(X)+2$.
(ii) If X is an LQEL-manifold of type δ and $\det(X) > 0$, then $\delta = \det(X) + 2$.

Proof In the hypotheses of (i), Proposition 4.4.7 yields

$$i(X) = \frac{n + \det(X) + 2}{2} > \frac{2n}{3}$$

so that X is a CC-manifold by Proposition 4.4.3. Proposition 4.2.1 yields $\delta \geq \operatorname{def}(X) + 2$ and also the remaining assertions of (i) and (ii). □

We recall that according to Hartshorne's Conjecture, if $n > \frac{2}{3}N$, then $X \subset \mathbb{P}^N$ should be a complete intersection and that complete intersections have no dual defect, see part (4) of Exercise 1.5.16. Thus, assuming Hartshorne's Conjecture, the following result yields the complete list of manifolds $X \subset \mathbb{P}^N$ such that $\dim(X^*) = \dim(X)$. The second part says that under the LQEL hypothesis the same results hold without any restriction.

Theorem 4.4.9 (Varieties with Small Duals, [51, Theorem 4.5], [103, Theorem 4.4]) *Let $X \subset \mathbb{P}^N$ be a manifold such that $\dim(X) = \dim(X^*)$.*

(i) If $N \geq \frac{3n}{2}$, then X is projectively equivalent to one of the following:

    (a) a smooth hypersurface $X \subset \mathbb{P}^{n+1}$, $n = 1, 2$;
    (b) a Segre variety $\mathbb{P}^1 \times \mathbb{P}^{n-1} \subset \mathbb{P}^{2n-1}$;
    (c) the Plücker embedding $\mathbb{G}(1, 4) \subset \mathbb{P}^9$;
    (d) the ten-dimensional spinor variety $S^{10} \subset \mathbb{P}^{15}$.

(ii) If X is an LQEL-manifold, then it is projectively equivalent either to a smooth quadric hypersurface $Q \subset \mathbb{P}^{n+1}$ or to a variety as in (b), (c), (d) above.

Proof Clearly $\operatorname{def}(X) = 0$ if and only if $X \subset \mathbb{P}^{n+1}$ is a hypersurface, giving case (a), respectively that of quadric hypersurfaces. From now on we suppose $\operatorname{def}(X) > 0$ and hence $n \geq 3$. By parts (i) and (ii) of Proposition 4.4.7, $\operatorname{def}(X) = n - 2$ and $N = 2n - 1$ if and only if we are in case (b); see also [51, Theorem 3.3, c)].

Thus, we may assume $0 < \operatorname{def}(X) \leq n - 4$, that is $N \leq 2n - 3$. Therefore $\delta \geq 4$ and X is a Fano manifold with $\operatorname{Pic}(X) \cong \mathbb{Z}\langle \mathcal{O}_X(1)\rangle$. Moreover, in case (i), $\operatorname{def}(X) = N - n - 1 > \frac{n-6}{3}$ by hypothesis.

Thus Proposition 4.4.8 yields that X is also a CC-manifold with $\delta \geq \operatorname{def}(X) + 2$. Taking into account also the last part of Proposition 4.4.8, from now on we can suppose that X is a CC-manifold with $\delta \geq \operatorname{def}(X) + 2 \geq 3$.

We have $n - \delta \leq N - 1 - n = \operatorname{def}(X) \leq \delta - 2$, that is $\delta \geq \frac{n}{2} + 1$. Zak's Linear Normality Theorem implies $SX = \mathbb{P}^N$, so that

$$N = \dim(SX) = 2n + 1 - \delta \leq \frac{3n}{2}.$$

Since $N \geq \frac{3n}{2}$, we get $N = \frac{3n}{2}$, $\delta = \frac{n}{2} + 1 = \operatorname{def}(X) + 2$ and n even. Therefore X is an LQEL-manifold of type $\delta = \frac{n}{2} + 1$ by Proposition 4.4.8. Corollary 4.3.2 concludes the proof, yielding cases (c) and (d). □

4.4.2   Bounds for the Dual Defect of a Manifold
and for the Secant Defect of an LQEL-Manifold

We defined inductively in Definition 4.2.7 some varieties naturally attached to an LQEL-manifold of type $\delta \geq 3$. Let $i_j = i(X^j), j = 1, \ldots, r_X - 1$, be the index of the Fano manifold X^j. Computing as in Theorem 4.2.8 we get:

$$i_j = \frac{n - \delta}{2^{j+1}} + \delta - 2j, \ 0 \leq j \leq r_X - 1. \tag{4.4}$$

Corollary 4.4.10 ([63, Theorem 3]) *Let $X \subset \mathbb{P}^N$ be an n-dimensional LQEL-manifold of type δ. If*

$$\delta > 2[\log_2 n] + 2 \ or \ \delta > \min_{k \in \mathbb{N}} \{ \frac{n}{2^{k-1} + 1} + \frac{2^k k}{2^{k-1} + 1} \},$$

then $N = n + 1$ and $X \subset \mathbb{P}^{n+1}$ is a quadric hypersurface.

Proof If $\delta > 2[\log_2 n] + 2$, then $n < 2^r$, where $r = [(\delta - 1)/2]$. By Theorem 4.2.8, 2^r divides $n - \delta$. This is possible only if $\delta = n$. Thus X is a hyperquadric. Now assume we have the second inequality. Note that for a fixed n, the minimum

$$\min_{k \in \mathbb{N}} \{ \frac{n}{2^{k-1} + 1} + \frac{2^k k}{2^{k-1} + 1} \}$$

is achieved for some $k \leq n/2$, so we may assume that for some $k \leq n/2$, we have

$$\delta > \frac{n}{2^{k-1} + 1} + \frac{2^k k}{2^{k-1} + 1} = 2k + \frac{n - 2k}{2^{k-1} + 1} \geq 2k,$$

so that $\delta \geq 2k + 1$. Now we can consider the variety $\mathscr{L}_k \subset \mathbb{P}^{\dim(\mathscr{L}_{k-1})-1}$. Note that $\dim(\mathscr{L}_k) = i(\mathscr{L}_{k-1}) - 2$ and

$$\dim(\mathscr{L}_{k-1}) = 2i_{k-1} - \delta(\mathscr{L}_{k-1}) = \frac{n - \delta}{2^{k-1}} + \delta - 2k + 2.$$

On the other hand, $\mathscr{L}_k \subset \mathbb{P}^{\dim(\mathscr{L}_{k-1})-1}$ is non-degenerate and it contains a hyperquadric of dimension $\delta - 2k$, which is strictly bigger than $(\dim(\mathscr{L}_{k-1} - 2)/2$ under our assumption on δ.

Now [198, Corollary I.2.20] implies that $\mathscr{L}_k \subset \mathbb{P}^{\dim(\mathscr{L}_{k-1})-1}$ is a hypersurface. Since it is a non-degenerate hypersurface by Theorem 4.2.8 and an LQEL-manifold, $\mathscr{L}_{k-1}$ is a quadric hypersurface. This easily implies the conclusion. □

We can now state a sharper bound for the secant defect of an *LQEL*-manifold, which can be reinterpreted as a Sharp Linear Normality Bound, see Theorem 5.1.6. The bound $\delta \leq \frac{n+8}{3}$ comes once again from Hartshorne's Conjecture on Complete Intersections applied to $\mathscr{L}_x \subset \mathbb{P}^{n-1}$, see Remark 5.3.

Corollary 4.4.11 (Bound for the Secant Defect of an *LQEL*-Manifold, [63]) *Let* $X \subset \mathbb{P}^N$ *be an LQEL-manifold of type δ, not a quadric hypersurface. Then*

$$\delta \leq \min_{k \in \mathbb{N}} \{ \frac{n}{2^{k-1}+1} + \frac{2^k k}{2^{k-1}+1} \} \leq \frac{n+8}{3}$$

and

$$N \geq \dim(SX) \geq 2n+1 - \min_{k \in \mathbb{N}} \{ \frac{n}{2^{k-1}+1} + \frac{2^k k}{2^{k-1}+1} \} \geq \frac{5}{3}(n-1).$$

Furthermore, $\delta = \frac{n+8}{3}$ if and only if $X \subset \mathbb{P}^N$ is projectively equivalent to one of the following:

i) a smooth four-dimensional quadric hypersurface $X \subset \mathbb{P}^5$;
ii) the ten-dimensional spinor variety $S^{10} \subset \mathbb{P}^{15}$;
iii) the E_6-variety $X \subset \mathbb{P}^{26}$ or one of its isomorphic projections in $\mathbb{P}^{25}$.

Proof We shall prove only the second part. If $\delta = \frac{n+8}{3}$, then $n - \delta = \frac{2n-8}{3}$. Suppose $\delta = 2r_X + 1$, so that $n - \delta = \frac{12r_X - 18}{3}$. By Theorem 4.2.8 we deduce that 2^{r_X} should divide $4r_X - 6$, which is not possible.

Suppose now $\delta = 2r_X + 2$, so that $n - \delta = \frac{12r_X - 12}{3} = 4(r_X - 1)$. Since 2^{r_X} has to divide $4(r_X - 1)$, we get $r_X = 1, 2, 3$ and, respectively, $n = 4, 10, 16$ with $\delta = 4, 6$, respectively 8. The conclusion follows from Theorems 4.3.2 and 4.3.3. □

Let us observe that Lazarsfeld and Van de Ven posed the question if for an irreducible smooth projective non-degenerate n-dimensional variety $X \subset \mathbb{P}^N$ with $SX \subsetneq \mathbb{P}^N$ the secant defect is bounded, see [125]. This question was motivated by the fact that for the known examples we have $\delta(X) \leq 8$, the bound being attained for the 16-dimensional Cartan variety $E_6 \subset \mathbb{P}^{26}$, which is an *LQEL*-variety of type $\delta = 8$. Based on these remarks and on the above results one could naturally formulate the following problem, see [63].

Question Is an *LQEL*-manifold $X \subset \mathbb{P}^N$ with $\delta > 8$ a smooth quadric hypersurface?

We now recall a bound for the dual defect of a manifold which, as always, comes from Hartshorne's Conjecture on Complete Intersections applied to $\mathscr{L}_x \subset \mathbb{P}^{n-1}$, see [106, Theorem 3.4, Corollary 3.5] and Remark 5.3.

Corollary 4.4.12 (Bound for the Dual Defect of a Manifold, Not a Scroll, [106, Corollary 3.5]) *Let $X^n \subset \mathbb{P}^N$ be a dual defective manifold and assume that X is not a scroll. Then*

$$\mathrm{def}(X) \leq \frac{n+2}{3}.$$

Moreover, $\mathrm{def}(X) = \frac{n+2}{3}$ *if and only if* $N = 15$, $n = 10$ *and* $X \subset \mathbb{P}^{15}$ *is projectively equivalent to the ten-dimensional spinorial variety* $S^{10} \subset \mathbb{P}^{15}$.

Chapter 5
Hartshorne Conjectures and Severi Varieties

5.1 Hartshorne Conjectures

Around the beginning of the 1970s there was a renewed interest in solving concrete problems and in finding applications of the new theories developed during the refoundation of Algebraic Geometry performed during the 1950s by Weil and Zariski and which culminated during the 1960s in the amazing contributions of Serre and of Grothendieck together with his school. To know the state of the art one can consult the beautiful book of Robin Hartshorne, [86]. In [86] several open problems were solved and many outstanding questions were discussed such as the set-theoretic complete intersection of curves in $\mathbb{P}^3$ (still open), the characterization of $\mathbb{P}^N$ among the smooth varieties with ample tangent bundle, which was solved by Mori in [136] and which cleared the path to the foundation of Mori theory in [137]. In related fields we only mention Deligne's proof of the Weil conjectures or later Faltings' proof of the Mordell conjecture.

The interplay between topology and algebraic geometry returned to flourish as seen in Chap. 3. Lefschetz's Theorem and the Barth–Larsen Theorem also suggested that smooth varieties whose codimension is small with respect to their dimension should satisfy very strong restrictions. To get a feeling for this we remark that a codimension two smooth complex subvariety of $\mathbb{P}^N$, $N \geq 5$, has to be simply connected. If $N \geq 6$, the transversal intersections of two hypersurfaces in $\mathbb{P}^N$ are the only known examples of smooth irreducible projective varieties of codimension 2.

Let us recall the definition of complete intersections and some of their notable properties.

Definition 5.1.1 (Complete Intersection) An equidimensional variety of dimension $n \geq 1$,

$$X^n \subset \mathbb{P}^N$$

© Springer International Publishing Switzerland 2016
F. Russo, *On the Geometry of Some Special Projective Varieties*,
Lecture Notes of the Unione Matematica Italiana 18,
DOI 10.1007/978-3-319-26765-4_5

is a *complete intersection* if there exist $N - n = \text{codim}(X)$ homogeneous polynomials

$$f_i \in K[X_0, \ldots, X_N]_{d_i}$$

of degree $d_i \geq 1$, generating the homogeneous ideal $I(X) \subset K[X_0, \ldots, X_N]$, that is if

$$I(X) = <f_1, \ldots, f_{N-n}> .$$

Let us recall that since $f_1, \ldots, f_{N-n}$ form a regular sequence in $K[X_0, \ldots, X_N]$, the homogeneous coordinate ring

$$S(X) = \frac{K[X_0, \ldots, X_N]}{I(X)}$$

has depth $n + 1$ so that $X^n \subset \mathbb{P}^N$ is an arithmetically Cohen–Macaulay variety by definition.

Thus a complete intersection $X^n \subset \mathbb{P}^N$ is connected if $n > 0$ and moreover $H^i(\mathscr{O}_X(m)) = 0$ for every i such that $0 < i < n$ and for every $m \in \mathbb{Z}$. Furthermore, a complete intersection $X \subset \mathbb{P}^N$ is projectively normal, meaning that the restriction morphisms

$$H^0(\mathscr{O}_{\mathbb{P}^N}(m)) \to H^0(\mathscr{O}_X(m))$$

are surjective for every $m \geq 0$.

Moreover, by the so-called Grothendieck–Lefschetz Theorem on the Picard group of Complete Intersections we have

$$\text{Pic}(X) \simeq \mathbb{Z} < \mathscr{O}_X(1) >,$$

as soon as $n \geq 3$, see [86, Chap. IV, Corollary 3.2]. This result was proved by Severi for smooth hypersurfaces of dimension $n \geq 3$ in [177] and extended to arbitrary smooth complete intersections by Fano in [58]. Notwithstanding, the previous statement is usually attributed to Lefschetz who stated and proved it in [126], see the historical note in [86, Chap. IV, Sect. 4].

By Lefschetz's Theorem on Hyperplane sections, see Chap. 3, complete intersections defined over $K = \mathbb{C}$ are simply connected, as soon as $n \geq 2$, and have the same cohomology $H^i(X, \mathbb{Z})$ of the projective spaces containing them for $i < n$, see [86].

An equidimensional variety $X^n \subset \mathbb{P}^N$ is a *scheme-theoretic intersection* of $m \geq N - n = \text{codim}(X)$ hypersurfaces $H_i = V(g_i) \subset \mathbb{P}^N$, $g_i \in K[X_0, \ldots, X_N]_{d_i}$ homogeneous of degree $d_i \geq 1$, if, as schemes,

$$X = V(g_1, \ldots, g_m).$$

This means that the ideal sheaf $\mathscr{I}_X \subset \mathscr{O}_{\mathbb{P}^N}$ is locally generated by the $m \geq$ codim$(X) = N - n$ elements $g_1, \ldots, g_m$. Equivalently, one can say that $X^n \subset \mathbb{P}^N$ is equal to $H_1 \cap \ldots \cap H_m$ and that at each point $p \in X$

$$T_p X = T_p H_1 \cap \ldots \cap T_p H_m$$

holds.

An equidimensional variety $X^n \subset \mathbb{P}^N$ is a *scheme-theoretic complete intersection* if it is the scheme-theoretic intersection of $N - n$ hypersurfaces.

One immediately sees, for example via the Koszul complex, that from the algebraic point of view the last condition is equivalent to the fact that the homogeneous ideal $I(X) \subset K[X_0, \ldots, X_N]$ is generated precisely by codim$(X) = N - n$ homogeneous elements. Thus a projective variety is a complete intersection if and only if it is a scheme-theoretic complete intersection, a useful observation used repeatedly later in this chapter.

On the basis of some empirical observations, inspired by the Theorem of Barth and Larsen but, according to Fulton and Lazarsfeld, also *"on the basis of few examples"*, Hartshorne was led to formulate the following conjectures.

Before stating them we wish to quote from the Introduction of the book [113]:

Algebraic geometry is a mixture of the ideas of two Mediterrean cultures. It is the superposition of the Arab science of the lightning calculation of the solutions of equations over the Greek art of position and shape. This tapestry was originally woven by on European soil and is still being refined under the influence of international fashion.

Algebraic geometry studies the delicate balance between the geometrically plausible and the algebraic possible. Whenever one side of this mathematical teeter–tooter outweighs the other, one immediately loses interest and runs off in search of a more exciting amusement.

George R. Kempf

Hartshorne's Complete Intersection Conjecture, [87]:

Let $X^n \subset \mathbb{P}^N$ be a smooth irreducible non-degenerate projective variety of dimension $n = \dim(X) \geq 1$.

If $N < \frac{3}{2} \dim(X)$, or equivalently if codim$(X) < \frac{1}{2} \dim(X)$, then X is a complete intersection.

Let us quote R. Hartshorne, see [87]:

While I am not convinced of the truth of this statement, I think it is useful to crystallize one's idea, and to have a particular problem in mind.

Hartshorne immediately remarks that the conjecture is sharp, due to the examples of $\mathbb{G}(1, 4) \subset \mathbb{P}^9$ and of the spinorial variety $S^{10} \subset \mathbb{P}^{15}$, see Examples 4.2.5 and 4.2.6 for the definition/construction of these varieties.

Moreover, the examples of cones over curves in $\mathbb{P}^3$, not complete intersections, reveal the necessity of the non-singularity assumption.

Definition 5.1.2 (Hartshorne Variety) A smooth irreducible projective variety $X^n \subset \mathbb{P}^{\frac{3n}{2}}$ is called a *Hartshorne variety* if it is not a complete intersection.

It is not accidental that the previous examples of Hartshorne varieties are homogeneous. Indeed, a useful technique for constructing varieties of not too high codimension is the theory of algebraic groups, see for example [198, Chap. 3] or the appendix to [125].

One of the main difficulties of Hartshorne's Conjecture is a good translation in geometrical terms of the algebraic condition of being a complete intersection and more generally of dealing with the equations defining a variety.

Here is not the place to stress how many important results originated and still arise today from this open problem in the areas of vector bundles on projective space, of the study of defining equations of a variety, of k-normality and so on. The list of these achievements is so long that we prefer to avoid citations, being confident that everyone has met at some time a problem or a result related to this conjecture.

For a long time the Hartshorne Conjecture has been forgotten and today continues to be out of the interests of the main research groups in Algebraic Geometry around the world, although it remains almost as open as before.

We recall that a quadratic manifold is a non-degenerate manifold $X^n \subset \mathbb{P}^N$ which is the scheme-theoretic intersection of $m \geq N - n$ quadratic hypersurfaces. Let us state our contributions in the subject, whose proofs are postponed in Sect. 5.2.

Theorem 5.1.3 (Hartshorne's Conjecture for Quadratic Manifolds, [105, Theorem 3.8, part (4)]) *Let $X^n \subset \mathbb{P}^N$ be a quadratic manifold. If $\mathrm{codim}(X) < \frac{\dim(X)}{2}$, then X is a complete intersection.*

Theorem 5.1.4 (Classification of Quadratic Hartshorne Varieties, [105, Theorem 3.9]) *Let $X^n \subset \mathbb{P}^{\frac{3n}{2}}$ be a quadratic manifold. Then either $X \subset \mathbb{P}^N$ is a complete intersection or it is projectively equivalent to one of the following:*

1. $\mathbb{G}(1, 4) \subset \mathbb{P}^9$;
2. $S^{10} \subset \mathbb{P}^{15}$.

In the same survey paper Hartshorne posed another conjecture, motivated by the remark that complete intersections are linearly normal and by the behavior of some examples in low dimension.

Let us recall that linearly normal varieties have been introduced in Definition 3.4.3 as those varieties $X \subset \mathbb{P}^N$ which are not isomorphic external projections of a variety $Y \subset \mathbb{P}^M$ with $M > N$.

Conjecture 5.1.5 (Hartshorne's Linear Normality Conjecture, [87]) Let $X \subset \mathbb{P}^N$ be a smooth irreducible non-degenerate projective variety.

If $N < \frac{3}{2} \dim(X) + 1$, i.e. if $\mathrm{codim}(X) < \frac{1}{2} \dim(X) + 1$, then X is linearly normal.

By Proposition 1.3.5 and by the discussion after Definition 3.4.3 relating linear normality and isomorphic linear projections, we can equivalently reformulate this conjecture via secant varieties. Indeed, putting "$N = N + 1$", we get the following equivalent formulation of the Linear Normality Conjecture:

$$\text{If } N < \frac{3}{2}\dim(X) + 2, \text{ then } SX = \mathbb{P}^N.$$

We quote Hartshorne to understand his point of view on this second problem:

> Of course in settling this conjecture, it would be nice also to classify all nonlinearly normal varieties with $N = \frac{3n}{2} + 1$, so as to have a satisfactory generalization of Severi's theorem. As noted above, a complete intersection is always linearly normal, so this conjecture would be a consequence of our original conjecture, except for the case $N = \frac{3n}{2}$. My feeling is that this conjecture should be easier to establish than the original one.
>
> R. Hartshorne in [87].

Once again the bound is sharp as shown by the projected Veronese surface in $\mathbb{P}^4$. The conjecture on linear normality was proved by Zak at the beginning of the 1980s as an immediate consequence of Terracini's Lemma and of Theorem 3.2.1. We now reproduce Zak's original proof.

Theorem 5.1.6 (Zak's Theorem on Linear Normality) *Let $X^n \subset \mathbb{P}^N$ be a smooth non-degenerate projective variety. If $N < \frac{3}{2}n + 2$, then $SX = \mathbb{P}^N$. Or equivalently if $SX \subsetneq \mathbb{P}^N$, then $\dim(SX) \geq \frac{3}{2}n + 1$ and hence $N \geq \frac{3}{2}n + 2$.*

Proof Suppose $SX \subsetneq \mathbb{P}^N$. Then there exists a hyperplane H containing the general tangent space to SX, let us say $T_z SX$. Then by Corollary 1.4.7, the hyperplane H is tangent to X along $\Sigma_z(X)$, which by the generality of z has pure dimension

$$\delta(X) = 2n + 1 - \dim(SX).$$

Since $T(\Sigma_z(X), X) \subseteq H$, the non-degenerate variety $S(\Sigma_z(X), X) \supseteq X$ is not contained in H, yielding $T(\Sigma_z(X), X) \neq S(\Sigma_z(X), X)$. By Theorem 3.2.1 we get

$$2n + 1 - \dim(SX) + n + 1 = \dim(S(\Sigma_z(X), X)) \leq \dim(SX),$$

yielding

$$3n + 2 \leq 2\dim(SX)$$

and finally

$$N - 1 \geq \dim(SX) \geq \frac{3}{2}n + 1.$$

$\square$

As outlined by Hartshorne, the previous result naturally leads to the notion of a Severi variety, first introduced by F.L. Zak in [198].

Definition 5.1.7 (Severi Variety) A smooth non-degenerate irreducible projective variety $X^n \subset \mathbb{P}^{\frac{3}{2}n+2}$ such that $SX \subsetneq \mathbb{P}^{\frac{3}{2}n+2}$ is called *a Severi variety*.

As we shall see in Corollary 5.4.2, Severi varieties are *LQEL*-manifolds of type $\delta = n/2$ so that their classification follows from Theorem 4.3.3. We shall come back to this in Sect. 5.4 below.

5.2 Proofs of Hartshorne's Conjecture for Quadratic Manifolds and of the Classification of Quadratic Hartshorne Manifolds

We introduce some results which will be crucial for the proof of Hartshorne's Conjecture for quadratic manifolds. First we recall some notation introduced in Sects. 2.3 and 2.3.3.

Let

$$X = V(f_1, \ldots, f_m) \subset \mathbb{P}^N \tag{**}$$

be a projective manifold, let $x \in X$ be a general point, let $n = \dim(X)$ and let $c = \mathrm{codim}(X) = N - n$.

Thus (**) means precisely that $X \subset \mathbb{P}^N$ is scheme-theoretically the intersection of $m \geq 1$ hypersurfaces of degrees $d_1 \geq d_2 \geq \ldots \geq d_m \geq 2$, where $d_i = \deg(f_i)$. Moreover, it is implicitly assumed that m is minimal, i.e. none of the hypersurfaces contains the intersection of the others. Define, following [105], the integer

$$d := \min\{\sum_{i=1}^{c}(d_i - 1) \text{ for expressions (**) as above}\} \geq c.$$

With these definitions $X \subset \mathbb{P}^N$ (or more generally a scheme $Z \subset \mathbb{P}^N$) is called *quadratic* if it is scheme-theoretically an intersection of quadrics, that is we can assume $d_1 = 2$ in (**). Equivalently, $X \subset \mathbb{P}^N$ is quadratic if and only if $d = c$.

5.2.1 The Bertram–Ein–Lazarsfeld Criterion for Complete Intersections

The following characterization of Complete Intersections is stated in a slightly different way in [19, Corollary 4].

Theorem 5.2.1 (Bertram–Ein–Lazarsfeld Criterion, [19, Corollary 4]) *Let $X^n \subset \mathbb{P}^N$ be a manifold of dimension $n \geq 1$ which is scheme-theoretically defined* *by hypersurfaces of degrees $d_1 \geq d_2 \geq \ldots \geq d_m$. Let $c = N - n = \mathrm{codim}(X)$. Then $X \subset \mathbb{P}^N$ is a complete intersection of type $(d_1, \ldots, d_c)$ if and only if*

$$\mathscr{O}(-K_X) = \mathscr{O}(N + 1 - \sum_{i=1}^{c} d_i). \tag{5.1}$$

Proof If $X^n \subset \mathbb{P}^N$ is a complete intersection of type $(d_1, \ldots, d_c)$, then (5.1) holds by the Adjunction Formula.

Suppose now that (5.1) holds. Then there exist $g_i \in H^0(\mathbb{P}^N, \mathscr{I}_X(d_i))$, $i = 1, \ldots, c$, such that, letting $Q_i = V(g_i) \subset \mathbb{P}^N$, we obtain the complete intersection scheme

$$Y = Q_1 \cap \ldots \cap Q_c = X \cup X',$$

where X' (if non-empty) is either disjoint from X or meets X in a divisor, see the proof of [19, Corollary 4] for further details.

Since $n \geq 1$ the scheme Y, being a complete intersection, is also connected so that to show that $Y = X$ it is enough to prove that $X \cap X' = \emptyset$. To this end, observe that the g_i's define a morphism

$$\bigoplus_{i=1}^{c} \mathscr{O}_{\mathbb{P}^N}(-d_i) \to \mathscr{I}_X$$

which restricted to X yields a morphism

$$\alpha : \bigoplus_{i=1}^{c} \mathscr{O}_X(-d_i) \to \frac{\mathscr{I}_X}{\mathscr{I}_X^2}$$

of locally free sheaves of rank c. Since the g_i's generically generate $\mathscr{I}_X$ outside a codimension one subset of X, the map α is an injective homomorphism of coherent sheaves. Furthermore, $\mathrm{coker}(\alpha)$ is supported on $X \cap X'$. Computing first Chern classes, it follows that $X \cap X'$ is supported on a divisor D such that

$$\mathscr{O}_X(D) \simeq \det(\frac{\mathscr{I}_X}{\mathscr{I}_X^2}) \otimes \mathscr{O}_X(\sum_{i=1}^{c} d_i) \simeq \mathscr{O}_X(\sum_{i=1}^{c} d_i - N - 1) \otimes \mathscr{O}(-K_X).$$

In conclusion, under the hypothesis (5.1), α is an isomorphism and $X \cap X' = \emptyset$. Thus $X \subset \mathbb{P}^N$ is the scheme-theoretic intersection of c hypersurfaces of degree $d_1, \ldots, d_c$ and hence X is the complete intersection of these hypersurfaces. $\qquad \square$

5.2.2 Faltings' and Netsvetaev's Conditions for Complete Intersections

We now prove an interesting result due to Faltings, following the ideas and presentation in [144].

Theorem 5.2.2 (Faltings' Condition, [57, Korollar zu Satz 3]) *Let $X^n \subset \mathbb{P}^N$ be a manifold scheme-theoretically defined by hypersurfaces of degrees $d_1 \geq d_2 \geq \ldots \geq d_m$. If $m \leq \frac{N}{2}$, then $X \subset \mathbb{P}^N$ is the complete intersection of c hypersurfaces among the m defining it scheme-theoretically.*

Proof For $i = 1, \ldots, m$ let

$$\pi_i : \bigoplus_{k=1}^{m} \mathcal{O}_X(d_k) \to \mathcal{O}_X(d_i)$$

be the canonical projection and let

$$j_i : \mathcal{O}_X(d_i) \to \bigoplus_{k=1}^{m} \mathcal{O}_X(d_k)$$

be the canonical injection.

From the surjection

$$\beta : \bigoplus_{k=1}^{m} \mathcal{O}_X(-d_k) \to \frac{\mathcal{I}_X}{\mathcal{I}_X^2}$$

we deduce the exact sequence of locally free sheaves

$$0 \longrightarrow N_{X/\mathbb{P}^N} \overset{a}{\longrightarrow} \bigoplus_{k=1}^{m} \mathcal{O}_X(d_k) \overset{b}{\longrightarrow} \mathcal{K} \longrightarrow 0, \tag{5.2}$$

with $a = \beta^t$ and with $\mathcal{K}$ locally free sheaf of rank $m - c$.

Let V_1 be the locus of points of X such that $\pi_1 \circ a$ is not surjective and let W_1 be the locus of points of X such that $b \circ j_1$ is not injective. Since (5.2) is an exact sequence we deduce $V_1 \cap W_1 = \emptyset$. Moreover, if $V_1 \neq \emptyset$, then $\text{codim}(V_1) \leq c$. Analogously if $W_1 \neq \emptyset$, then $\text{codim}(W_1) \leq m - c$.

Suppose $W_1 \neq \emptyset$ and let $[W_1] \in \text{H}^i(X, \mathbb{C})$ with

$$i = 2\,\text{codim}(W_1) \leq 2(m + n - N) \leq 2n - N.$$

By Theorem 3.1.1 we can suppose that $[W_1]$ is cut out by the classes of hyperplanes, yielding $V_1 = \emptyset$. Otherwise $\text{codim}(W_1) + \text{codim}(V_1) \leq m \leq n$ since $m \leq N/2$

implies $n \geq N/2$ and a fortitori $n \geq m$. Thus $W_1 \cap V_1 \neq \emptyset$, yielding a contradiction. In conclusion we deduce that either $V_1 = \emptyset$ or $W_1 = \emptyset$.

If $V_1 = \emptyset$, let $N^{(2)} = \ker(\pi_1 \circ a)$, let $\mathcal{K}^{(2)} = \mathcal{K}$ and let

$$0 \longrightarrow N^{(2)} \xrightarrow{a_2} \bigoplus_{k=2}^{m} \mathcal{O}_X(d_k) \xrightarrow{b_2} \mathcal{K}^{(2)} \longrightarrow 0, \tag{5.3}$$

be the corresponding exact sequence of coherent sheaves.

If $W_1 = \emptyset$, let $N^{(2)} = N$, let $\mathcal{K}^{(2)} = \mathrm{coker}(b \circ j_1)$ and let

$$0 \longrightarrow N^{(2)} \xrightarrow{a_2} \bigoplus_{k=2}^{m} \mathcal{O}_X(d_k) \xrightarrow{b_2} \mathcal{K}^{(2)} \longrightarrow 0, \tag{5.4}$$

be the corresponding exact sequence of coherent sheaves.

We can define analogously V_2, respectively W_2, via $\pi_2 \circ a_2$, respectively $b_2 \circ j_2$. As before from (5.3) we deduce $V_2 \cap W_2 = \emptyset$. In this way we can construct the sets V_i, W_i and the coherent sheaves $N^{(i)}$, $\mathcal{K}^{(i)}$ for $i = 1, \ldots, m$. The number r of indices $j \in \{1, \ldots, m\}$ such that $V_j = \emptyset$ is greater than or equal to c. Indeed, on the contrary, we would have an injection of a locally free sheaf of rank $m - r$, $r < c$, into $\mathcal{K}$ which has rank $m - c$. In conclusion, $r = c$.

Thus, modulo a renumbering, we can suppose $V_1 = \ldots = V_c = \emptyset$ (and $W_{c+1} = \ldots = W_m = \emptyset$) and that there exists an isomorphism induced by a and the π_i's

$$N_{X/\mathbb{P}^N} \simeq \bigoplus_{k=1}^{c} \mathcal{O}_X(d_k).$$

From this it easily follows that $X^n \subset \mathbb{P}^N$ is the scheme-theoretic intersection of the selected c hypersurfaces among the initial $m \geq c$ and hence that $X^n \subset \mathbb{P}^N$ is the complete intersection of these c hypersurfaces.

We shall immediately state Netsvetaev's improvement of Faltings' condition. The proof follows the same path, see [144, Sect. 4.12], showing that under the previous hypothesis and with the notation introduced above we still have either $V_i = \emptyset$ or $W_i = \emptyset$ for every $i = 1, \ldots, m$. Instead of Barth's Theorem 3.1.1 Netsvetaev applies the following topological result due to Oka.

Theorem 5.2.3 (Oka's Theorem, See [146]) *Let $X \subset \mathbb{P}^N$ be an algebraic set defined set-theoretically by m equations. Then the restriction maps*

$$r : \mathrm{H}^i(\mathbb{P}^N_{\mathbb{C}}, \mathbb{C}) \to \mathrm{H}^i(X, \mathbb{C})$$

are isomorphisms for $i < N - m$.

Netsvetaev's condition has an apparently mysterious formulation but the conditions appearing in its statement are the ideal tools for the proof of Theorem 5.1.4.

Theorem 5.2.4 (Netsvetaev's Condition, see [144, Theorem 3.2]) *Let $X \subset \mathbb{P}^N$ be a manifold scheme-theoretically defined by hypersurfaces of degrees $d_1 \geq d_2 \geq \ldots \geq d_m$. Suppose one of the following conditions holds:*

1. *$m < N - \frac{2n}{3}$;*
2. *$n \geq \frac{3N}{4} - \frac{1}{2}$.*

If $m \leq n + 1$, then $X \subset \mathbb{P}^N$ is the complete intersection of c hypersurfaces among the m defining it scheme-theoretically.

The last two results are among the most important sufficient conditions to ensure that a smooth manifold $X \subset \mathbb{P}^N$ is a complete intersection. Notwithstanding they did not seem to have any direct or obvious relation with Hartshorne's Conjecture on Complete Intersections in arbitrary codimension. The main applications were obviously to varieties defined by a small number of equations with respect to codimension, yielding restricted forms of the conjecture on Complete Intersections, see [57, Sect. 3]. On the other hand, when these results are applied to $\mathscr{L}_x \subset \mathbb{P}^{n-1}$ for a quadratic manifold of small codimension but without any assumption on the number $m \geq c$ of quadratic equations defining X, they reveal their deepness and powerfulness as we shall see in the next subsection.

We remark that an obvious sufficient condition for $\mathscr{L}_x \neq \emptyset$ is $d \leq n - 1$, see Proposition 2.3.8. On the other hand, $c \leq d$ by definition, while the hypothesis in Hartshorne's Conjecture reads as $c \leq \frac{n-1}{2}$ and finally for quadratic manifolds $d = c$. In conclusion quadratic manifolds satisfying the hypothesis in Hartshorne's Conjecture are ruled by a large family of lines and the linear system of quadrics in the second fundamental form at a general point satisfies Faltings' condition in Theorem 5.2.2.

5.2.3 Proofs of the Main Results

All the necessary tools to provide simple proofs of Theorem 5.1.3 and of Theorem 5.1.4 have now been introduced and we can proceed quickly.

Proof of Theorem 5.1.3 Since $c \leq \frac{n-1}{2}$, the Barth–Larsen Theorem 3.1.1 yields $\text{Pic}(X) \simeq \mathbb{Z} < O_X(1) >$. Since $\mathscr{L}_x \subset \mathbb{P}^{n-1}$ is not empty for $x \in X$ general, then X is a Fano manifold of index $i(X) = \dim(\mathscr{L}_x) + 2$ by Proposition 2.3.9. Since $X \subset \mathbb{P}^N$ is a quadratic manifold we deduce from Proposition 2.3.8 that $\mathscr{L}_x \subset \mathbb{P}^{n-1}$ is scheme-theoretically defined by $m \leq c \leq \frac{n-1}{2}$ quadrics in the second fundamental form of X at x. Faltings' Theorem yields that $\mathscr{L}_x \subset \mathbb{P}^{n-1}$ is the complete intersection of $s \leq c$ quadrics in the second fundamental form of X at x.

Thus we have

$$\dim(\mathscr{L}_x) \geq n - 1 - s \geq n - 1 - c \geq n - 1 - \frac{n-1}{2} = \frac{n-1}{2}$$

and $i(X) = \dim(\mathscr{L}_x) + 2 \geq \frac{n+3}{2}$. Proposition 2.3.9 implies that $\mathscr{L}_x \subset \mathbb{P}^{n-1}$ is non-degenerate while Proposition 2.3.5 assures that there are c linearly independent quadrics vanishing on $\mathscr{L}_x$. Since $\mathscr{L}_x \subset \mathbb{P}^{n-1}$ is a non-degenerate complete intersection of s quadrics we deduce $s = \mathrm{h}^0(\mathscr{I}_{\mathscr{L}_x}(2)) \geq c$, yielding $s = c$, $\dim(\mathscr{L}_x) = n - 1 - c$ and $i(X) = n + 1 - c$. In conclusion, $X \subset \mathbb{P}^N$ is a complete intersection by Theorem 5.2.1. □

Proof of Theorem 5.1.4 Suppose $X \subset \mathbb{P}^N$ is not a complete intersection. Reasoning as in the previous proof we know that $\mathscr{L}_x \subset \mathbb{P}^{n-1}$ is a manifold scheme-theoretically defined by c linearly independent quadratic equations. Moreover, $\dim(\mathscr{L}_x) \geq n - 1 - c$ with equality holding if and only if X is a complete intersection. Thus we can assume $\dim(\mathscr{L}_x) \geq n - c = \frac{n}{2} > \frac{n-1}{2}$ and that $\mathscr{L}_x \subset \mathbb{P}^{n-1}$ is non-degenerate by Proposition 2.3.9.

Netsvetaev's Condition in Theorem 5.2.4 applied to $\mathscr{L}_x \subset \mathbb{P}^{n-1}$ gives the inequalities

$$\frac{3n - 6}{4} \leq \dim(\mathscr{L}_x) < \frac{3n - 5}{4}.$$

We may assume $n \geq 6$ (otherwise $\mathscr{L}_x$ is a complete intersection), so we deduce that $n = 4r + 2$ and $\dim(\mathscr{L}_x) = 3r$, for a suitable r. If $r > 2$ (equivalently $n > 10$), we would deduce $\dim(\mathscr{L}_x) \geq 2(n - 1 - \dim(\mathscr{L}_x)) + 1$ and $\mathscr{L}_x$ would be a complete intersection by Theorem 5.1.3. In conclusion, $n = 6$ or $n = 10$. In the first case $i(X) = \dim(\mathscr{L}_x) + 2 = 5$ and we get case (1) by the classification of del Pezzo manifolds, see [65]. In the second case, $i(X) = 8$, leading to case (2) by [138]. □

It is worth remarking that there exist Hartshorne varieties different from the ones described in the previous theorem, see for example [52, Proposition 1.9].

5.3 Speculations on Hartshorne's Conjecture

From the point of view of the defining equations quadratic manifolds represent the simplest possible case. This case, although important and meaningful, did not shed new light on the general case of Hartshorne's Conjecture, which remains as intriguing as before. In any case, we would like to point out in this section that the applications of Hartshorne's Conjecture to $\mathscr{L}_x \subset \mathbb{P}^{n-1}$ can be used as a method to produce natural geometrical bounds on X itself and to pose some new natural questions on Fano manifolds of high index.

Before applying this method we shall analyze some of the results obtained until now from this perspective. The classification of *LQEL*-manifolds of type $\delta = \frac{n}{2}$ was well known classically for $n = 2, 4$. Thus suppose $n \geq 6$ and that $X^n \subset \mathbb{P}^N$ is a *LQEL*-manifold of type $\delta = \frac{n}{2} \geq 3$. In particular, $N \geq \dim(SX) = \frac{3n}{2} + 1$.

Then, by Theorem 4.2.3, $\mathscr{L}_x \subset \mathbb{P}^{n-1}$ is a non-degenerate *QEL*-manifold of dimension $\frac{3n}{4} - 2$ and type $\delta = \frac{n}{2} - 2 \geq 1$, which is not a complete intersection since

$$h^0(\mathscr{I}_{\mathscr{L}_x}(2)) \geq \dim(|II_{x,X}|) + 1 = N - n \geq \frac{n}{2} + 1 > \frac{n}{4} + 1 = \mathrm{codim}(\mathscr{L}_x, \mathbb{P}^{n-1}).$$

Moreover, we have that

$$\frac{3n}{4} - 2 = \dim(\mathscr{L}_x) > \frac{2(n-1)}{3} \text{ if and only if } n > 16.$$

In conclusion the existence of an *LQEL*-manifold of type $\delta = \frac{n}{2}$ and of dimension $n > 16$ would have produced a counterexample to Hartshorne's Conjecture on Complete Intersections!

Obviously the interesting fact in the proof of Theorem 4.3.3 is that we can prove directly that for $n \geq 6$ necessarily $n = 8$ or $n = 16$ without invoking Hartshorne's Conjecture. To the best of our knowledge, these sharp connections have been overlooked before and they were first pointed out in [160, Remark 3.3].

Analogously if $X^n \subset \mathbb{P}^N$ is an *LQEL*-manifold the bound $\delta \leq \frac{n+3}{3}$, which for $\delta = n/2$ implies $n \leq 16$, is equivalent to the condition $\dim(\mathscr{L}_x) \leq \frac{2(n-1)}{3}$. This is precisely the negation of the condition in Hartshorne's Conjecture for $\mathscr{L}_x \subset \mathbb{P}^{n-1}$. Indeed, reasoning as above, one deduces that for an arbitrary *LQEL*-manifold we have

$$h^0(\mathscr{I}_{\mathscr{L}_x}(2)) > \mathrm{codim}(\mathscr{L}_x, \mathbb{P}^{n-1}),$$

unless $\delta = n$, which implies that $X \subset \mathbb{P}^{n+1}$ is a quadric hypersurface. Thus the variety $\mathscr{L}_x \subset \mathbb{P}^{n-1}$ of an *LQEL*-manifold with $n > \delta > \frac{n+8}{3}$ would have been a counterexample to Hartshorne's Conjecture. Corollary 4.4.11 shows that this is impossible and once again the interesting fact is that one proves directly the previous bound on δ via the Divisibility Theorem without assuming Hartshorne's Conjecture.

Similarly the bound $\mathrm{def}(X) \leq \frac{n+2}{3}$ in Corollary 4.4.12 is equivalent to the hypothesis in Hartshorne's Conjecture for the manifold $\mathscr{L}_x \subset \mathbb{P}^{n-1}$ containing the *contact lines* appearing on a dual defective manifold.

The previous analysis reveals that the application of Hartshorne's Conjecture to the Hilbert scheme of lines passing through a general point of a prime Fano variety of high index is an efficient way to postulate some new problems, which are either true or would produce a counterexample to Hartshorne's Conjecture.

The validity of Hartshorne's Conjecture for quadratic manifolds also supports the possibility that this conjecture could be valid at least for $\mathscr{L}_x \subset \mathbb{P}^{n-1}$ under some hypothesis on $X^n \subset \mathbb{P}^N$ like small codimension or under other geometrical restrictions (prime Fano and/or small degree, etc, etc). Now we let this principle act in practice.

From this point of view, taking into account that the quadrics in the second fundamental form of $X^n \subset \mathbb{P}^N$ at a general point $x \in X$ vanish on $\mathscr{L}_x \subset \mathbb{P}^{n-1}$, one could pose this first basic question, which is a kind of restricted Hartshorne Conjecture. In this context we put a restriction on the variety but not on the defining equations.

Conjecture 5.3.1 (HCL = Hartshorne's Conjecture for $\mathscr{L}_x \subset \mathbb{P}^{n-1}$, [105]) Assume that $X^n \subset \mathbb{P}^N$ is covered by lines with $\dim(\mathscr{L}_x) \geq \frac{n-1}{2}$ and let $T = \langle \mathscr{L}_x \rangle$ be the span of $\mathscr{L}_x$ in $\mathbb{P}^{n-1}$. If $\dim(\mathscr{L}_x) > 2 \operatorname{codim}(\mathscr{L}_x, T)$, then $\mathscr{L}_x \subset \mathbb{P}^{n-1}$ is a complete intersection.

Since for smooth complete intersections $X^n \subset \mathbb{P}^N$, the variety $\mathscr{L}_x \subset \mathbb{P}^{n-1}$ is also a smooth complete intersection, see Example 2.3.11, one may ask if for manifolds covered by lines the converse also holds, illustrating a kind of recovery principle from $\mathscr{L}_x$ to X.

Conjecture 5.3.2 If $X \subset \mathbb{P}^N$ is covered by lines and $\mathscr{L}_x \subset \mathbb{P}^{n-1}$ is a (say smooth irreducible non-degenerate) complete intersection, then X is a complete intersection.

We now specialize to some *restricted* form of the Complete Intersection Conjecture.

Conjecture 5.3.3 (HCF = Hartshorne's Conjecture for Fano Manifolds, [105]) If $X^n \subset \mathbb{P}^N$ is a Fano manifold and if $n \geq 2c + 1$, then $X^n \subset \mathbb{P}^N$ is a complete intersection.

Manifolds of (very) small degree are known to be complete intersections, cf. [14]. Ionescu proved in [101] that for a manifold $X^n \subset \mathbb{P}^N$ with $\deg(X) \leq n - 1$ there are only the following two possibilities: either $X \simeq \mathbb{G}(1, 4) \subset \mathbb{P}^9$ or $c \leq \frac{n-1}{2}$ and $X^n \subset \mathbb{P}^N$ is a prime Fano manifold. Therefore Conjecture 5.3.3 would yield the following optimal result, which obviously can be stated independently.

Conjecture 5.3.4 (Barth–Ionescu Conjecture, [14, 101]) If $\deg(X) \leq n - 1$, then X is a complete intersection, unless it is projectively equivalent to $\mathbb{G}(1, 4) \subset \mathbb{P}^9$.

Note that the Segre embedding $\mathbb{P}^1 \times \mathbb{P}^{n-1} \subset \mathbb{P}^{2n-1}$ has degree n and it is not a complete intersection if $n \geq 3$.

Hartshorne's Conjecture combined with the previous Conjectures can be applied to $\mathscr{L}_x \subset \mathbb{P}^{n-1}$, with $X^n \subset \mathbb{P}^N$ a prime Fano manifold of high index, to derive some characterizations of Fano Complete Intersections of high index. This is also justified by the remark that prime Fano manifolds of high index tend to be complete intersections, see [65, 138], and also quadratic manifolds. Let us recall these results.

Proposition 5.3.5 ([105, Proposition 3.6])

1. Let $X \subset \mathbb{P}^N$ be a Fano manifold with $\operatorname{Pic}(X) \simeq \mathbb{Z}\langle \mathscr{O}(1) \rangle$ and of index $i(X) \geq \frac{2n+5}{3}$. If the (HCL) and the (HCF) are true, then X is a complete intersection.
2. The same conclusion holds assuming only the (HCF), but asking instead that $i(X) \geq \frac{3(n+1)}{4}$.

Thus we have found two different statements pointing out two explicit bounds on the index of a prime Fano manifold which so far have never been stated, to the best of our knowledge.

We now put our results in the perspective of being natural generalizations of some known properties of quadratic manifolds. Mumford in his seminal series of lectures [140] called attention to the fact that many interesting embedded manifolds are scheme-theoretically defined by quadratic equations.

As an application of the main vanishing theorem in [19, Corollary 2] it is proved that for a quadratic manifold $X \subset \mathbb{P}^N$ the following implications hold:

a) If $n \geq c - 1$, then X is projectively normal;
b) If $n \geq c$, then X is projectively Cohen–Macaulay.

The natural continuation is contained in the next result.

Theorem 5.3.6 ([105, Theorem 3.8]) *Assume that $X \subset \mathbb{P}^N$ is a quadratic manifold.*

1. *If $n \geq c$, then X is Fano.*
2. *If $n \geq c + 1$, then X is covered by lines. Moreover, $\mathscr{L}_x \subset \mathbb{P}^{n-1}$ is scheme-theoretically defined by c independent quadratic equations.*
3. *If $n \geq c + 2$, then the following conditions are equivalent:*

   (i) *$X \subset \mathbb{P}^N$ is a complete intersection;*
   (ii) *$\mathscr{L}_x \subset \mathbb{P}^{n-1}$ is a complete intersection;*
   (iii) *$\dim(\mathscr{L}_x) = n - 1 - c$.*

Proof We use the notation in (**). Moreover, denote by $a := \dim(\mathscr{L}_x)$. Since X is quadratic, we have $d = c$ and that $N^*_{X/\mathbb{P}^N}(2)$ is spanned by global sections. Therefore, $\det(N^*_{X/\mathbb{P}^N})(2c)$ is also spanned and $\det(N^*_{X/\mathbb{P}^N})(2c + 1)$ is ample. From $n \geq c$ it follows that $N + 1 \geq 2c + 1$. We get the ampleness of $-K_X = \det(N^*_{X/\mathbb{P}^N})(N + 1)$, thus proving (1).

From Proposition 2.3.8 we deduce the first part in (2) and that $\mathscr{L}_x \subset \mathbb{P}^{n-1}$ is defined scheme-theoretically by at most c quadratic equations. Since $n \geq c + 1$, we must have $\delta \geq n - c + 1 > 0$. From Proposition 2.3.5 we infer that the quadratic equations defining $\mathscr{L}_x$ are independent, proving (2). This also shows that, when $\mathscr{L}_x$ is a complete intersection, it has codimension c in $\mathbb{P}^{n-1}$. From Barth's Theorem (3.1.1) we deduce that $X^n \subset \mathbb{P}^N$ is a prime Fano manifold. The equivalence between conditions (i), (ii) and (iii) of part (3) is now clear, the implication (iii)$\Rightarrow$ (i) coming from Theorem 5.2.1. $\square$

In Proposition 2.3.8 we saw that if $d \leq n - 1$ then $\mathscr{L}_x \subset \mathbb{P}^{n-1}$ is set-theoretically defined by d equations. If one is able to prove that these equations define $\mathscr{L}_x$ scheme-theoretically, Faltings' condition could help to prove that $\mathscr{L}_x$ is the complete intersection of some of these equations. Expressing the hypothesis in Faltings' condition one could pose the following problem, which is a weaker form of Hartshorne's Conjecture since in general $d \geq c$.

Conjecture 5.3.7 (Scheme-Theoretic Hartshorne Conjecture) Let

$$X^n = V(f_1, \ldots, f_m) \subset \mathbb{P}^N$$

be a manifold which is scheme-theoretically defined as the intersection of the hypersurfaces $V(f_i)$ of degree d_i with $d_1 \geq d_2 \geq \ldots \geq d_m \geq 2$.
   Let

$$d := \min\{\sum_{i=1}^{c}(d_i - 1) \text{ for expressions as above}\} \geq c.$$

If

$$d \leq \frac{n-1}{2},$$

then $X \subset \mathbb{P}^N$ is a complete intersection.

5.4   A Refined Linear Normality Bound and Severi Varieties

Now we are in position to provide a slight refinement of Zak's Linear Normality Theorem, [198, Theorem 2.8]. The proof is essentially identical to Zak's original argument, but it reveals the importance of the tangential invariants defined in Sect. 3.3 to which we refer for the notation.

   The bound obtained also strengthens the bound of Landsberg for smooth varieties involving $\widetilde{\gamma}(X) = \gamma(X) - \delta(X)$, which equals $\dim(F_v)$ in Landsberg's notation, see [119] and also [107, 3.15].

Theorem 5.4.1 *Let $X \subset \mathbb{P}^N$ be an irreducible non-degenerate variety such that $SX \subsetneq \mathbb{P}^N$. Let $b = \dim(\mathrm{Sing}(X))$, $\xi = \xi(X)$ and $\delta = \delta(X)$. Then:*

$$\dim(SX) \geq \frac{3}{2}n + \frac{1-b}{2} + \frac{\xi - \delta}{2}; \tag{5.5}$$

$$N \geq \frac{3}{2}n + 1 + \frac{1-b}{2} + \frac{\xi - \delta}{2}; \tag{5.6}$$

$$n \leq \frac{1}{3}(2N + b - (\xi - \delta)) - 1. \tag{5.7}$$

In particular, if $X \subset \mathbb{P}^N$ is also smooth, then

$$\dim(SX) \geq \frac{3}{2}n + 1 + \frac{\xi - \delta}{2}; \tag{5.8}$$

$$N \geq \frac{3}{2}n + 2 + \frac{\xi - \delta}{2} \tag{5.9}$$

and

$$n \leq \frac{2(N-2)}{3} - \frac{\xi - \delta}{3}. \tag{5.10}$$

Proof If $\xi \leq b + 1$, then

$$\dim(SX) > \frac{3}{2}n + \frac{1-b}{2} + \frac{\xi - \delta}{2},$$

so that we can assume $\xi \geq b + 2$ and hence $n \geq b + 3$.

Fix a general $p \in SX$ and consider a general hyperplane $H \subset \mathbb{P}^N$ containing T_pSX. By definition H is tangent to X along $\Xi_p \setminus \mathrm{Sing}(X)$. Consider a general $L = \mathbb{P}^{N-b-1}$ and set $\widehat{X} = X \cap L$, $\widehat{\Xi}_p = \Xi_p \cap L$ and $\widehat{H} = H \cap L$. The variety $\widehat{X} \subset \mathbb{P}^{N-b-1} = L$ is smooth, irreducible and non-degenerate of dimension $n - b - 1 \geq 2$, while the hyperplane $\widehat{H} = \mathbb{P}^{N-b-2}$ is tangent to $\widehat{X}$ along the variety $\widehat{\Xi}_p$, whose dimension is $\xi - b - 1 \geq 1$.

Since $\widehat{X} \subset L$ is non-degenerate and contained in $S(\widehat{\Xi}_p, \widehat{X})$, we get $S(\widehat{\Xi}_p, \widehat{X}) \neq T(\widehat{\Xi}_p, \widehat{X})$, where $T(\widehat{\Xi}_p, \widehat{X}) := \cup_{y \in \widehat{\Xi}_p} T_y\widehat{X} \subseteq \widehat{H}$. Therefore by applying Theorem 3.2.1 to $S(\widehat{\Xi}_p, \widehat{X})$ we deduce that

$$\dim(SX) - b - 1 = \dim(SX \cap L) \geq \dim(S\widehat{X}) \geq \dim(S(\widehat{\Xi}_p, \widehat{X})) =$$

$$= \dim(\widehat{\Xi}_p) + \dim(\widehat{X}) + 1 = (\xi - b - 1) + (n - b - 1) + 1.$$

Hence

$$2n + 1 - \delta \geq n + \xi - b = n + \delta + \widetilde{\xi} - b,$$

where $\widetilde{\xi} = \xi - \delta$, that is

$$\delta \leq \frac{n + b + 1 - \widetilde{\xi}}{2}.$$

Thus

$$\dim(SX) = 2n + 1 - \delta \geq \frac{3}{2}n + \frac{1-b}{2} + \frac{\widetilde{\xi}}{2} = \frac{3}{2}n + \frac{1-b}{2} + \frac{\xi - \delta}{2}.$$

Since $SX \subsetneq \mathbb{P}^N$, we deduce $N \geq \dim(SX) + 1$, which combined with the above estimates yields $n \leq \frac{1}{3}(2N + b - (\xi - \delta)) - 1$. The other assertions are now obvious. $\qquad\square$

Theorem 5.4.1 implies that a Severi variety $X \subset \mathbb{P}^{\frac{3}{2}n+2}$ has $\xi(X) = \gamma(X) = \delta(X) = \frac{n}{2}$ and that $SX \subsetneq \mathbb{P}^{\frac{3}{2}n+2}$ is a hypersurface.

With these powerful tools at hand and via Scorza's Lemma we can immediately prove the next consequence, in a way different from [198, IV.2.1, IV.3.1, IV.2.2].

Corollary 5.4.2 *Let $X \subset \mathbb{P}^{\frac{3}{2}n+2}$ be a Severi variety. Then*

1. *$X \subset \mathbb{P}^{\frac{3}{2}n+2}$ is an LQEL-variety of type $\delta = \frac{n}{2}$.*
2. *The image of a general tangential projection of X, $\pi_x(X) = W_x \subset \mathbb{P}^{\frac{n}{2}+1}$, is a smooth quadric hypersurface.*
3. *Given three general points $x, y, z \in X$, let $Q_{x,z}$, respectively $Q_{y,z}$, be the smooth quadrics passing through x and z, respectively y and z. Then $Q_{x,z} \cap Q_{y,z} = z$, the intersection being transversal.*

Proof As we observed above, for a Severi variety we have $\xi(X) = \gamma(X) = \delta(X) = \frac{n}{2}$ due to Theorem 5.4.1. The conclusion of the first part follows from Scorza's Lemma, Theorem 3.3.3.

The proof of Theorem 5.4.1 yields $SX = S(\Sigma_q, X)$ for a general $q \in SX$. From Terracini's Lemma we get $T_x X \cap T_w \Sigma_q = \emptyset$ for a general $x \in X$ and for a general $w \in \Sigma_q$. Thus $\dim(\pi_x(\Sigma_q)) = \frac{n}{2}$ for $q \in SX$ general. Since $\pi_x(X) = W_x$ has dimension $\frac{n}{2}$, we deduce that $\pi_x(\Sigma_q) = W_x$ for $q \in SX$ general. Therefore the variety $W_x \subset \mathbb{P}^{\frac{n}{2}+1}$ is a quadric hypersurface, being a hypersurface and also a non-degenerate linear projection of a quadric hypersurface. The smoothness of W_x follows from $0 = \widetilde{\xi}(X) = \mathrm{def}(W_x)$. In particular, the restriction of π_x to Σ_q is an isomorphism. Scorza's Lemma also yields $\Sigma_p = \pi_x^{-1}(\pi_x(z))$. Take $q \in < y, z >$ general and consider Σ_q. From the previous analysis $\pi_{x|\Sigma_q} : \Sigma_q \to W_x$ is an isomorphism so that Σ_q intersects Σ_p only at z, the intersection being transversal. $\square$

Clearly the dimension n of a Severi variety $X \subset \mathbb{P}^{\frac{3}{2}n+2}$ is even so that the first case to be considered is $n = 2$. These are smooth surfaces in $\mathbb{P}^5$ such that $SX \subsetneq \mathbb{P}^5$. They were completely classified in the classical and well-known theorem of Severi, [176], see Theorem 3.4.1. This justifies the name given by Zak to such varieties. By Theorem 5.1.6, it follows that $SX \subset \mathbb{P}^{\frac{3}{2}n+2}$ is necessarily a hypersurface, that is $\dim(SX) = \frac{3}{2}n + 1$.

In Exercise 1.5.11 we showed that the Segre variety $X = \mathbb{P}^2 \times \mathbb{P}^2 \subset \mathbb{P}^8$ is an example of a Severi variety of dimension 4. Indeed, $N = 8 = \frac{3}{2} \cdot 4 + 2$ and SX is a cubic hypersurface, see *loc. cit*. By the classical work of Scorza, last page of [163], it turns out that $\mathbb{P}^2 \times \mathbb{P}^2$ is the only Severi variety of dimension 4. We provided a short, geometrical, self-contained and elementary proof in Theorem 4.3.3.

The realization of the Grassmann variety of lines in $\mathbb{P}^5$ Plücker embedded, $X = \mathbb{G}(1,5) \subset \mathbb{P}^{14}$, as the variety given by the pfaffians of the general antisymmetric 6×6 matrix, yields that $\mathbb{G}(1,5)$ is a Severi variety of dimension 8 such that its secant variety is a degree 3 hypersurface whose equation is the pfaffian of the 6×6 antisymmetric matrix, see for example [85, p. 112 and p. 145] for the last assertion.

A less trivial example is a variety studied by Elie Cartan and also by Room. It is a homogeneous complex variety of dimension 16, $X \subset \mathbb{P}^{26}$, associated to the representation of E_6 and for this reason called an E_6-*variety*, or *Cartan variety* by Zak. It has been shown by Lazarsfeld and Zak that its secant variety is a degree 3 hypersurface, see, for example, [125] and [198, Chap. 3].

There is a unitary way to look at these 4 examples, by realizing them as *Veronese surfaces over the composition algebras over* K, $K = \overline{K}$, char(K)=0, [198, Chap. 3]. Let $\mathscr{U}_0 = K$, $\mathscr{U}_1 = K[t]/(t^2 + 1)$, $\mathscr{U}_2 =$ quaternion algebra over K, $\mathscr{U}_3 =$ Cayley algebra over K. Let $\mathscr{I}_i$, $i = 0, \ldots, 3$, denote the Jordan algebra of Hermitian (3×3)-matrices over $\mathscr{U}_i$, $i = 0, \ldots, 3$. A matrix $A \in \mathscr{I}_i$ is called *Hermitian* if $\overline{A}^t = A$, where the bar denotes the natural involution in $\mathscr{U}_i$ (for $\mathscr{U}_0$ there is no involution and we have symmetric matrices). Let

$$X_i = \{[A] \in \mathbb{P}(\mathscr{I}_i) \; : \; \mathrm{rk}(A) = 1\} \subset \mathbb{P}(\mathscr{I}_i).$$

Then

$$N_i = \dim(\mathbb{P}(\mathscr{I})) = 3 \cdot 2^i + 2, \quad n_i = \dim(X_i) = 2^{i+1} = 2\dim_K(\mathscr{U}_i),$$

and

$$SX = \{[A] \in \mathbb{P}(\mathscr{I}_i) \; : \; \mathrm{rk}(A) \leq 2\} = V(\det(A)) \subset \mathbb{P}(\mathscr{I}_i)$$

is a degree 3 hypersurface. By definition $X_i \subset \mathbb{P}(\mathscr{I}_i)$ is a Severi variety of dimension 2^{i+1}, projectively equivalent to one of the above examples.

A theorem of Jacobson states that over a fixed algebraically closed field K, modulo isomorphism, there exist only four simple rank three Jordan algebras, which are exactly the algebras $\mathscr{I}_i$'s. Equivalently there are only three composition algebras over such a field K. Thus this construction gives the four known examples.

A highly non-trivial and notable result is the classification of Severi varieties, which was first proved by Zak in [196] and which now is an immediate consequence of Corollary 4.3.3. Via the previous construction the next theorem is equivalent to Jacobson's Classification Theorem recalled above.

Theorem 5.4.3 (Zak's Classification of Severi Varieties, [27, 119, 125, 196, 198], [160, Corollary 3.2] or Theorem 4.3.3 Here) *Let $X \subset \mathbb{P}^{\frac{3}{2}n+2}$ be a Severi variety of dimension n, defined over an algebraically closed field K of characteristic 0. Then X is projectively equivalent to one of the following:*

1. *the Veronese surface $v_2(\mathbb{P}^2) \subset \mathbb{P}^5$;*
2. *the Segre fourfold $\mathbb{P}^2 \times \mathbb{P}^2 \subset \mathbb{P}^8$;*
3. *the Grassmann variety $\mathbb{G}(1,5) \subset \mathbb{P}^{14}$;*
4. *the E_6-variety $X \subset \mathbb{P}^{26}$.*

Proof By Corollary 5.4.2 a Severi variety $X^n \subset \mathbb{P}^{\frac{3}{2}n+2}$ is a *QEL*-manifold of type $\delta = \frac{n}{2}$ so that the conclusion follows from Corollary 4.3.3.   □

Since in Corollary 4.3.3 we explicitly proved only that the dimension of a Severi variety is equal to 2, 4, 8 or 16 in the next subsection we shall provide a direct proof that a Severi Variety of dimension 2^r, $r \in \{1, 2, 3, 4\}$, is as stated in Theorem 5.4.3.

5.5 Reconstruction of Severi Varieties of Dimension 2, 4, 8 and 16

We propose an elementary approach to the classification of Severi varieties of dimension 2, 4, 8 or 16. At this point it should be clear that the classification of Severi varieties in dimension 2, 4, 8 and 16 is a straightforward consequence of the above results. As we said above for dimension 2 and 4 it is classical and elementary. Due to the relevance of this classification, we shall reproduce here a short argument which will also be an interesting detour into higher-dimensional projective and birational geometry.

Let us recall the following picture of the known Severi varieties in dimension $n = 2, 4, 8$ and 16. Let $Y \subset \mathbb{P}^{n-1} \subset \mathbb{P}^n$ be either $\emptyset$, respectively $\mathbb{P}^1 \sqcup \mathbb{P}^1 \subset \mathbb{P}^3$, $\mathbb{P}^1 \times \mathbb{P}^3 \subset \mathbb{P}^7$ Segre embedded, $S^{10} \subset \mathbb{P}^{15}$ the ten dimensional spinor variety. On $\mathbb{P}^{n-1}$ we take coordinates $x_0, \ldots, x_{n-1}$ and on $\mathbb{P}^n$ coordinates $x_0, \ldots, x_n$. Let $Q_1, \ldots, Q_{\frac{n}{2}+2}$ be the quadratic forms in the variables $x_0, \ldots, x_{n-1}$ defining $Y \subset \mathbb{P}^{n-1}$. The subvariety $Y \subset \mathbb{P}^n$ is scheme-theoretically defined by the $\frac{3n}{2} + 3$ quadratic forms: Q_i, $x_n x_j$, $i = 1, \ldots, \frac{n}{2} + 2$, $j = 0, \ldots, n$. More precisely, these quadric hypersurfaces form the linear system of quadrics on $\mathbb{P}^n$ vanishing along Y, that is $|H^0(\mathscr{I}_{Y,\mathbb{P}^n}(2))|$. Let

$$\phi_{|H^0(\mathscr{I}_{Y,\mathbb{P}^n}(2))|} : \mathbb{P}^n \dashrightarrow \mathbb{P}^{\frac{3n}{2}+2}.$$

For $Y = \emptyset$, clearly $\phi(\mathbb{P}^2) = \nu_2(\mathbb{P}^2) \subset \mathbb{P}^5$. For $Y = \mathbb{P}^1 \sqcup \mathbb{P}^1 \subset \mathbb{P}^3$ we get $\phi(\mathbb{P}^4) = \mathbb{P}^2 \times \mathbb{P}^2 \subset \mathbb{P}^8$, a particular form of a result known to C. Segre, see [170], found by him when he first studied what are now called *Segre varieties*. For $Y = \mathbb{P}^1 \times \mathbb{P}^3 \subset \mathbb{P}^7$, one obtains $\phi(\mathbb{P}^8) = \mathbb{G}(1, 5) \subset \mathbb{P}^{14}$ Plücker embedded, a particular case of a general result of Semple, see [162, 173]. For $Y = S^{10} \subset \mathbb{P}^{15}$, Zak has shown in [125, 198], Chap. III, that $\phi(\mathbb{P}^{16}) = E_6 \subset \mathbb{P}^{26}$ is the Cartan, or E_6, variety.

The birational inverse of ϕ, $\phi^{-1} : X \dashrightarrow \mathbb{P}^n$, with $X \subset \mathbb{P}^{\frac{3}{2}n+2}$ one of the Severi variety described above, is given by the linear projection from the linear space

$$\mathbb{P}_p^{\frac{n}{2}+1} = \langle \Sigma_p \rangle,$$

$p \in SX$ a general point, onto a skew $\mathbb{P}^n$.

We show that, more generally and a priori, a Severi variety of dimension n can be birationally projected from $\mathbb{P}_p^{\frac{n}{2}+1}$, $p \in SX$ general, onto $\mathbb{P}^n$. This was originally proved in [198, IV.2.4 f)] and we provide a proof for the reader's convenience and with the desire of being as self-contained as possible.

Proposition 5.5.1 *Let* $X \subset \mathbb{P}^{\frac{3}{2}n+2}$ *be a Severi variety. Let* $p \in SX$ *be a general point, let* $\Sigma_p \subset \mathbb{P}_p^{\frac{n}{2}+1}$ *be its entry locus and let*

$$\pi = \pi_{L_p} : X \dashrightarrow \mathbb{P}^n$$

be the projection from $L_p = \langle \Sigma_p \rangle = \mathbb{P}_p^{\frac{n}{2}+1}$. *Let* $\widetilde{X} = Bl_{\Sigma_p} X \xrightarrow{\alpha} X$, *let* E *be the exceptional divisor and let* F *be the strict transform of* $H_p = T_p SX \cap X$ *on* $\widetilde{X}$. *Let* $\widetilde{\pi} : \widetilde{X} \to \mathbb{P}^n$ *be the resolution of* π_L *and let* $\widetilde{\pi}(F) = T_p SX \cap \mathbb{P}^n = \mathbb{P}^{n-1} \subset \mathbb{P}^n$. *By definition of* $\widetilde{\pi}$ *we have* $\widetilde{\pi}^{-1}(\mathbb{P}^{n-1}) = E \cup F$. *Then:*

i) *if* $\dim(\widetilde{\pi}^{-1}(z)) > 0$, $z \in \mathbb{P}^n$, *then* $\widetilde{\pi}^{-1}(z) \subseteq E \cup F$;
ii) *the morphism* $\widetilde{\pi}$ *is birational and defines an isomorphism between* $\widetilde{X} \setminus (E \cup F)$ *and* $\mathbb{P}^n \setminus \mathbb{P}^{n-1}$. *In particular, the locus of indetermination of* $\widetilde{\pi}^{-1}$ *is a subscheme* $Y \subset \mathbb{P}^{n-1}$.

Proof Let $w \in \mathbb{P}^n \setminus \mathbb{P}^{n-1}$, let $W = \widetilde{\pi}^{-1}(w))$ and suppose $\dim(W) > 0$. Then

$$W \cap (E \cup F) = \emptyset,$$

so that $\alpha(W) = W'$ is positive-dimensional and it does not cut L_p. Thus W' contains a positive-dimensional variety $M \subset X$ such that $L_p \cap M = \emptyset$ and such that $\pi(M) = w$. This contradicts the fact that a linear projection, when it is defined everywhere, is a finite morphism. The first part is proved.

To prove part (ii) let us remark that if $p \in \langle x, y \rangle$, $x, y \in X$ general points, then

$$L_p = \langle \Sigma_p \rangle = \langle T_x X, y \rangle \cap \langle T_y X, x \rangle, \tag{5.11}$$

by Terracini's Lemma (see also the proof of Scorza's Lemma). The projection from the linear space $\langle T_x X, y \rangle$ can be regarded as the composition of the tangential projection $\pi_x : X \dashrightarrow W_x \subset \mathbb{P}^{\frac{n}{2}+1}$ and the projection of the smooth quadric hypersurface W_x from the point $\pi_x(y)$. Thus the projection from $\langle T_x X, y \rangle$, $\pi_{x,y} : X \dashrightarrow \mathbb{P}^{\frac{n}{2}}$ is dominant and for a general point $z \in X$ we get

$$\langle T_x X, y, z \rangle \cap X \setminus (\langle T_x X, y \rangle \cap X) = \pi_{x,y}^{-1}(\pi_{x,y}(z)) = Q_{x,z} \setminus (Q_{x,z} \cap \langle T_x X, y \rangle), \tag{5.12}$$

where as always $Q_{x,z}$ is the entry locus of a general point on $\langle x, z \rangle$. Similarly

$$\langle T_y X, x, z \rangle \cap X \setminus (\langle T_y X, x \rangle \cap X) = \pi_{y,x}^{-1}(\pi_{y,x}(z)) = Q_{y,z} \setminus (Q_{y,z} \cap \langle T_x X, y \rangle). \tag{5.13}$$

By definition of projection we have that

$$\pi^{-1}(\pi(z)) = \langle L_p, z \rangle \cap (X \setminus \Sigma_p). \tag{5.14}$$

By the generality of z, we get

$$\pi^{-1}(\pi(z)) = \langle L_p, z \rangle \cap (X \setminus H_p). \tag{5.15}$$

By Terracini's Lemma the linear spaces $\langle T_x X, y \rangle$ and $\langle T_y X, x \rangle$ are contained in $T_p SX$, so that $\langle T_x X, y \rangle \cap X$ and $\langle T_y X, x \rangle \cap X$ are contained in H_p. By combining (5.11), (5.12), (5.13) and (5.15) we finally get

$$z \subseteq \pi^{-1}(\pi(z)) \subseteq Q_{x,z} \cap Q_{y,z} = z,$$

where the last equality is scheme-theoretical by Corollary 5.4.2 and by the generality of $x, y, z \in X$. □

Proposition 5.5.2 *Let $X^n \subset \mathbb{P}^{\frac{3n}{2}+2}$ be a Severi variety with $n \in \{2, 4, 8, 16\}$. Then $X^n \subset \mathbb{P}^{\frac{3n}{2}+2}$ is projectively equivalent to one of the following:*

1. *the Veronese surface $v_2(\mathbb{P}^2) \subset \mathbb{P}^5$;*
2. *the Segre fourfold $\mathbb{P}^2 \times \mathbb{P}^2 \subset \mathbb{P}^8$;*
3. *the Grassmann variety $\mathbb{G}(1, 5) \subset \mathbb{P}^{14}$;*
4. *the Cartan (or E_6) variety $X \subset \mathbb{P}^{26}$.*

Proof For $n = 2$ one can apply Proposition 4.3.1 (or Corollary 3.4.1) to get case (1).

Assume $n = 4$, so that $\delta = 2$. The base locus scheme of $|II_{x,X}|$, which is a linear system of dimension 3, is a smooth not necessarily irreducible curve in $\mathbb{P}^3$ with one apparent double point by Theorem 4.2.3. It immediately follows that $\mathscr{L}_x \subset \mathbb{P}^3$ is the union of two skew lines and that $\mathscr{L}_x \subset \mathbb{P}^3$ coincides with the base locus scheme B_x of $|II_{x,X}|$.

Suppose $n = 8$ and $\delta = 4$. By Theorem 4.2.3, the variety $\mathscr{L}_x \subset \mathbb{P}^7$ is a smooth, irreducible, non-degenerate, QEL-manifold of dimension 4 and such that $S\mathscr{L}_x = \mathbb{P}^7$. Furthermore, by Theorem 4.2.3 part 1), there are at least six quadric hypersurfaces vanishing on $\mathscr{L}_x \subset \mathbb{P}^7$. By restricting to a general $\mathbb{P}^3 \subset \mathbb{P}^7$, the usual Castelnuovo Lemma yields $\deg(\mathscr{L}_x) \leq 4$ and hence $\deg(\mathscr{L}_x) = 4$ since $\mathscr{L}_x \subset \mathbb{P}^7$ is non-degenerate. Thus $\mathscr{L}_x \subset \mathbb{P}^7$ is projectively equivalent to the Segre variety $\mathbb{P}^1 \times \mathbb{P}^3 \subset \mathbb{P}^7$ and clearly $\mathscr{L}_x$ is the base locus scheme B_x of $|II_{x,X}|$.

Suppose $n = 16$. Then by Theorem 4.2.3, $\mathscr{L}_x \subset \mathbb{P}^{15}$ is a Mukai variety of dimension 10 and type $\delta = 6 > n/2 = 5$. Thus by Corollary 4.3.2, $\mathscr{L}_x \subset \mathbb{P}^{15}$ is projectively equivalent to $S^{10} \subset \mathbb{P}^{15}$.

From now on we suppose $n \geq 4$ so that a general entry locus is not a divisor on X. Let $p \in SX \setminus X$ be a general point and let $\mathbb{P}_p^{\frac{n}{2}+1}$ be the locus of secant lines through p. Take a $\mathbb{P}^n$ disjoint from $\mathbb{P}_p^{\frac{n}{2}+1}$ and let $\varphi : X \dashrightarrow \mathbb{P}^n$ be the projection from $\mathbb{P}_p^{\frac{n}{2}+1}$. By Proposition 5.5.1 the map φ is birational and an isomorphism on $X \setminus (T_p SX \cap X)$. Let $Y \subset T_p SX \cap \mathbb{P}^n = \mathbb{P}^{n-1}$ be the base locus scheme of $\varphi^{-1} : \mathbb{P}^n \dashrightarrow X \subset \mathbb{P}^{\frac{3n}{2}+2}$, that is of φ^{-1} composed with the inclusion $i : X \to \mathbb{P}^{\frac{3n}{2}+2}$.

Take a general point $y \in \Sigma_p$. A general smooth conic through y cuts Σ_p transversally so that it is mapped onto a line by φ. By Theorem 4.2.2 there is an irreducible family of dimension $\frac{3n}{2} - 2$ of such conics through y. By varying $y \in \Sigma_p$ the projected lines form a $(2n - 2)$-dimensional family of lines on $\mathbb{P}^n$, which is then a part of the whole family of lines in $\mathbb{P}^n$. This means that φ^{-1} is given by a linear system of quadric hypersurfaces vanishing on the subscheme $Y \subset \mathbb{P}^{n-1}$. Moreover, since $X \subset \mathbb{P}^{\frac{3n}{2}+2}$ is linearly normal, φ^{-1} is given by $|H^0(\mathscr{I}_{Y,\mathbb{P}^n}(2))|$.

Consider a general point $q \in \mathbb{P}^n \setminus \mathbb{P}^{n-1}$, which we can write as $\varphi(x)$ with $x \in X$ general. Consider the family of lines through $\varphi(x)$ and parametrized by the not necessarily irreducible variety $Y_{\mathrm{red}} \subset \mathbb{P}^{n-1}$. The image via φ^{-1} of these lines are lines passing through x. Indeed, these lines cannot be contracted by Proposition 5.5.1, they cut Y and the restriction of φ^{-1} to such a line is given by a sublinear system of $|\mathscr{O}_{\mathbb{P}^1}(2)|$ with a base point. Thus we get a morphism $\alpha_x : Y_{\mathrm{red}} \to \mathscr{L}_x$, since $\mathscr{L}_x$ is isomorphic to the Hilbert scheme of lines through x. Moreover, the birational map φ^{-1} is an isomorphism near $\varphi(x)$, so that the morphism α_x is one-to-one.

A general line through a general point $x \in X$ is sent into a line passing through $\varphi(x)$, because it does not cut the center of projection. Since φ^{-1} is given by a linear system of quadric hypersurfaces, a general line through $\varphi(x)$ cuts Y in one point, proving that $\alpha_x : Y_{\mathrm{red}} \to \mathscr{L}_x$ is dominant and hence surjective. Thus $\alpha_x : Y_{\mathrm{red}} \to \mathscr{L}_x$ is an isomorphism by Zariski's Main Theorem. Moreover, the variety $Y_{\mathrm{red}} \subset \mathbb{P}^{n-1}$ is projectively equivalent to $\mathscr{L}_x$. Thus $Y_{\mathrm{red}} \subset \mathbb{P}^{n-1}$ has homogeneous ideal generated by $\frac{n}{2} + 2$ quadratic equations and therefore it coincides with $Y \subset \mathbb{P}^{n-1}$.

Therefore the previous analysis yields that $Y \simeq \mathscr{L}_x \subset \mathbb{P}^{n-1}$ is projectively equivalent to $\mathbb{P}^1 \sqcup \mathbb{P}^1 \subset \mathbb{P}^3$, $\mathbb{P}^1 \times \mathbb{P}^3 \subset \mathbb{P}^7$, respectively $S^{10} \subset \mathbb{P}^{15}$. The conclusion follows from the birational representation of the known Severi varieties recalled above.   □

Chapter 6
Varieties n-Covered by Curves of a Fixed Degree and the XJC Correspondence

In this chapter, for computational reasons which will become immediately clear, we shall put $\dim(X) = r + 1$, $r \geq 0$, and $n \geq 2$ will indicate the cardinality of finite sets of at least two points on the projective irreducible variety X. As before we shall use the coherent notation $X^{r+1} \subset \mathbb{P}^N$.

The theory of rationally connected varieties is quite recent and was formalized by Campana, Kollár, Miyaoka, Mori, although these varieties were intensively studied from different points of view by classical algebraic geometers such as Darboux, C. Segre, Scorza, Bompiani.

An important result in this theory, see [116, Theorem IV.3.9], asserts that through n general points of a smooth rationally connected complex variety X there passes an irreducible rational curve, which can be taken also to be smooth as soon as $\dim(X) \geq 3$.

From this one deduces that for $\dim(X) \geq 3$ a fixed smooth curve of arbitrary genus can be embedded into X in such a way that it passes through n arbitrary fixed general points, see *loc. cit.* When a (rationally connected) variety X is embedded in some projective space $\mathbb{P}^N$ (or more generally when a polarization or an arbitrary Cartier divisor is fixed on X), one can consider the property of being generically n-(rationally) connected by (rational) curves of a fixed degree δ (with respect to the fixed polarization or to an arbitrary Cartier divisor, let us say D for fixing notation). A variety $X^{r+1} \subset \mathbb{P}^N$ such that through $n \geq 2$ general points there passes an irreducible curve $C \subset X$ of degree $\delta \geq n - 1$ will be indicated by $X^{r+1}(n, \delta) \subset \mathbb{P}^N$ (or simply by $X^{r+1}(n, \delta)$).

This stronger condition depends on the embedding, on the number $n \geq 2$, on the degree $\delta \geq 1$ and natural constraints for the existence of such varieties immediately appear. We shall see that there exists a bound on the embedding dimension N (or on $\dim(|D|)$) depending on $\dim(X)$, on the number of points $n \geq 2$ and on the degree of the curves $\delta \geq n - 1$, see Theorem 6.2.3. Furthermore, the boundary examples

© Springer International Publishing Switzerland 2016
F. Russo, *On the Geometry of Some Special Projective Varieties*,
Lecture Notes of the Unione Matematica Italiana 18,
DOI 10.1007/978-3-319-26765-4_6

are rational varieties which are n-connected by smooth rational curves of degree δ in such a way that there exists a unique curve of the family passing though n general points, see Theorem 6.3.3.

In Theorem 6.3.2 we shall also describe another result generalizing in various directions Theorem 3.4.4 and which at the end will appear as another embedded (or projective) incarnation of Mori's famous characterization of projective spaces as the unique projective manifolds having ample tangent bundle. We point out that in this characterization the maximal projective variety is not assumed to be a priori smooth.

Another consequence of the bound in Theorem 6.2.3 is that under the same hypothesis we have $D^{r+1} \leq \delta^{r+1}/(n-1)^r$ if D is nef, see Theorem 6.3.4 (recall that $\dim(X) = r + 1$). This is a generalization of a result usually attributed to Fano in the case $n = 2$, see for example [116, Proposition V.2.9].

Several extremal $X^{r+1}(n, \delta)$ (for arbitrary $n \geq 2$, $r \geq 1$ and $\delta \geq n - 1$) can be constructed via the theory of Castelnuovo varieties, as briefly recalled in Sect. 6.2.3. The main result of [158] ensures that these examples *of Castelnuovo type* are the only extremal ones except possibly when $n > 2$, $r > 1$ and $\delta = 2n - 3$.

The first open case, that is the classification of extremal varieties $X = X^{r+1}(3, 3) \subset \mathbb{P}^{2r+3}$, is considered in Sect. 6.4, where it is proved that these varieties are in one-to-one correspondence, modulo projective transformations, with quadro-quadric Cremona transformations on $\mathbb{P}^r$, [154, Theorem 5.2] and Theorem 6.4.5 here. One of the key steps to establish this result is to prove a priori the equality $B_x = \mathcal{L}_x$ as schemes for a general $x \in X$. This has some very interesting consequences like Corollary 6.4.6 or the classification of smooth varieties $X^{r+1}(3, 3) \subset \mathbb{P}^{2r+3}$ showing that there are two infinite series: smooth rational normal scrolls and $\mathbb{P}^1 \times Q^r$ Segre embedded; and four isolated examples appearing for $r = 5, 8, 14$ and 26, whose variety of lines through a general point is one of the four Severi varieties. We end the chapter with the so-called *XJC*-correspondence and its applications in various contexts.

6.1 Preliminaries and Definitions

As stated above, in this chapter we shall suppose that $X \subset \mathbb{P}^N$ has dimension $r + 1$, $r \geq 0$, for computational reasons.

The next result, which seems to go back to Veronese in [189], is well known and reveals the first instance of the problem we wish to consider in arbitrary dimension.

Lemma 6.1.1 *Let C be an irreducible projective curve.*

1. **(Veronese Lemma)** *If $C \subset \mathbb{P}^N$ is non-degenerate, then $\delta = \deg(C) \geq N$.*
   Moreover, the following conditions are equivalent:

   a. *$N = \delta$ and $C \subset \mathbb{P}^\delta$ is a rational normal curve of degree δ;*
   b. *for a general $x \in C$ there exists a hyperplane $H_x \subset \mathbb{P}^N$ such that $H_x \cap X = \delta \cdot x$ as schemes.*

2. *The following conditions are equivalent:*

(a') *C is a smooth rational curve;*
(b') *there exists a Cartier divisor D of degree $\delta \geq 1$ on C such that* dim $(|D|) = \delta$;
(c') $\mathcal{O}_C(\delta \cdot x_1) \simeq \mathcal{O}_C(\delta \cdot x_2)$ *for some $\delta \geq 1$ and for $x_1, x_2 \in C$ general points.*

The first part is a classical fact about the osculating behavior of projective curves, the second one is the abstract incarnation of the same statement. It is important to emphasize that, in both cases, a priori the curve is not assumed to be non-singular, a property shown only a posteriori via the maximality condition.

Now we start to discuss the possible generalizations to higher dimensions. First, we need to introduce some notation.

Definition 6.1.2 (Varieties n-Covered by Curves of Degree δ) Fixed $n \geq 2, \delta \geq n - 1$ and an embedding $X^{r+1} \subset \mathbb{P}^N$

$X^{r+1} \subset \mathbb{P}^N$ is said to be $\qquad$ $\exists\, C = C_{p_1,\dots,p_n} \subseteq X$ irreducible curve
n–covered by curves $\qquad \Longleftrightarrow \qquad$ through $n \geq 2$ general points $p_1, \dots, p_n \in X$
of degree δ $\qquad\qquad$ with $\deg(C) = \delta$.

In this case we shall use the notation:

$$X = X^{r+1}(n, \delta) \subset \mathbb{P}^N.$$

6.1.1 Examples and Reinterpretation of Known Results

We shall also assume $X^{r+1} \subseteq \mathbb{P}^N$ non-degenerate.

Example 6.1.3 Let us reinterpret some facts proved before from this point of view. The easiest case is the following:

$$X = X^{r+1}(2, 1) \subset \mathbb{P}^N \iff N = r + 1, X = \mathbb{P}^{r+1}.$$

One implication is obvious. To prove the relevant one we can observe that fixing a general point $p \in X$ and a general point $q \in X$ the line $<p, q>$ is contained in X by definition. Thus $q \in T_pX$ and the generality of q yields $X = T_pX = \mathbb{P}^{r+1}$.

Example 6.1.4 The next case is the following:
$X = X^{r+1}(3, 2) \subset \mathbb{P}^N \qquad\qquad N = r + 2$
non-degenerate $\qquad \Longleftrightarrow \qquad X^{r+1} \subset \mathbb{P}^{r+2}$ quadric hypersurface.

Clearly a quadric hypersurface of dimension $r + 1$ is a $X^{r+1}(3, 2)$. If we take a general point $p \in X^{r+1}(3, 2) \subset \mathbb{P}^N$ and we project X from p we obtain the variety $\pi_p(X) = X^{r+1}(2, 1) \subset \mathbb{P}^{N-1}$ since conics through p are projected into lines. Thus $N - 1 = r + 1$ and $\deg(X) = \deg(\pi_p(X)) + 1 = 2$.

Example 6.1.5 Reasoning as above via successive projections from $n - 2$ general points one proves the following projective characterization of varieties of minimal degree:

$$X = X^{r+1}(n, n-1) \subset \mathbb{P}^N \quad \Longleftrightarrow \quad \begin{aligned} &N = r + n - 1, \\ &X^{r+1} \subset \mathbb{P}^{r+n-1}, \\ &\deg(X) = n - 1 = \text{codim}(X) + 1 \\ &(\textit{minimal degree variety}). \end{aligned}$$

non-degenerate

6.2 Bounding the Embedding Dimension

The previous definitions/examples suggest the following problem. Letting $X = X^{r+1}(n, \delta) \subset \mathbb{P}^N$, one could ask if there exists a sharp (or optimal) universal function $\overline{\pi}(\dim(X), n, \delta)$ such that

$$N \leq \overline{\pi}(\dim(X), n, \delta) - 1.$$

This bound should extend to arbitrary dimension the bound $N \leq \delta = \overline{\pi}(1, n, \delta) - 1$ in Lemma 6.1.1 and also generalize the previous examples.

For simplicity, since $\dim(X) = r+1$, we shall write $\overline{\pi}(\dim(X), n, \delta) = \overline{\pi}(r, n, \delta)$. Assuming a uniform minimal bound existed, one could also ask if

$$X^{r+1} \subset \mathbb{P}^{\overline{\pi}(r,n,\delta)-1},$$

can be classified, generalizing to higher dimension the content of Veronese Lemma 6.1.1. In particular, one would know if such varieties are rational and n-covered by rational curves, as the examples recalled above.

Remark 6.2.1 The first problem has been completely solved in [158] and in [154]. More precisely, in [158] all the results were proved under the assumption that a general curve in the n-covering family is rational. Furthermore, Pirio and Trépreau also classify all the extremal varieties with $\delta \neq 2n - 3$, see Theorem 6.2.8, showing they are of a very particular type dubbed by them *Castelnuovo type*. The assumption on the rationality of the general curve n-covering X has been removed in [154] where it was also proved that under the maximality condition $N = \overline{\pi}(r, n, \delta) - 1$ the curve in the n-covering family is rational and the variety itself is rational, see Theorem 6.3.3 for the most general form of this statement. We now present the contents of these two papers after surveying some classical results which can be considered to have been the first steps towards the solution of both problems.

6.2.1 Previously Known Versions

Let us begin with $\delta = 2$ (or $n = 2$). Recall that $\delta \geq n - 1$ so we have $n \in \{2, 3\}$. Example 6.1.4 can be rephrased by $\overline{\pi}(r, 3, 2) = r + 3$ and with the characterization of irreducible quadric hypersurfaces as the only $X^{r+1}(3, 2) \subset \mathbb{P}^{\overline{\pi}(r,3,2)-1}$.

It was known to C. Segre and Scorza that $\overline{\pi}(r, 2, 2) \leq \binom{r+1+2}{2}$, see [166]. Indeed, we can use an argument similar to that used in the proof Proposition 2.3.5. Let the notation be as in Sect. 2.3.2. If $X = X^{r+1}(2, 2) \subset \mathbb{P}^N$, then $\widetilde{\pi}_x(E) = W_x \subset \mathbb{P}^{N-r-2}$ since a conic $C_{x,y} \subset X$, passing through x and a general point y, is contracted to the general point $\widetilde{\pi}_x(y) = \widetilde{\pi}_x([t_x C_{x,y}])$. The restriction of $\widetilde{\pi}_x$ to E is given by a sublinear system of $|\mathcal{O}_{\mathbb{P}^r}(2)|$, yielding

$$N - r - 1 \leq \binom{r+2}{2}.$$

Thus $N \leq \binom{r+2}{2} + r + 1 = \binom{r+1+2}{2} - 1$, as claimed.

Moreover, the Veronese embedding $v_2(\mathbb{P}^{r+1}) \subset \mathbb{P}^{\binom{r+1+2}{2}-1}$ shows that $\overline{\pi}(r, 2, 2) \geq \binom{r+1+2}{2}$ so that $\overline{\pi}(r, 2, 2) = \binom{r+1+2}{2}$. Scorza also proved that an extremal $X^{r+1}(2, 2) \subset \mathbb{P}^{\binom{r+1+2}{2}-1}$ is projectively equivalent to $X = v_2(\mathbb{P}^{r+1}) \subset \mathbb{P}^{\binom{r+1+2}{2}-1}$, see Theorem 3.4.4 for a proof under the smoothness assumption (a $X^{r+1}(2, 2)$ is obviously secant defective) and Theorem 6.3.2 for a significant generalization.

Obviously a smooth $X^{r+1}(2, 2) \subset \mathbb{P}^N$ is nothing but a CC-manifold, so that these manifolds were completely classified in [104, Theorem 2.1], recalled here as Theorem 4.4.1. Thus the problem previously stated has a positive solution in the first easiest cases or for $\delta = n - 1$, see Example 6.1.5.

Let us consider the case $\delta = 3$ so that $n \in 2, 3, 4$. Example 6.1.5 provides $\overline{\pi}(r, 4, 3) = r + 3$ and that the corresponding extremal varieties are varieties of minimal degree equal to 3. Let us consider $\overline{\pi}(r, 3, 3)$. The tangential projection of a $X^{r+1}(3, 3) \subset \mathbb{P}^N$ is a $X^s(2, 1) = \mathbb{P}^s \subseteq \mathbb{P}^{N-r-2}$ with $s \leq r + 1$. Since $X^{r+1} \subset \mathbb{P}^N$ is assumed to be non-degenerate $s = N - r - 2$ and

$$N - r - 2 = s \leq r + 1,$$

proving that $N \leq 2r + 3$. The construction of $X^{r+1}(3, 3) \subset \mathbb{P}^{2r+3}$ for every $r \geq 1$ in Sect. 6.4 will show that $\overline{\pi}(r, 3, 3) = 2r + 4$.

6.2.2 Looking for the Function $\overline{\pi}(r, n, \delta)$ via Projective Geometry

We introduce some definitions. Let $x \in X_{\mathrm{reg}} \subset \mathbb{P}^N$. For any $\ell \in \mathbb{N}$, we denote by

$$T_x^\ell X \subset \mathbb{P}^N$$

the ℓth-order osculation space of X at x. If

$$\psi : (\mathbb{C}^{r+1}, 0) \to (X, x),$$
$$u \quad \mapsto \quad \psi(u)$$

is a regular local parametrization of X at $x = \psi(0)$, then $T_x^\ell X$ can be defined as the projective subspace

$$\overline{\langle \partial^{|\alpha|} \psi(0)/\partial u^\alpha \mid \alpha \in \mathbb{N}^{r+1}, |\alpha| \le \ell \rangle} \subset \mathbb{P}^N.$$

By definition $\dim (T_x^\ell X) \le \binom{r+1+\ell}{r+1} - 1$ and in general it is expected that equality holds at general points of $X \subset \mathbb{P}^N$ as soon as $N \ge \binom{r+1+\ell}{r+1} - 1$. In this case, we shall say that the *osculation of order ℓ of X at x is regular*.

This space can also be defined more abstractly as the linear subspace spanned by the ℓth order infinitesimal neighborhood of X at x and also generalized to the case of arbitrary Cartier divisors on X.

Indeed, for every integer $\ell \in \mathbb{N}$, let $\mathscr{P}_X^\ell(D)$ denote the ℓth principal part bundle (or ℓth jet bundle) of $\mathscr{O}_X(D)$. For every linear subspace $V \subseteq H^0(X, \mathscr{O}_X(D))$ we have a natural homomorphism of sheaves

$$\phi^\ell : V \otimes \mathscr{O}_X \to \mathscr{P}_X^\ell(D), \tag{6.1}$$

sending a section $s \in V$ to its ℓth jet $\phi_x^\ell(s)$ evaluated at $x \in X$, that is $\phi_x^\ell(s)$ is represented in local coordinates by the Taylor expansion of s at x, truncated after the order ℓ. Taking a smooth point $x \in X \subset \mathbb{P}^N = \mathbb{P}(V)$ (Grothendieck's notation) and $\mathscr{O}_X(D) = \mathscr{O}_X(1)$, it is easily verified that $T_x^\ell X = \mathbb{P}(\text{Im}(\phi_x^\ell))$.

Let us remark that, if $T_x^\ell X \subsetneq \mathbb{P}^N$ and if $H \supseteq T_x^\ell X$ is a hyperplane, then the hyperplane section $H \cap X$ has a point of multiplicity at least $\ell + 1$ at x. The next result uses this observation to calculate the linear span of a curve.

Lemma 6.2.2 *For an irreducible curve $C \subset \mathbb{P}^N$ of degree δ, for non-negative integers $a_1, \ldots, a_\kappa$, with $\kappa > 0$ fixed and such that $\sum_{i=1}^\kappa (a_i + 1) \ge \delta + 1$ and for $x_1, \ldots, x_\kappa \in C$ pairwise distinct smooth points, one has:*

$$\langle C \rangle = \langle T_{x_i}^{a_i} C \mid i = 1, \ldots, \kappa \rangle \tag{6.2}$$

Proof Otherwise there would exist a hyperplane $H \supsetneq \langle C \rangle$ containing

$$\langle T_{x_i}^{a_i} C \mid i = 1, \ldots, \kappa \rangle$$

and

$$\delta = \deg(C) = \deg(H \cap C) \ge \sum_{i=1}^\kappa (a_i + 1),$$

contrary to our assumption.

Let $X = X^{r+1}(n, \delta) \subset \mathbb{P}^N$ and let Σ be a fixed n-covering family of irreducible curves of degree δ on X. If $x_1, \ldots, x_{n-1}$ are distinct general points on X one can consider the subfamily

$$\Sigma_{x_1, \ldots, x_{n-1}} = \{C \in \Sigma \mid x_i \in C \text{ for } i = 1, \ldots, n-1\}.$$

Since Σ is n-covering, the family $\Sigma_{x_1, \ldots, x_{n-1}}$ covers X and we can also assume that the general curve in this family is non-singular at $x_1, \ldots, x_{n-1}$.

Let $\{a_1, \ldots, a_{n-1}\}$ be a set of $n-1$ non-negative integers such that

$$\sum_{i=1}^{n-1}(a_i + 1) \geq \delta + 1.$$

From (6.2), after remarking that $T_{x_i}^{a_i} C \subset T_{x_i}^{a_i} X$ for every $i = 1, \ldots, n-1$, we deduce that

$$\langle C \rangle \subseteq \langle T_{x_i}^{a_i} X \mid i = 1, \ldots, n-1 \rangle$$

for a general $C \in \Sigma_{x_1, \ldots, x_{n-1}}$.

Since the elements of $\Sigma_{x_1, \ldots, x_{n-1}}$ cover X, one obtains

$$\langle X \rangle = \langle T_{x_i}^{a_i} X \mid i = 1, \ldots, n-1 \rangle. \tag{6.3}$$

Therefore for these varieties we deduce that

$$\dim(\langle X \rangle) + 1 \leq \sum_{i=1}^{n-1} \binom{r+1+a_i}{r+1}. \tag{6.4}$$

Now following [158] we shall try to minimize this bound in a uniform way depending only on r, n, δ.

We shall use the following notation. Let r, n and δ be positive integers such that $n - 1 \leq \delta$. Let

$$\rho = \left\lfloor \frac{\delta}{n-1} \right\rfloor,$$

$$\epsilon = \delta - \rho(n-1) \in \{0, \ldots, n-2\},$$

$$m = \epsilon + 1 = \delta - \rho(n-1) + 1 \in \{1, \ldots, n-1\}$$

and

$$m' = n - 1 - m \geq 0.$$

Observe that $m + m' = n - 1$ and

$$\delta + 1 = \rho(n-1) + \epsilon + 1 = \rho(n-1) + m = (\rho + 1)m + \rho m'.$$

Taking $a_1 = \ldots = a_m = \rho + 1$ and $a_{m+1} = \ldots = a_{n-1} = \rho$ we get from (6.4):

$$\overline{\pi}(r, n, \delta) := m \binom{r + \rho + 1}{r + 1} + m' \binom{r + \rho}{r + 1}. \tag{6.5}$$

Combining (6.5) with (6.4) we obtain the desired result.

Theorem 6.2.3 (Pirio–Trépreau Bound, [158]) *Let $X = X^{r+1}(n, \delta) \subset \mathbb{P}^N$ be a non-degenerate variety. Then*

$$N + 1 \leq \overline{\pi}(r, n, \delta) := m \binom{r + \rho + 1}{r + 1} + m' \binom{r + \rho}{r + 1}, \tag{6.6}$$

where $\rho = \lfloor \frac{\delta}{n-1} \rfloor$, $m = \delta - \rho(n-1) + 1$, $m' = n - 1 - m \geq 0$.

We abused the notation by defining $\overline{\pi}(r, n, \delta)$ as in (6.5) since Theorem 6.2.3 only proves

$$\overline{\pi}(r, n, \delta) \leq m \binom{r + \rho + 1}{r + 1} + m' \binom{r + \rho}{r + 1},$$

according to our previous definitions. The construction of examples of $X^{r+1}(n, \delta) \subset \mathbb{P}^N$ with $N = m \binom{r+\rho+1}{r+1} + m' \binom{r+\rho}{r+1}$ for every r, n, δ in Sect. 6.2.3 will prove that this notation is coherent with the previous definitions. In particular, the next definition is well posed.

Definition 6.2.4 (Extremal Variety $\overline{X}^{r+1}(n, \delta) \subset \mathbb{P}^{\overline{\pi}(r,n,\delta)-1}$)   An irreducible non-degenerate projective variety

$$X = X^{r+1}(n, \delta) \subset \mathbb{P}^{\overline{\pi}(r,n,\delta)-1}$$

will be denoted by

$$\overline{X}^{r+1}(n, \delta) \subset \mathbb{P}^{\overline{\pi}(r,n,\delta)-1}$$

or simply by

$$\overline{X}^{r+1}(n, \delta).$$

Remark 6.2.5 Since $\dim(T_x^\ell X) \le \binom{r+1+\ell}{r+1} - 1$ for any point $x \in X_{\text{reg}}$ and for any integer $\ell \in \mathbb{N}$, for a non-degenerate $\overline{X}^{r+1}(n, \delta) \subset \mathbb{P}^{\overline{\pi}(r,n,\delta)-1}$ the following properties hold:

(i) the osculation of order ρ of X at a general point $x \in X$ is regular, that is

$$\dim\left(T_x^\rho X\right) = \binom{r+1+\rho}{r+1} - 1 ; \tag{6.7}$$

(ii) if $x_1, \ldots, x_{n-1}$ are general points of X, then

$$\langle X \rangle = \left(\bigoplus_{i=1}^{m} T_{x_i}^\rho X \right) \oplus \left(\bigoplus_{j=1}^{m'} T_{x_{m+j}}^{\rho-1} X \right) = \mathbb{P}^{\overline{\pi}(r,n,\delta)-1}. \tag{6.8}$$

6.2.3 Relation to the Castelnuovo–Harris Bound

The function $\overline{\pi}(r, n, \delta)$ is related to the Castelnuovo–Harris function bounding the geometric genus of an irreducible variety as we now show. Moreover, Castelnuovo varieties allow the construction of a quasi exhaustive class of $\overline{X}^{r+1}(n, \delta)$'s.

Let $V^r \subset \mathbb{P}^{r+n-1}$ be an irreducible non-degenerate variety and let $d = \deg(V) > 1$. By hypothesis $n - 1 = \text{codim}(X)$.

The *geometric genus* of V is $g(V) = h^0(K_{\tilde{V}})$ where $\tilde{V} \to V$ is a resolution of singularities of V. Let

$$\pi(r, n, d) = \sum_{\sigma \ge 0} \binom{\sigma + r - 1}{\sigma}\left(d - (\sigma + r)(n-1) - 1\right)^+, \tag{6.9}$$

where for $k \in \mathbb{N}$ we define $k^+ = \max\{0, k\}$. If

$$d = \delta + r(n-1) + 2,$$

one easily verifies that

$$\overline{\pi}(r, n, \delta) = \pi(r, n, \delta + r(n-1) + 2),$$

see [158] for all the details.

Theorem 6.2.6 (Castelnuovo–Harris Bound, [84]) *The bound*

$$g(V) \le \pi(r, n, d) \tag{6.10}$$

holds for the geometric genus of $V \subset \mathbb{P}^{n+r-1}$. In particular, $g(V) = 0$ if $d < r(n-1) + 2$.

An irreducible variety $V \subset \mathbb{P}^{n+r-1}$ as above and such that $g(V) = \pi(r, n, d) > 0$ is called a *Castelnuovo variety*. Note that in this case necessarily $d \geq r(n-1) + 2$.

The classification of projective curves of maximal genus was obtained by Castelnuovo in 1889. More recently, Harris proved the following result, see [84] and [32], the last paper filling out some imprecisions in the first one.

Proposition 6.2.7 *Let $V \subset \mathbb{P}^{n+r-1}$ be a Castelnuovo variety of dimension $r \geq 1$ and codimension at least 2. The linear system $|\mathscr{I}_V(2)|$ cuts out a variety of minimal degree $Y \subset \mathbb{P}^{n+r-1}$ of dimension $r + 1$.*

Let us recall that we saw in Example 6.1.5 that varieties $X^{r+1} \subset \mathbb{P}^{r+n-1}$ of minimal degree $\deg(X) = \text{codim}(X) + 1 = n - 1$ are exactly the extremal varieties $\overline{X}^{r+1}(n, n-1) \subset \mathbb{P}^{r+n-1}$.

Thus a Castelnuovo variety $V \subset \mathbb{P}^{n+r-1}$ of dimension r is a divisor in the variety of minimal degree $Y \subset \mathbb{P}^{n+r-1}$ cut out by $|\mathscr{I}_V(2)|$. This property has been used by Harris to describe Castelnuovo varieties, see [84] for details. We outline the description in [154, Sect. 4] as follows:

- if $p : \tilde{Y} \to Y$ denotes a desingularization (obtained for instance by blowing-up the vertex of the cone Y when it is singular), Harris determines the class $[\tilde{V}]$ of $\tilde{V}$ (the strict transform of V in $\tilde{Y}$ via p) in the Picard group of $\tilde{Y}$.
- Assuming (to simplify) that $\tilde{V}$ is smooth, consider $K_{\tilde{Y}} + \tilde{V}$. By adjunction theory, there is a short exact sequence of sheaves:

$$0 \to \mathscr{O}(K_{\tilde{Y}}) \to \mathscr{O}(K_{\tilde{Y}} + \tilde{V}) \to \mathscr{O}(K_{\tilde{V}}) \to 0.$$

- Since $h^0(\tilde{Y}, K_{\tilde{Y}}) = h^1(\tilde{Y}, K_{\tilde{Y}}) = 0$ (because $\tilde{Y}$ is smooth and rational), the map

$$H^0(\tilde{Y}, K_{\tilde{Y}} + \tilde{V}) \to H^0(\tilde{V}, K_{\tilde{V}}) \tag{6.11}$$

is an isomorphism. Thus it induces rational maps $\phi_V = \phi_{|K_V|}$ and $\Phi_V = \Phi_{|L_V|} \circ p^{-1}$ such that the following diagram of rational maps is commutative:

$$
\begin{array}{ccc}
V & \overset{\phi_V}{\dashrightarrow} & \mathbb{P}^{\pi(r,n,d)-1} \\
\Big\uparrow & & \Big\| \\
Y & \underset{\Phi_V}{\dashrightarrow} & \mathbb{P}^{\pi(r,n,d)-1}
\end{array}
\tag{6.12}
$$

- Let X_V be (the closure of) the image of Φ_V.
- It is an irreducible non-degenerate subvariety in $\mathbb{P}^{\pi(r,n,d)-1}$ and $\dim(X_V) = \dim(Y) = r + 1$. Moreover, one proves that the image by Φ_V of a generic one-dimensional linear section of Y, that is of a rational normal curve of degree

$n - 1$ passing through n general points of Y, is a rational normal curve of degree $\delta = d - r(n-1) - 2$ contained in X_V.

- Thus $X_V \subset \mathbb{P}^{\pi(r,n,d)-1} = \mathbb{P}^{\overline{\pi}(r,n,\delta)-1}$ is an example of $\overline{X}^{r+1}(n,\delta)$. These examples are called *of Castelnuovo type* in [158].

One of the main contributions in [158] is the following intriguing result.

Theorem 6.2.8 (Pirio–Trépreau Theorem, [158]) *Let $\overline{X}^{r+1}(n,\delta) \subset \mathbb{P}^{\overline{\pi}(r,n,\delta)-1}$. If $\delta \neq 2n - 3$, then X is of Castelnuovo type.*

It follows that varieties $\overline{X}^{r+1}(n,\delta)$ not of Castelnuovo type can exist only for $\delta = 2n - 3$, the first significant case being $n = 3$ (for $n = 2$ we have only $\mathbb{P}^{r+1}$). As we shall see in Sect. 6.6 there exist a lot of examples of $\overline{X}^{r+1}(3,3) \subset \mathbb{P}^{2r+3}$ not of Castelnuovo type showing that the previous theorem is sharp. Moreover, in [156] all the examples of $\overline{X}^{r+1}(3,3) \subset \mathbb{P}^{2r+3}$ not of Castelnuovo type are classified, see Sect. 6.6. For $n \geq 4$ some sporadic examples are known to exist for some n but no classification has been obtained until now.

6.3 Rationality of $\overline{X}^{r+1}(n,\delta)$ and of the General Curve of the n-Covering Family

The following simple remark will play a central role several times in our analysis.

Lemma 6.3.1 ([154, Lemma 2.1]) *Let $\phi : X \dashrightarrow X'$ be a dominant rational map between proper varieties of the same dimension, let Σ be an irreducible n-covering family of irreducible curves on X and let Σ' be the induced n-covering family on X'. If X' is projective, if the restriction of ϕ to a general curve $C \in \Sigma$ induces a morphism birational onto its image and if through n-general points of X' there passes a unique curve $C' \in \Sigma'$, then the same is true for Σ on X and moreover ϕ is a birational map.*

We are in a position to present the study of the case $\overline{X}^{r+1}(2,\delta)$. This case was classically considered by Bompiani in [20], where he has essentially provided details only for surfaces. Under the assumption that the general two-covering curve is smooth and rational, this result was also obtained by Ionescu in [100, Theorem 2.8]. It is also a wide generalization of Theorem 3.4.4.

Theorem 6.3.2 ([154, Theorem 2.2]) *An irreducible projective variety*

$$X = \overline{X}^{r+1}(2,\delta) \subset \mathbb{P}^{\binom{r+1+\delta}{r+1}-1}$$

is projectively equivalent to the Veronese manifold $v_\delta(\mathbb{P}^{r+1})$. In particular, every curve in the two-covering family is a rational normal curve of degree δ in the given embedding and there exists a unique such curve passing through two points of X.

Proof By definition $\rho = \delta$ so that by (6.7), for $x \in X$ general we have

$$\dim(T_x^{\delta-1}X) = \binom{r+\delta}{r+1} - 1 \quad \text{and} \quad T_x^\delta X = \mathbb{P}^{\binom{r+1+\delta}{r+1}-1}.$$

Let $x \in X$ be a fixed general point and let $T = T_x^{\delta-1}X = \mathbb{P}^{\binom{r+\delta}{r+1}-1}$. Let

$$p_T : X \dashrightarrow \mathbb{P}^{\overline{\pi}(r-1,2,\delta)-1}$$

be the restriction to X of the projection from T. The rational map p_T is given by the linear system $|D_x|$ cut on X by hyperplanes containing T so that the corresponding hyperplane sections have a point of multiplicity δ at $x \in X$. A general irreducible curve of degree δ passing through x is thus contracted by p_T. Let $X_T = p_T(X) \subset \mathbb{P}^{\overline{\pi}(r-1,2,\delta)-1}$.

We claim that X_T is projectively equivalent to $v_\delta(\mathbb{P}^r) \subset \mathbb{P}^{\overline{\pi}(r-1,2,\delta)-1}$. Indeed, let $\pi : \mathrm{Bl}_x(X) \to X$ be the blow-up of X at x, let $E = \mathbb{P}^r$ be the exceptional divisor and let

$$p_T' = p_T \circ \pi : \mathrm{Bl}_x(X) \dashrightarrow X_T$$

be the induced rational map. The restriction of p_T' to E is a rational dominant map from $\mathbb{P}^r$ to $X_T \subset \mathbb{P}^{\overline{\pi}(r-1,2,\delta)-1}$ given by a sublinear system of $|\mathcal{O}_{\mathbb{P}^r}(\delta)|$ of dimension $\overline{\pi}(r+1,2,\delta)-1$ so that it is given by the complete linear system $|\mathcal{O}_{\mathbb{P}^r}(\delta)|$. Note that since $T_x^\delta X = \mathbb{P}^{\overline{\pi}(r,2,\delta)-1}$, the restriction of the strict transform of the linear system of hyperplane sections containing $T_x^{\delta-1}X$ to E is not zero. Thus the restriction of p_T' to E induces an isomorphism between E and X_T given by $|\mathcal{O}_{\mathbb{P}^r}(\delta)|$, proving the claim.

Moreover, since a general curve $C \in \Sigma$ is not contracted by p_T, we have that $p_T(C)$ is a curve on X_T of degree $\delta' \leq \delta$. Thus $p_T(C)$ is a smooth rational curve of degree δ, $T \cap C = \emptyset$, the rational map p_T is defined along C and it gives an isomorphism between C and $p_T(C)$.

By solving the indeterminacies of p_T', we can suppose that there exists a smooth variety $\widetilde{X}$, a birational morphism $\phi : \widetilde{X} \to X$ and a morphism $\widetilde{p}_T : \widetilde{X} \to X_T \simeq v_\delta(\mathbb{P}^r)$ such that $p_T \circ \phi = \widetilde{p}_T$.

Let $\phi^*(|D_x|) = F_x + |\widetilde{D}_x|$ with $|\widetilde{D}_x|$ base point free and let $|\overline{D}_x| = \widetilde{p}_T^*(|\mathcal{O}_{\mathbb{P}^r}(1)|)$. Then $\widetilde{D}_x \sim \delta \overline{D}_x$ and $\dim(|\overline{D}_x|) \geq r$. Furthermore, for the strict transform of a general curve C_x in Σ passing through x we have $(\widetilde{D}_x \cdot C_x) = 0$ and $(F_x \cdot C_x) = \delta$ while for the strict transform of a general curve $C \in \Sigma$ we have $(\widetilde{D}_x \cdot C) = \delta$. Thus $(\overline{D}_x \cdot C_x) = 0$ and $(\overline{D}_x \cdot C) = 1$.

Letting $T' = T_{x'}^{\delta-1}X$ with $x' \in X$ general and performing the same analysis we can also suppose that on $\widetilde{X}$ the rational map $p_{T'} \circ \phi = \widetilde{p}_{T'}$ is defined and that there exists a linear system $|\overline{D}_{x'}|$ such that $\dim(|\overline{D}_{x'}|) \geq r$, $(\overline{D}_{x'} \cdot C_{x'}) = 0$ for general $C_{x'} \in \Sigma_{x'}$ and $(\overline{D}_{x'} \cdot C) = 1$ for general $C \in \Sigma$. Since a general curve $C_{x'}$ in $\Sigma_{x'}$ does not pass through x we have $|\overline{D}_x| \neq |\overline{D}_{x'}|$. On the other hand for $x \in X$ general, the linear systems $|\overline{D}_x|$ vary in the same linear system $|D|$ on $\widetilde{X}$ since $\widetilde{X}$ is rationally connected.

Thus $\dim(|D|) \geq r+1$, $(D \cdot C) = 1$ for the strict transform of a general curve C in Σ and C does not intersect the base locus of $|D|$ by the previous analysis. Let $s+1 = \dim(|D|)$ and let $\psi = \psi_{|D|} : \widetilde{X} \dashrightarrow \widetilde{X}' \subseteq \mathbb{P}^{s+1}$ be the associated rational map. Since $\psi(C)$ is a line passing through two general points of X', we deduce $X' = \mathbb{P}^{s+1}$ and $r = s$. Moreover, by Lemma 6.3.1 the rational map ψ is birational. Hence there exists a birational map $\varphi = \phi \circ \psi^{-1} : \mathbb{P}^{r+1} \dashrightarrow X$ sending a general line in $\mathbb{P}^{r+1}$ onto a general curve of degree δ in Σ. Composing φ with the inclusion $X \subset \mathbb{P}^{\overline{\pi}(r,2,\delta)-1}$ we get a birational map from $\mathbb{P}^{r+1}$ given by a sublinear system of $|\mathcal{O}_{\mathbb{P}^{r+1}}(\delta)|$ of dimension $\binom{r+1+\delta}{r+1} - 1$, that is φ is given by the complete linear system $|\mathcal{O}_{\mathbb{P}^{r+1}}(\delta)|$. In conclusion, $X \subset \mathbb{P}^{\overline{\pi}(r,2,\delta)-1}$ is projectively equivalent to the Veronese manifold $v_\delta(\mathbb{P}^{r+1})$. $\square$

The rationality and the smoothness of the general member Σ of the two-covering family of a $\overline{X}(r+1,2,\delta)$ could also have been deduced differently. Indeed, in the previous proof we saw that the linear system of hyperplane sections having a point of multiplicity greater than or equal to δ at a general $x \in X$ cuts a general $C \in \Sigma_x$ in the Cartier divisor δx. By varying x on C we see that this property holds for the general point of C. Thus the smoothness and rationality of a general element of Σ are consequences of the abstract Veronese Lemma 6.1.1.

Thus via the Veronese Lemma one could also prove Theorem 6.3.2 above differently, following the steps of Mori's characterization of projective spaces given in [136] because in this case the family of smooth rational curves Σ_x is easily seen to be proper.

The main result on extremal varieties is the maximal generalization to the higher dimensional case of Veronese Lemma contained in the following theorem, see [154, Theorem 2.4] for a proof obtained by reducing to $n = 2$ and then by using Theorem 6.3.2 and Lemma 6.3.1.

Theorem 6.3.3 ([154, Theorem 2.4]) *Let X be an irreducible proper variety of dimension $r+1$ and let D be a Cartier divisor on X. Suppose that through $n \geq 2$ general points of X there passes an irreducible curve C such that $(D \cdot C) = \delta \geq n-1$. Then:*

(i) *$h^0(X, \mathcal{O}_X(D)) \leq \overline{\pi}(r,n,\delta)$;*
(ii) *equality holds in (i) if and only if $\phi_{|D|}$ maps X birationally onto a*

$$\overline{X}^{r+1}(n,\delta) \subseteq \mathbb{P}^{\overline{\pi}(r,n,\delta)-1}.$$

In this case the general deformation of C does not intersect the indeterminacy locus of $\phi_{|D|}$.

(iii) *If equality holds in (i), then*

  a. *the variety X is rational;*
  b. *the general deformation $\overline{C}$ of C is a smooth rational curve and through n general points of X there passes a unique smooth rational curve $\overline{C}$ such that $(D \cdot \overline{C}) = \delta$.*

In particular:

1. *a $\overline{X}^{r+1}(n,\delta) \subset \mathbb{P}^{\overline{\pi}(r,n,\delta)-1}$ is rational, the general curve of the n-covering family is a rational normal curve of degree δ and through n general points of X there passes a unique rational normal curve of degree δ;*
2. *a $\overline{X}^{r+1}(n,\delta) \subset \mathbb{P}^{\overline{\pi}(r,n,\delta)-1}$ is a linear birational projection of $v_\delta(\mathbb{P}^{r+1})$. Equivalently, a $\overline{X}^{r+1}(n,\delta)$ is the birational image of $\mathbb{P}^{r+1}$ via the birational map given by a linear system of hypersurfaces of degree δ and dimension $\overline{\pi}(r,n,\delta) - 1$.*

6.3.1   Bound for the Top Self Intersection of a Nef Divisor

We cannot resist including the proof of a sharp bound for the top self intersection of a nef divisor. The computations pass through some formulas from calculus which one might not expect to be related to the problems considered until now and which one would probably never think of.

The bound (6.13) below generalizes a result usually attributed to Fano, who proved it for $n = 2$. The reader can consult the modern reference [116, Proposition V.2.9] for the case $n = 2$ of Fano's result and also the several applications given in *loc. cit.*, e.g. to the boundedness of the number of components of families of smooth Fano varieties of a fixed dimension.

Theorem 6.3.4 ([154, Theorem 3.1]) *Let X be a proper irreducible variety of dimension $r + 1$, let D be a nef Cartier divisor on X and suppose that through $n \geq 2$ general points there passes an irreducible curve C such that $(D \cdot C) = \delta \geq n - 1$. Then*

$$D^{r+1} \leq \frac{\delta^{r+1}}{(n-1)^r} \, . \tag{6.13}$$

In particular, if $X = X^{r+1}(n,\delta) \subset \mathbb{P}^N$, then

$$\deg(X) \leq \frac{\delta^{r+1}}{(n-1)^r} \, . \tag{6.14}$$

Proof By the Asymptotic Riemann–Roch Theorem, we know that

$$h^0(\mathcal{O}_X(\ell D)) = D^{r+1}\frac{\ell^{r+1}}{(r+1)!} + O(\ell^r)$$

so that

$$D^{r+1} = \lim_{\ell \to +\infty} \frac{(r+1)! h^0(\mathcal{O}_X(\ell D))}{\ell^{r+1}} .$$ (6.15)

Since X is *n*-covered by a family of irreducible curves having intersection with D equal to δ, X is also *n*-covered by a family of irreducible curves having intersection $\delta\ell$ with ℓD for any $\ell > 0$. Theorem 6.3.3 yields

$$h^0(\mathcal{O}_X(\ell D)) \leq \overline{\pi}(r, n, \delta\ell)$$

for every positive integer ℓ. From (6.15) we deduce

$$D^{r+1} \leq \liminf_{\ell \to +\infty} \frac{(r+1)!\,\overline{\pi}(r, n, \delta\ell)}{\ell^{r+1}} .$$ (6.16)

Let $\rho_\ell = \lfloor \frac{\delta\ell}{n-1} \rfloor$ for $\ell > 0$. The definition of $\overline{\pi}(r, n, \delta\ell)$ in (6.5) implies that

$$D^{r+1} \leq \liminf_{\ell \to +\infty} \frac{(n-1)\,(r+1+\rho_\ell)!}{\ell^{r+1}\rho_\ell!} .$$

Using Stirling's formula for $k \to +\infty$:

$$k! \sim \sqrt{2\pi k}\,(\frac{k}{e})^k,$$

we deduce, for $\ell \to +\infty$,

$$\frac{(n-1)\,(r+1+\rho_\ell)!}{\ell^{r+1}\rho_\ell!} \sim \frac{(n-1)\,\sqrt{r+1+\rho_\ell}\,(\frac{r+1+\rho_\ell}{e})^{r+1+\rho_\ell}}{\ell^{r+1}\sqrt{\rho_\ell}\,(\frac{\rho_\ell}{e})^{\rho_\ell}}$$

$$\sim \frac{(n-1)\,(r+1+\rho_\ell)^{r+1}}{\ell^{r+1}e^{r+1}}\,\left(1 + \frac{r+1}{\rho_\ell}\right)^{\rho_\ell} .$$

Since $\rho_\ell \to +\infty$ if $\ell \to +\infty$ and recalling that

$$\lim_{x \to +\infty} \left(1 + \frac{r+1}{x}\right)^x = e^{r+1},$$

we obtain

$$\frac{(n-1)\,(r+1+\rho_\ell)!}{\ell^{r+1}\rho_\ell!} \sim \frac{(n-1)\,\rho_\ell^{r+1}}{\ell^{r+1}} .$$

But $\rho_\ell \sim \frac{\delta \ell}{n-1}$ if $\ell \to +\infty$ hence we finally get

$$D^{r+1} \le \liminf_{\ell \to +\infty} \frac{(n-1)\left(\frac{\delta \ell}{n-1}\right)^{r+1}}{\ell^{r+1}} = \frac{\delta^{r+1}}{(n-1)^r} .$$

$\square$

6.4 Quadro-Quadric Cremona Transformations and $\overline{X}^n(3,3) \subset \mathbb{P}^{2n+1}$

We begin this section by constructing a huge series of examples of $\overline{X}^n(3,3) \subset \mathbb{P}^{2n+1}$ not of Castelnuovo type via quadro-quadric Cremona transformations. In this section we shall suppose, as in the whole chapter, that all varieties are defined over the complex field.

Let $n \ge 2$ and let $f : \mathbb{P}^{n-1} \dashrightarrow \mathbb{P}^{n-1}$ be a quadro-quadric, or $(2,2)$ Cremona transformation, meaning that f and f^{-1} are given by linear systems of quadric hypersurfaces. By definition there exists an $N(\mathbf{x}) \in \mathbb{C}[\mathbf{x}]_3$ such that

$$(f^{-1} \circ f)(\mathbf{x}) = N(\mathbf{x})\mathbf{x} \quad \forall \, \mathbf{x} \in \mathbb{C}^n. \tag{6.17}$$

There are only two possibilities: either the linear system of quadrics has a fixed component so that f reduces to a projective transformation or f has no fixed component. In the first case f will be called a *fake* $(2,2)$ *Cremona transformation* and for $n = 2$ this is the only case occurring.

Let us consider the following affine embedding

$$\mu_f : \mathbb{C}^n \longrightarrow \mathbb{P}\big(\mathbb{C} \oplus \mathbb{C}^n \oplus \mathbb{C}^n \oplus \mathbb{C}\big) = \mathbb{P}^{2n+1} \tag{6.18}$$

$$\mathbf{x} \longmapsto \big[1 : \mathbf{x} : f(\mathbf{x}) : N(\mathbf{x})\big].$$

Let

$$X_f = \overline{\big[1 : \mathbf{x} : f(\mathbf{x}) : N(\mathbf{x})\big]} \subset \mathbb{P}^{2n+1},$$

which is a non-degenerate irreducible n-dimensional subvariety of $\mathbb{P}^{2n+1}$ containing $0_f = \mu_f(\mathbf{0}) = [1 : \mathbf{0} : \mathbf{0} : 0]$.

Proposition 6.4.1 *Let the notation be as above. Then* $X_f = X^n(3,3) \subset \mathbb{P}^{2n+1}$, *i.e.* X_f *is three-covered by twisted cubics.*

Moreover, if f *is a fake* $(2,2)$ *Cremona transformation, then* X_f *is a smooth rational normal scroll so that* X_f *is of Castelnuovo type.*

Proof Let the notation be as above. Consider $\mathbb{C}^n$ as the hyperplane $\mathbb{P}^n \setminus V(x_0)$ so that $(x_0 : x_1 \ldots : x_n)$ are projective coordinates on $\mathbb{P}^n$ and $\mu_f : \mathbb{P}^n \dashrightarrow X_f$ is a rational map defined on $\mathbb{C}^n$. If f is a fake $(2,2)$ Cremona transformation, let $H = \mathbb{P}^{n-2} \subset V(x_0)$ be the base locus of f. Then $X_f \subset \mathbb{P}^{2n+1}$ is a smooth rational normal scroll since the pencil of hyperplanes in $\mathbb{P}^n$ passing through H is mapped to a pencil of linear spaces of dimension $n-1$ on X_f.

From now on we shall suppose that f is not fake, yielding $n \geq 3$. Consider three general points $p_1, p_2, p_3 \in \mathbb{P}^n$ and let $\Pi \subset \mathbb{P}^n$ be their linear span. We claim that the line $L = \Pi \cap V(x_0)$ determines a plane Π' cutting the base locus scheme of f in a length three subscheme $\mathscr{P}$ spanning Π.

Indeed, $D = f(L) \subset \mathbb{P}^{n-1}$ is a conic cutting the base locus scheme of f^{-1} in a length three subscheme $\mathscr{P}'$ spanning a plane $\overline{\Pi}$ since $f^{-1}(D) = L$. Then taking $\Pi' = f^{-1}(\overline{\Pi})$ the claim is proved. The length six scheme $\{p_1, p_2, p_3, \mathscr{P}\}$ spans the three-dimensional linear space $\langle \Pi, \Pi' \rangle$ so that it determines a unique twisted cubic $C \subset \mathbb{P}^{n-1}$ containing it.

The birational map $\mu_f : \mathbb{P}^n \dashrightarrow X_f$ is given by a linear system of cubic hypersurfaces having points of multiplicity at least 2 along its base locus scheme $V(x_0, N(x)) \subset \mathbb{P}^n$. Then $\mu_f(C) \subset X_f$ is a twisted cubic passing through the three general points $\mu_f(p_i)$, $i = 1, 2, 3$, showing that $X_f = \overline{X}^n(3,3)$. $\qquad\square$

One immediately verifies that the projective equivalence class of X_f does not depend on f but only on its linear equivalence class, that is composition on the left and on the right by a projective transformation of $\mathbb{P}^{n-1}$.

Remark 6.4.2 It is worthwhile to point out the physical (quantum mechanical?) phenomenon appearing in the previous geometric construction. Given three points in $\mathbb{P}^n$ spanning a plane the Cremona transformation f on $\mathbb{P}^{n-1}$ produced other three points spanning a plane in $\mathbb{P}^{n-1}$ such that the six resulting points span a $\mathbb{P}^3$ and lie on a unique twisted cubic curve.

We shall now survey in detail the contents of [154, Sect. 5]. Let us recall some facts, which were proved in the previous sections or which are easy consequences of them.

Lemma 6.4.3 *Let $X = \overline{X}^n(3,3) \subset \mathbb{P}^{2n+1}$ and let $x \in X$ be a general point. Then the tangential projection $\pi_x : X \dashrightarrow \mathbb{P}^n$ from the tangent space $T_x X$ is birational. In particular, X is a rational variety, $SX = \mathbb{P}^{2n+1}$ and X is not a cone.*

Proof The family of twisted cubics passing through x is a two-covering family and a general twisted cubic in this family projects from $T_x X$ onto a general line in $\mathbb{P}^n$. Thus the birationality of π_x follows from Lemma 6.3.1 while the last part is a consequence of Terracini's Lemma, see Proposition 1.4.10. $\qquad\square$

By the previous description a general twisted cubic passing through $x \in X$ general is mapped by π_x birationally onto a general line in $\mathbb{P}^n$. The birational map

$$\phi = \pi_x^{-1} : \mathbb{P}^n \dashrightarrow X \subset \mathbb{P}^{2n+1}$$

is thus given by a linear system of cubic hypersurfaces mapping a general line of $\mathbb{P}^n$ birationally onto a twisted cubic passing through x and mapping a general cubic hypersurface in this linear system birationally onto a general hyperplane section of X.

Let $\overline{\alpha}_x : \mathrm{Bl}_x X \to X$ be the blow-up of X at x and let $E = \mathbb{P}^{n-1}$ be the exceptional divisor of $\overline{\alpha}_x$. Let $\tilde{\pi}_x : \mathrm{Bl}_x X \dashrightarrow \mathbb{P}^n$ be the induced rational map. The restriction of $\tilde{\pi}_x$ to E is defined by $|II_{x,x}|$, see Definition 2.3.4. Recall that by definition, see (2.18), $B_x \subset E = \mathbb{P}^{n-1}$ is the base locus scheme of $|II_{x,x}|$.

We claim that $E' = \tilde{\pi}_{x|E}(E) = \mathbb{P}^{n-1} \subset \mathbb{P}^n$ is a hyperplane and that the restriction of $\tilde{\pi}_x$ to E is birational onto its image. Indeed, if $\dim(E') < n-1$, then a general line in $\mathbb{P}^n$ would not cut E' and its image by ϕ would not pass through x. If $\deg(E') \geq 2$, then a general line $l \subset \mathbb{P}^n$ would cut E' at $\deg(E')$ distinct points where ϕ is defined. From $\phi(E') = x$ we would deduce that $\phi(l)$ is singular at x, in contrast with the fact that $\phi(l)$ is a twisted cubic. From this picture the birationality of the restriction of $\tilde{\pi}_x$ to E also immediately follows.

Therefore $\dim(|II_{x,x}|) = n - 1$ and $\tilde{\pi}_{x|E} : E \dashrightarrow E'$ is a Cremona transformation not defined along B_x. Moreover, since $\phi(E') = x$, the restriction of the linear system of cubic hypersurfaces defining ϕ to E' is constant and given by a cubic hypersurface $C'_x \subset E' = \mathbb{P}^{n-1}$. One can assume that $E' \subset \mathbb{P}^n$ is cut out by $x_0 = 0$. Let $\mathbf{x} = (x_1, \ldots, x_n) \in \mathbb{C}^n$ and let $N(\mathbf{x})$ be a cubic equation for $C'_x \subset E'$. Let us choose homogeneous coordinates $(y_0 : \cdots : y_{2n+1})$ on $\mathbb{P}^{2n+1}$ such that $x = (0 : \cdots : 0 : 1)$ and $T_x X = V(y_0, \ldots, y_n)$.

The map $\phi : \mathbb{P}^n \dashrightarrow X \subset \mathbb{P}^{2n+1}$ is given by $2n + 2$ cubic polynomials $g_0, \ldots, g_{2n+1}$. We can suppose that x_0 does not divide g_{2n+1} and that x_0 divides g_j for every $j = 0, \ldots, 2n$. Moreover, x_0^2 divides $g_0, \ldots g_n$ since the hyperplane sections of the form $\lambda_0 g_0 + \cdots + \lambda_n g_n = 0$ correspond to hyperplane sections of X containing $T_x X$ and hence having at least a double point at x. Modulo a change of coordinates on $\mathbb{P}^n$ we can thus suppose $g_i = x_0^2 x_i$ for every $i = 0, \ldots, n$ and that $g_{2n+1} = x_0 g + N$, with $g = g(\mathbf{x})$ quadratic polynomial. The hyperplane sections of X passing through x and not containing $T_x X$ are smooth at x so that we can also suppose $g_{n+1+j} = x_0 f_j$ with $j = 0, \ldots, n-1$ and $f_j = f_j(\mathbf{x})$ quadratic polynomials. We can also suppose $g = 0$, or equivalently $g \in \langle f_0, \ldots, f_{n-1} \rangle$. Otherwise $X \subset \mathbb{P}^{2n+1}$ would be the birational projection on the hyperplane $V(y_{2r+4}) = \mathbb{P}^{2n+1} \subset \mathbb{P}^{2n+2}$ from the external point $(0 : \ldots : 0 : 1 : -1) \in \mathbb{P}^{2n+2}$ of the variety $X' \subset \mathbb{P}^{2n+2}$ having the parametrization $\tilde{\phi} : \mathbb{P}^n \dashrightarrow X' \subset \mathbb{P}^{2n+2}$ given by the following homogenous cubic polynomials: $\tilde{g}_i = g_i$ for $i = 0, \ldots, 2n$; $\tilde{g}_{2n+1} = x_0 g$ and $\tilde{g}_{2n+2} = N$, which is also 3-covered by twisted cubics, contrary to (6.6).

By blowing-up the point x on $\mathbb{P}^{2n+1}$ it immediately follows that the restriction of $\tilde{\pi}_x^{-1}$ to E' is given by $(f_0 : \cdots : f_{n-1})$. Hence

$$\psi_x := \tilde{\pi}_{x|E} : E \dashrightarrow E'$$

is a quadro-quadric Cremona transformation. In conclusion we can suppose that the rational map ϕ is given by the $2n + 2$ cubic polynomials

$$x_0^3, x_0^2 x_1, \ldots, x_0^2 x_n, x_0 f_0, \ldots, x_0 f_{n-1}, N \tag{6.19}$$

and that the base locus of ψ_x^{-1}, $B_x' \subset \mathbb{P}^{n-1} = E'$, is $V(f_0, \ldots, f_{n-1}) \subset \mathbb{P}^{n-1}$, where in this case the polynomials $f_i(\mathbf{x})$ are considered as polynomials in the variables $x_1, \ldots, x_n$.

In practice, we have reversed the previous construction of varieties of type X_f, associating to each $X = \overline{X}^n(3,3) \subset \mathbb{P}^{2n+1}$ a Cremona transformation $\psi = \psi_x :$ $\mathbb{P}^{n-1} \dashrightarrow \mathbb{P}^{n-1}$ such that $X_\psi = X$. First, we consider the case in which ψ_x is fake for $x \in X$ general.

Theorem 6.4.4 ([154, Theorem 5.2]) *Let* $X = \overline{X}^n(3,3) \subset \mathbb{P}^{2n+1}$ *and let the notation be as above. Let* $x \in X$ *be general and let* $\psi_x : \mathbb{P}^r \dashrightarrow \mathbb{P}^r$ *be the associated Cremona transformation. Then the following conditions are equivalent:*

(a) ψ_x *is a fake quadro-quadric Cremona transformation;*
(b) X *is projectively equivalent to a smooth rational normal scroll;*
(c) *the affine parametrization deduced from (6.19) is either*

$$\left(1 : x_1 : \ldots : x_n : x_1^2 : x_1 x_2 : \ldots : x_1 x_n : x_1^2 x_2 \right)$$

or

$$\left(1 : x_1 : \ldots : x_n : x_1^2, x_1 x_2 : \ldots : x_1 x_n : x_1^3 \right);$$

(d) *the projection from* $T = T_x X$ *of a general twisted cubic included in* X *is a conic.*

In particular, a $\overline{X}^n(3,3)$ *is of Castelnuovo type if and only if* ψ_x *is fake so that for every* $n \geq 3$ *there exists an* $\overline{X}^n(3,3) \subset \mathbb{P}^{2n+1}$ *not of Castelnuovo type.*

Theorem 6.4.5 ([154, Theorem 5.2]) *If* $X = \overline{X}^n(3,3) \subset \mathbb{P}^{2n+1}$ *is not a rational normal scroll as above, then:*

1. *the linear system defining* $\phi : \mathbb{P}^n \dashrightarrow X \subset \mathbb{P}^{2n+1}$ *consists of the cubic hypersurfaces in* $\mathbb{P}^n$ *having double points along* $B_x' \subset E' \subset \mathbb{P}^n$;
2. *the scheme* $B_x \subset \mathbb{P}^n$ *is equal (as a scheme) to* $\mathcal{L}_x \subset \mathbb{P}^{n-1}$. *Moreover,* $B_x' \subset E' = \mathbb{P}^{n-1}$ *and* $B_x = \mathcal{L}_x \subset E = \mathbb{P}^{n-1}$ *are projectively equivalent so that* ψ_x *and its inverse have the same base loci, modulo this identification;*
3. *if* X *is also smooth, then* $B_x = \mathcal{L}_x$ *and* B_x' *are smooth schemes.*

Proof By the discussion above on $\mathbb{P}^n \setminus E'$ the map ϕ has an affine expression

$$\phi(\mathbf{x}) = \left(1 : x_1 : \cdots : x_n : f_0(\mathbf{x}) : \cdots : f_{n-1}(\mathbf{x}) : N(\mathbf{x}) \right).$$

Let $(y_0 : \cdots : y_{2n+1})$ be a system of homogeneous coordinates on $\mathbb{P}^{2n+1}$ as above. Then the equations defining $X \subset \mathbb{P}^{2n+1}$ in the affine space $\mathbb{A}^{2n+1} = \mathbb{P}^{2n+1} \setminus V(y_0)$ are $y_i = x_i$, $i = 1, \ldots, n$; $y_{n+1+j} = f_j(\mathbf{x})$, $j = 0, \ldots, n-1$ and $y_{2n+1} = N(\mathbf{x})$, that is, letting $\mathbf{y} = (y_1, \ldots, y_n)$, we get the equations $y_{n+1+j} = f_j(\mathbf{y})$ for $j = 0, \ldots, n-1$ and $y_{2n+1} = N(\mathbf{y})$.

Let $p = \phi(1 : \mathbf{p}) = (1 : \mathbf{p} : f_0(\mathbf{p}) : \cdots : f_r(\mathbf{p}) : f(\mathbf{p}))$ be a general point of X, with $\mathbf{p} = (p_1, \ldots, p_n) \in \mathbb{C}^n$. In particular, $(0 : \mathbf{p})$ is a general point on E'. A tangent direction at $p \in X$ corresponds to the image via $d\phi_q$ of the tangent direction to some line passing through $q = (1 : \mathbf{p}) \in \mathbb{P}^n$. We shall parametrize lines through q via points $(0 : \mathbf{y}) \in E'$ so that such a line, denoted by $L_\mathbf{y}$, admits $t \mapsto \mathbf{p} + t\mathbf{y}$ as an affine parametrization. Then for $i = 0, \ldots, n-1$, one has

$$f_i(\mathbf{p} + t\mathbf{y}) = f_i(\mathbf{p}) + 2tf_i^1(\mathbf{p}, \mathbf{y}) + t^2 f_i(\mathbf{y}), \tag{6.20}$$

where f_i^1 stands for the bilinear form associated to the quadratic form f_i. Moreover,

$$N(\mathbf{p} + t\mathbf{y}) = N(\mathbf{p}) + tN(\mathbf{p}, \mathbf{y}) + t^2 N(\mathbf{y}, \mathbf{p}) + t^3 f(\mathbf{y}), \tag{6.21}$$

where $N(\mathbf{p}, \mathbf{y}) = dN_\mathbf{p}(\mathbf{y})$ is quadratic in $\mathbf{p}$ and linear in $\mathbf{y}$.

Clearly, the base locus of the second fundamental form at $p = \phi(1 : \mathbf{p})$ is the scheme

$$B_p = V\big(f_0(\mathbf{y}), \ldots, f_{n-1}(\mathbf{y}), f(\mathbf{y}, \mathbf{p})\big) = V\big(f_0(\mathbf{y}), \ldots, f_{n-1}(\mathbf{y})\big) \subset \mathbb{P}^{n-1},$$

where the second equality of schemes follows from the equality

$$\dim(< f_0, \ldots, f_{n-1} >) = n$$

combined with the fact that $\dim(|II_{X,p}|) = \dim(|II_{X,x}|) = n$ by the generality of $p \in X$. In particular, we deduce that for $\mathbf{z} \in B_p$ we have $N(\mathbf{z}, \mathbf{p}) = 0$. Because $(0 : \mathbf{p})$ is general in E', this implies $dN_\mathbf{z} = 0$ (since $N(\mathbf{z}, \mathbf{p}) = dN_\mathbf{z}(\mathbf{p})$ for every $\mathbf{p}$) on one hand, and gives $N(\mathbf{z}) = 0$ on the other hand (since $0 = N(\mathbf{z}, \mathbf{z}) = 3N(\mathbf{z})$ after specializing $\mathbf{p} = \mathbf{z}$). The previous facts show that the cubic $C_x' = V(N(\mathbf{x})) \subset \mathbb{P}^n$ has double points along B_x' and part (1) is proved. From these facts it also immediately follows that the closure of the image of the line $L_\mathbf{z}$ (for $\mathbf{z} \in B_p$) via the map ϕ is a line included in X and passing through p, proving (2). Put more intrinsically, the equation of $\mathscr{L}_p$, the Hilbert scheme of lines contained in X and passing through p in its natural embedding into $\mathbb{P}((t_p X)^*)$, is the scheme defined by the equations $f_j(\mathbf{x}), f(\mathbf{x}, \mathbf{p}), N(\mathbf{x})$ and we proved that the ideal generated by these polynomials coincides with the ideal generated by the f_j's which defines B_p as a scheme.

To prove (3) we recall that for a smooth variety $X^n \subset \mathbb{P}^N$ the scheme $\mathscr{L}_x \subset \mathbb{P}^{n-1}$, when non-empty, is a smooth scheme for $x \in X$ general, see for example Proposition 2.2.1.   □

As far as we know, the next result has not been noticed before.

Corollary 6.4.6 ([154, Corollary 5.3]) *Let* $\varphi = (\varphi_0, \ldots, \varphi_r) : \mathbb{P}^r \dashrightarrow \mathbb{P}^r$ *be a Cremona transformation of bidegree* $(2,2)$ *with* $r \geq 2$. *Let* B, *respectively* B', *be the base locus of* φ, *respectively of* φ^{-1}. *Then* B *and* B' *are projectively equivalent.*

We apply the previous correspondence between quadro-quadric Cremona transformations and $\overline{X}(3,3)$'s to the case of smooth varieties or equivalently to quadro-quadric Cremona transformations whose base locus scheme is smooth.

Proposition 6.4.7 ([53], [154, Proposition 5.6]) *Let* $\varphi : \mathbb{P}^r \dashrightarrow \mathbb{P}^r$ *be a Cremona transformation of type* $(2,2)$ *whose base locus* $B \subset \mathbb{P}^r$ *is smooth. Then one of the following holds:*

1. $r \geq 2$, $B = Q^{r-2} \sqcup p$ *with* Q^{r-2} *a smooth quadric hypersurface and* $p \notin \langle Q^{r-2} \rangle$;
2. $r = 5$ *and* B *is projectively equivalent to the Veronese surface* $v_2(\mathbb{P}^2)$;
3. $r = 8$ *and* B *is projectively equivalent to the Segre variety* $\mathbb{P}^2 \times \mathbb{P}^2$;
4. $r = 14$ *and* B *is projectively equivalent to the Grassmann variety* $\mathbb{G}(1,5)$;
5. $r = 26$ *and* B *is projectively equivalent to the 16-dimensional* E_6 *variety.*

Proof If B is reducible, then one obtains case (1), see [154, Proposition 5.2] for the simple details.

Suppose from now on that B is irreducible. Consider $\pi_1 : \mathrm{Bl}_B(\mathbb{P}^r) \to \mathbb{P}^r$, the blow-up of $\mathbb{P}^r$ along B and $\pi_2 : \mathrm{Bl}_{B'}(\mathbb{P}^r) \to \mathbb{P}^r$, the blow-up of $\mathbb{P}^r$ along B'. We deduce the following diagram of birational maps:

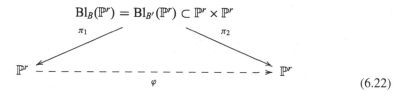

$$\mathrm{Bl}_B(\mathbb{P}^r) = \mathrm{Bl}_{B'}(\mathbb{P}^r) \subset \mathbb{P}^r \times \mathbb{P}^r$$

(6.22)

where π_i are naturally identified with the restriction of the projections on each factor. Let $E_1 = \pi_1^{-1}(B)$ and $E_2 = \pi_2^{-1}(B')$ be the π_i-exceptional Cartier divisors, $i = 1, 2$.

Then (6.22) shows that for general $q \in \pi_1(E_1') \setminus B$ there exists a linear space $\mathbb{P}_q^{r-1-\dim(B')}$ passing through q and cutting X along a quadric hypersurface of dimension $r - 2 - \dim(B)$. If $\varphi(q) = q'$, then naturally $\mathbb{P}^{r-1-\dim(B')} = \mathbb{P}((N_{B'/\mathbb{P}^r})_x^*)$. This immediately implies that $\pi_1(E_1')$ is the variety of secant lines to B and that $B \subset \mathbb{P}^r$ is a *QEL*-manifold of type $\delta(B) = \frac{1}{2}\dim(B)$, see [160, Proposition 4.2]. Indeed, $r - 2 - \dim(B) = \delta(B) = 2\dim(B) + 1 - \dim(\pi_1(E_1'))$ yields $\dim(B) = \frac{2}{3}(r - 2)$. Thus $B \subset \mathbb{P}^r$ is a *QEL*-manifold of type $\delta = \dim(B)/2$ which is also a Severi variety. The classification of Severi varieties due to Zak, see [196, 198] and also Corollary 4.3.3 here, assures us that we are in one of the cases (2)–(5). $\square$

The classification of arbitrary $\overline{X}^n(3,3) \subset \mathbb{P}^{2n+1}$ is difficult due to the existence of a lot of singular examples and it is equivalent to the classification of all quadro-quadric Cremona transformations. In contrast, Proposition 6.4.7 and Theorem 6.4.5 provide a surprisingly simple proof of the classification of smooth $\overline{X}^n(3,3) \subset \mathbb{P}^{2n+1}$.

Theorem 6.4.8 ([154, Theorem 5.7]) *Let $X = \overline{X}^n(3,3) \subset \mathbb{P}^{2n+1}$ be smooth. Then one of the following holds, modulo projective equivalence:*

(i) *X is a smooth rational normal scroll;*
(ii) *X is the Segre embedding $\mathbb{P}^1 \times Q^{n-1} \subset \mathbb{P}^{2n+1}$ with Q^{n-1} a smooth hyperquadric and with $n \geq 2$;*
(iii) *$r = 5$ and X is the Lagrangian Grassmannian $LG_3(\mathbb{C}^6) \subset \mathbb{P}^{13}$;*
(iv) *$r = 8$ and X is the Grassmannian $\mathbb{G}(2,5) \subset \mathbb{P}^{19}$;*
(v) *$r = 14$ and X is the Orthogonal Grassmannian $OG_6(\mathbb{C}^{12}) \subset \mathbb{P}^{31}$;*
(vi) *$r = 26$ and X is the E_7-variety in $\mathbb{P}^{55}$.*

Proof If the associated Cremona transformation is equivalent to a projective transformation we are in case (i) by Theorem 6.4.4. Otherwise, by Theorem 6.4.5, the associated Cremona transformation ψ_x is of type $(2,2)$ with smooth base locus. Let $\phi : \mathbb{P}^n \dashrightarrow X \subset \mathbb{P}^{2r+3}$ be the birational representation of X given by the linear system of cubic hypersurfaces having double points along B'_x. Then B'_x is projectively equivalent to a variety as in cases (1)–(5) of Proposition 6.4.7 so that X is as in cases (ii)–(vi) by a well-known description of the corresponding varieties, see for example [139].   ∎

Remark 6.4.9 Let $X = \overline{X}^n(3,3) \subset \mathbb{P}^{2n+1}$ be smooth and let $\Sigma \subset \text{Hilb}^{3t+1}(X)$ be the family of smooth cubics three-covering X. Then $\dim(\Sigma) = 3(n-1)$ and if $[C] \in \Sigma$ is a general curve, then Corollary 2.1.13 reads as

$$3n - 3 = \dim(\Sigma) = h^0(N_{C/X}) = -K_X \cdot C + n - 3,$$

yielding $-K_X \cdot C = 2n$.

Suppose that $L \subset X$ is a line such that $[3L] \equiv [C]$, which is always the case if $X^n \subset \mathbb{P}^{2n+1}$ is a prime Fano manifold. Under this hypothesis we have: $n \geq 3$ by Theorem 6.4.4; $-K_X \cdot L = 2n/3$; $\mathscr{L}_x$ equidimensional of dimension $\dim(\mathscr{L}_x) = 2(n-1)/3 \geq (n-1)/2$. Thus $\mathscr{L}_x$ is irreducible and $\delta(\mathscr{L}_x) = (n-2)/3 \geq 1$.

Since the restriction of $\widetilde{\pi}_x$ to E is birational onto the image and since $\delta(\mathscr{L}_x) \geq 1$, we deduce $S\mathscr{L}_x \subsetneq \mathbb{P}^{n-1}$ (otherwise $|II_{x,X}|$ would contract the secant lines to $\mathscr{L}_x$ passing through a general point of E).

In conclusion, for $n \geq 3$ and under the hypothesis $3[L] \equiv [C]$ we proved a priori that $\mathscr{L}_x \subset \mathbb{P}^{n-1}$ is a Severi variety. From this and from the classification of Severi varieties, one concludes that the only prime Fano manifolds which are also $\overline{X}^n(3,3) \subset \mathbb{P}^{2n+1}$ are the examples in (iii),.....,(vi) above.

This shows once again, if needed, the powerfulness of the method of studying $\mathscr{L}_x$ for special projective manifolds via the modern tools of deformation theory and it reveals another appearance of the definition of Severi variety.

6.5   A Digression on Power Associative Algebras and Some Involutive Cremona Transformations

In this section we shall present some applications of the theory of (power associative) algebras to the study and classification of Cremona transformations. The idea of constructing interesting examples via the inversion map of (associative) algebras surely dates back to Scorza and his school in Catania around the 1920s, see for example [25, 26, 180]. Let us quote G. Scorza once again, illustrating his point of view on the theory of (associative) algebras:

> Pure si tratta di argomenti che non può rassegnarsi a vedere scarsamente sconosciuti e inadeguatamente apprezzati chi ne abbia *sentita* la svelta e alta eleganza, o ne abbia riconosciuta alla prova la snella pieghevolezza alle applicazioni più varie.
> .....[...] Tanto avrei voluto mostrare con prove concrete l' importanza centrale della teoria delle algebre per la matematica moderna.
>
> G. Scorza, Introduzione a Corpi Numerici e Algebre, 1921, [167]

The monograph *Corpi numerici e Algebre* published in Messina by Edizioni Principato in 1921, see [167], was the first complete treatise on the theory of associative algebras (Dickson's celebrated book *Algebras and their arithmetics* appeared only two years later, published by Chicago University Press). In the Introduction Scorza also remarks that at that time:

> La teoria delle algebre non associative manca ancora di euritmia.

A notable account of the contributions of Scorza and his school in the theory of algebras and on the history of the beginning of these studies in Italy (and abroad) can be found in [21]. In 1937 the applications of the theory of algebras seem to be still considered exotic, at least in Italy. Indeed, Scorza in his last written expository paper points out what will be our *motto* for the rest of the section:

> [...] se non mi inganno, quanto sarà detto sarà sufficiente per far vedere quali profonde armonie soggiaciono a questo mondo delle algebre a prima vista così turbolento.
>
> G. Scorza, La teoria delle algebre e le sue applicazioni, 1937, see [168]

Another apparently turbulent world, hiding deep harmonies, is surely that of Cremona transformations.

Let $\phi : \mathbb{P}^n \dashrightarrow \mathbb{P}^n$ be a Cremona transformation given by a linear system without fixed component in $|\mathcal{O}_{\mathbb{P}^n}(d_1)|$. The base locus scheme of the linear system is called *the base locus scheme of ϕ*. Let $\phi^{-1} : \mathbb{P}^n \dashrightarrow \mathbb{P}^n$ be its inverse, given by a linear system without fixed component in $|\mathcal{O}_{\mathbb{P}^n}(d_2)|$. Then (d_1, d_2) is called the *type* of ϕ and quadro-quadric Cremona transformations are those of type $(2, 2)$ by definition.

Two Cremona transformations $\phi_i : \mathbb{P}^n \dashrightarrow \mathbb{P}^n$, $i = 1, 2$, are said to be *linearly equivalent* if there exist two projective transformations $\omega_i : \mathbb{P}^n \to \mathbb{P}^n$ such that $\phi_2 = \omega_2 \circ \phi_1 \circ \omega_1$.

Corollary 6.4.6 shows that any Cremona transformation of type $(2, 2)$ is linearly equivalent to an involution. Thus one can naturally ask if a birational involution of type $(2, 2)$ is linearly equivalent to a *canonical form*. Let us look at some examples of well-known birational involutions in their "canonical forms" to better understand the problem, to justify the expectation and to construct some significant examples.

Example 6.5.1 (Standard Involution of $\mathbb{P}^n$) The *standard (or elementary) involution of $\mathbb{P}^n$* is the map $\phi : \mathbb{P}^n \dashrightarrow \mathbb{P}^n$ given by

$$\phi(x_0 : \ldots : x_n) = (x_1 x_2 x_3 \cdots x_{n-1} x_n : x_0 x_2 x_3 \cdots x_{n-1} x_n : \ldots : x_0 x_1 x_2 \cdots x_{n-2} x_{n-1}). \tag{6.23}$$

From

$$\phi^2(\mathbf{x}) = (x_0 x_1 \cdots x_{n-1} x_n)^{n-1} \mathbf{x} \tag{6.24}$$

we deduce that ϕ is a birational involution of $\mathbb{P}^n$ of type (n, n).

For an arbitrary birational involution $\phi : \mathbb{P}^n \dashrightarrow \mathbb{P}^n$ of type (d, d), we have $\phi^2(\mathbf{x}) = F(\mathbf{x})\mathbf{x}$ with $F(\mathbf{x})$ homogeneous polynomial of degree $d^2 - 1$. For the standard involution $F(\mathbf{x})$ has a very particular form, being the $(n - 1)^{\text{th}}$-power of a polynomial of degree $n + 1$. Let us consider another example which generalizes considerably the previous one.

Example 6.5.2 (Birational Involutions via Laplace Formulas, See Also [26]) Let $A \in M_{n \times n}(\mathbb{K})$, $\mathbb{K}$ an arbitrary field, and let $A^\# \in M_{n \times n}(\mathbb{K})$ be the matrix defined by the Laplace formula:

$$A \cdot A^\# = \det(A) \cdot I_{n \times n} = A^\# \cdot A, \tag{6.25}$$

that is $A^\#$ is the transpose of the matrix of cofactors of A. The following formulas are immediate consequences of (6.25) and the Binet formula:

$$\det(A^\#) = \det(A)^{n-1}, \tag{6.26}$$

$$(A^\#)^\# = \det(A)^{n-2} \cdot A. \tag{6.27}$$

Via (6.27) we can define a birational involution

$$\phi = [\#] : \mathbb{P}(M_{n \times n}(\mathbb{K})) \dashrightarrow \mathbb{P}(M_{n \times n}(\mathbb{K}))$$

given by

$$\phi([A]) = [A^\#].$$

Thus ϕ is a birational involution of type $(n-1, n-1)$ on $\mathbb{P}(M_{n \times n}(\mathbb{K}))$. For invertible matrices $[A^\#] = [A^{-1}]$ so that ϕ can be considered as a projective realization of the inversion on $M_{n \times n}(\mathbb{K})$, which is a rational map of homogeneous degree -1 on $M_{n \times n}(\mathbb{K})$.

Let

$$U = \{ \text{diagonal matrices} \} \subset M_{n \times n}(\mathbb{K}).$$

Since for $A \in U$ we have $A^{\#} \in U$, the map $\phi = [\#]$ naturally defines by restriction the birational map $\psi : \mathbb{P}(U) \dashrightarrow \mathbb{P}(U)$ given by $\psi([A]) = [A^{\#}]$.

For $A \in U \setminus \mathbf{0}$, letting $x_i = a_{i,i}$ and identifying U with $\mathbb{K}^n$ we see that ψ is nothing but the standard involution on $\mathbb{P}^{n-1} = \mathbb{P}(U)$ with homogeneous coordinates $(x_1 : x_2 : \ldots : x_n)$. Hence the standard involution derives from Laplace formulas and (6.24) is an incarnation of (6.27). In conclusion, the standard involution is a projective realization of the inversion (or adjoint) formula for diagonal matrices.

In the previous case U is a subalgebra of $M_{n \times n}(\mathbb{K})$ but there are instances where, apparently, this is not the case. Indeed, let

$$W = \{ A \in M_{n \times n}(\mathbb{K}) : A = A^t \} \subset M_{n \times n}(\mathbb{K}).$$

Although W is not a subalgebra of $M_{n \times n}(\mathbb{K})$ for $n \geq 2$, one immediately verifies that for $A \in W$ we have $A^{\#} \in W$. Therefore ϕ induces by restriction a birational involution $\varphi : \mathbb{P}(W) \dashrightarrow \mathbb{P}(W)$, which at a first glance seems not to be the projective realization of an inversion map for a multiplication on W. We shall see below that also in this case we can define a structure of (non-associative) algebra on W for which ϕ realizes an inversion (or better a cofactor) map.

The inversion in the non-associative setting requires particular care since the inverse of an element in the usual sense is not necessarily unique, see Example 6.5.17 below. Before developing the theory of (power associative) algebras from scratch we shall look at other well-known birational involutions on $\mathbb{P}^n = \mathbb{P}(V)$ from the perspective of formula (6.27) (or as projective realizations of an inversion map for a suitable structure of algebra on V).

Example 6.5.3 (Non-standard Quadratic Involutions of $\mathbb{P}^2$) Let $\mathbf{A} = \mathbb{C} \times \frac{\mathbb{C}[\epsilon]}{(\epsilon^2)}$. Then $\mathbf{A}$ is a commutative and associative $\mathbb{C}$-algebra. For $\mathbf{x} \in \mathbf{A}$ we shall use the notation $\mathbf{x} = (x_0, x_1, x_2) = (x_0, x_1 + x_2\epsilon)$. In particular, $\mathbf{e} = (1, 1, 0) = (1, 1 + 0\epsilon)$ is the unit element of $\mathbf{A}$. If we define

$$\mathbf{x}^{\#} = (x_1^2, x_0 x_1, -x_0 x_2) = (x_1^2, x_0 x_1 - x_0 x_2 \epsilon),$$

then we deduce the following formulas, similar to (6.25) and (6.27):

$$\mathbf{x} \cdot \mathbf{x}^{\#} = (x_0, x_1 + x_2\epsilon) \cdot (x_1^2, x_0 x_1 - x_0 x_2 \epsilon) = (x_0 x_1^2)\mathbf{e},$$

$$(\mathbf{x}^{\#})^{\#} = (x_0 x_1^2)\mathbf{x}.$$

This suggests that one can define

$$\det(\mathbf{x}) := x_0 x_1^2$$

so that an element $\mathbf{x} \in \mathbf{A}$ is invertible if and only if $\det(\mathbf{x}) \neq 0$. In this case one has

$$\mathbf{x}^{-1} = \frac{\mathbf{x}^{\#}}{\det(\mathbf{x})} = (\frac{1}{x_0}, \frac{1}{x_1} - \frac{x_2}{x_1^2}\epsilon).$$

The associated map $\phi : \mathbb{P}^2 = \mathbb{P}(\mathbf{A}) \dashrightarrow \mathbb{P}(\mathbf{A}) = \mathbb{P}^2$ is given by

$$\phi(x_0 : x_1 : x_2) = (x_1^2 : x_0 x_1 : -x_0 x_2)$$

so that it is the non-standard quadro-quadric Cremona transformation of the plane given by the net of conics passing through $p_1 = (1 : 0 : 0)$ and $p_2 = (0 : 0 : 1)$ and tangent in p_2 to the line of equation $x_0 = 0$. We note that no algebraic geometer would have written $-x_0 x_2$ in the last component of the expression of ϕ above but this is crucial to define inversion on $\mathbf{A}$. This remark points out a particularity of this approach to the *right canonical form* of birational involutions associated to inversions in algebras.

The most degenerate case of a quadro-quadric non-standard Cremona transformation in the plane is given by the net of conics osculating at a point. This example also has a peculiar canonical form coming from inversion in the commutative and associative algebra $\mathbf{B} = \frac{\mathbb{C}[\epsilon]}{(\epsilon^3)}$. For $\mathbf{x} \in \mathscr{B}$ we shall use the notation $\mathbf{x} = (x_0, x_1, x_2) = x_0 + x_1\epsilon + x_2\epsilon^2$. Thus $\mathbf{e} = (1, 0, 0) = 1 + 0\epsilon + 0\epsilon^2$ is the unit element of $\mathbf{B}$. If we define

$$\mathbf{x}^{\#} = (x_0^2, -x_0 x_1, x_1^2 - x_0 x_2) = x_0^2 - x_0 x_1\epsilon + (x_1^2 - x_0 x_2)\epsilon^2,$$

then we get the following formula, once again very similar to (6.25) and (6.27):

$$\mathbf{x} \cdot \mathbf{x}^{\#} = x_1^3 \mathbf{e},$$

$$(\mathbf{x}^{\#})^{\#} = x_1^3 \mathbf{x}.$$

Once again one can define

$$\det(\mathbf{x}) := x_1^3$$

so that an element $\mathbf{x} \in \mathbf{B}$ is invertible if and only if $\det(\mathbf{x}) \neq 0$. Also in this case one deduces

$$\mathbf{x}^{-1} = \frac{\mathbf{x}^{\#}}{\det(\mathbf{x})} = \frac{x_0}{x_1} - \frac{x_0}{x_1^2}\epsilon + (\frac{1}{x_1} - \frac{x_0 x_2}{x_1^3})\epsilon^2.$$

We end this informal discussion by describing a well-known series of elementary quadro-quadric transformation in $\mathbb{P}^n$, which modulo linear equivalence are birational involutions of type $(2, 2)$ by Corollary 6.4.6. A priori it is not clear at all if they admit a canonical form coming from inversion in some algebra of dimension $n + 1$ where formulas analogous to the Laplace formulas hold.

Example 6.5.4 (Elementary Quadro-Quadric Cremona Transformations) Let $Q^n \subset \mathbb{P}^{n+1}$ be an irreducible quadric hypersurface. Let $p_1 \in Q_{\mathrm{reg}}$, let

$$p_2 \in Q_{\mathrm{reg}} \setminus (T_{p_1} Q \cap Q)$$

and let $\pi_{p_i} : Q \dashrightarrow \mathbb{P}^n$ be the birational projection from p_i onto a hyperplane in $\mathbb{P}^n$, $i = 1, 2$. Then

$$\pi_{p_2} \circ \pi_{p_1}^{-1} : \mathbb{P}^n \dashrightarrow \mathbb{P}^n$$

is a Cremona transformation of type $(2, 2)$. Indeed, a general line $L \subset \mathbb{P}^n$ is sent by $\pi_{p_1}^{-1}$ into the conic $C = <L, p_1> \cap Q$ passing through p_1 but not through p_2. Thus $\pi_{p_2} \circ \pi_{p_1}^{-1}(L)$ is the conic $\pi_{p_2}(C)$, proving the claim.

The algebraic theory sketched in the next section will allow us to construct interesting examples and to show that all quadro-quadric Cremona transformations of $\mathbb{P}^n$ are linearly equivalent to an inversion (or cofactor) map coming from particular classes of non-associative algebras. In particular, they have a canonical form coming from suitable Laplace formulas for a multiplication on $\mathbb{C}^{n+1}$.

6.5.1 Power Associative Algebras, Jordan Algebras and Generalizations of Laplace Formulas

For simplicity the basic definitions will be introduced in the case of (finite-dimensional) $\mathbb{C}$-algebras although most of the notions are valid on any ground field and in arbitrary dimension.

Definition 6.5.5 (Basic Definitions) Let V be a complex vector space. A *multiplication or product* on V is a bilinear map

$$* : V \times V \to V$$

which to $\mathbf{u}, \mathbf{v} \in V$ associates the vector $\mathbf{u} * \mathbf{v}$.

Then $\mathbf{A} = (V, *)$ is called a $\mathbb{C}$-*algebra* with product $*$ and the dimension of $\mathbf{A}$ is the dimension of V as complex vector space. Thus, besides the usual axioms on the additive structure and multiplication by scalars on V, we have introduced a bilinear "product" of vectors in V.

An element $\mathbf{e} \in V$ is said to be a *unity* in $\mathbf{A}$ if for every $\mathbf{x} \in V$ we have

$$\mathbf{e} * \mathbf{x} = \mathbf{x} = \mathbf{x} * \mathbf{e}.$$

If there exists a unit element $\mathbf{e}$, it is clearly unique and in this case we shall use the notation $\mathbf{A} = (V, *, \mathbf{e})$.

The algebra $\mathbf{A} = (V, *)$ is said to be *commutative* if for every $\mathbf{x}, \mathbf{y} \in V$ we have

$$\mathbf{x} * \mathbf{y} = \mathbf{y} * \mathbf{x}.$$

The algebra $\mathbf{A} = (V, *)$ is said to be *associative* if for every $\mathbf{x}, \mathbf{y}, \mathbf{z} \in V$ we have

$$(\mathbf{x} * \mathbf{y}) * \mathbf{z} = \mathbf{x} * (\mathbf{y} * \mathbf{z}).$$

For $\mathbf{x} \in \mathbf{A}$ we define

$$\mathbf{x}^2 = \mathbf{x} * \mathbf{x}$$

and inductively, for every $n \in \mathbb{N}$,

$$\mathbf{x}^n = \mathbf{x}^{n-1} * \mathbf{x}.$$

The algebra $\mathbf{A} = (V, *)$ is said to be *power associative* if for every $\mathbf{x} \in V$ and for every $m, n \in \mathbb{N}$ we have

$$\mathbf{x}^{m+n} = \mathbf{x}^m * \mathbf{x}^n.$$

For our purposes power associative algebras are general enough to include a lot of significant examples and also to extend the key formulas (6.25) and (6.27).

Jordan algebras, introduced below, are a subclass of power associative algebras and are among the most studied non-associative algebras together with the more widely known Lie algebras, see [108, 134] for an introduction, for the history of the subject and for a precise account of the theory of Jordan algebras. This theory is relatively new and non-classical in a sense masterfully summarized by McCrimmon:

I am unable to prove Jordan algebras were known to Archimedes, or that a complete theory has been found in the unpublished papers of Gauss. Their first appearance in recorded history seems to be in the early 1930's when the theory bursts forth full-grown from the mind, not of Zeus, but of Pascual Jordan, John von Neumann, and Eugene Wigner in their 1934 paper, *On an algebraic generalization of the quantum mechanical formalism* [110].
.... the physicists largely abandoned Jordan algebras, and the algebraists took over: A.A. Albert, N. Jacobson, and others developed a complete theory of finite-dimensional Jordan algebras over arbitrary fields of characteristic $\neq 2$. It would be most unfair to picture

lions finishing a kill and leaving the remains to the jackals. For one thing, physicists have not finished feeding on the carcass.[......].

For another thing, these Jordan algebras (especially the 27-dimensional exceptional algebra) which arose in an unsuccessful attempt to find a new algebra suitable for quantum mechanics turned out to have unforeseen applications to Lie groups and algebras, geometry, and analysis.

.........[....] Jordan algebras arose as an attempt to capture the algebraic essence of Hermitian matrices (or symmetric elements of an associative algebra with involution). They are close enough to associative algebras to remain tractable, yet they let in just enough generality to include $\text{Herm}_3(\mathbb{O})$, thereby yielding connections with exceptional structures in many branches of mathematics. As a result, Jordan methods have proved useful tools in a variety of settings.

K. McCrimmon, excerpts from *Jordan algebras and their applications*, [132] 1978

Definition 6.5.6 (Jordan Algebra) A $\mathbb{C}$-algebra with unity $\mathbf{J} = (V, *, \mathbf{e})$ is called a *Jordan algebra* if it is a commutative and if the *Jordan identity*

$$\mathbf{x}^2 * (\mathbf{x} * \mathbf{y}) = \mathbf{x} * (\mathbf{x}^2 * \mathbf{y}) \tag{6.28}$$

holds for every $\mathbf{x}, \mathbf{y} \in \mathbf{J}$.

Here we shall also assume that $\mathbf{J}$ is finite dimensional.

By specializing the identity above with $\mathbf{y} = \mathbf{x}$ and taking into account commutativity, one deduces that

$$\mathbf{x}^2 * \mathbf{x}^2 = (\mathbf{x}^2 * \mathbf{x}) * \mathbf{x} \tag{6.29}$$

holds for every $\mathbf{x} \in \mathbf{J}$. By definition $(\mathbf{x}^3) * \mathbf{x} = \mathbf{x}^4$ so that (6.29) yields $\mathbf{x}^2 * \mathbf{x}^2 = \mathbf{x}^4$. Albert showed that, arguing by induction, one can deduce from (6.29) that a Jordan algebra is power associative, see [59, Proposition II.1.2] for a different and elementary direct proof.

Albert's criterion for power associativity has different levels of generality, see [5–7], and it cannot be extended to arbitrary positive characteristic. We state it in its simplest form suitable for our purposes although the criterion is valid also for non-associative rings of characteristic zero (and also for large characteristic). One can consult the previous references for the precise statements of these generalizations.

Theorem 6.5.7 (Albert Criterion for Power Associativity, [5]) *Let* $\mathbb{K}$ *be a field and let* $\mathbf{A}$ *be a* $\mathbb{K}$*-algebra of characteristic zero. Then* $\mathbf{A}$ *is power associative if and only if the following two identities hold for every* $\mathbf{x} \in \mathbf{A}$:

$$\mathbf{x}^2 * \mathbf{x} = \mathbf{x} * \mathbf{x}^2 \ and \quad \mathbf{x}^2 * \mathbf{x}^2 = (\mathbf{x}^2 * \mathbf{x}) * \mathbf{x}.$$

In particular, a commutative $\mathbb{K}$*–algebra* $\mathbf{A}$ *of characteristic zero is power associative if and only if* $\mathbf{x}^2 * \mathbf{x}^2 = \mathbf{x}^3 * \mathbf{x}$ *for every* $\mathbf{x} \in \mathbf{A}$.

Example 6.5.8 (Some Examples of Jordan Algebras)

1. Let $\mathbf{A} = (V, *, \mathbf{e})$ be an associative algebra with unity. Denote by $\mathbf{A}^+ = (V, \bullet, \mathbf{e})$ the algebra with the symmetrized (or Jordan) product

$$\mathbf{x} \bullet \mathbf{y} = \frac{1}{2}(\mathbf{x} * \mathbf{y} + \mathbf{y} * \mathbf{x}).$$

Then $\mathbf{A}^+$ is a Jordan algebra. Note that $\mathbf{A}^+ = \mathbf{A}$ if $\mathbf{A}$ is commutative. Jordan algebras of the form $\mathbf{A}^+$ with $\mathbf{A}$ associative are said to be *special*.

2. Let $q : W \to \mathbb{C}$ be a quadratic form on the vector space W. For $(\lambda, \mathbf{w}), (\lambda', \mathbf{w}') \in \mathbb{C} \times W$, the product

$$(\lambda, \mathbf{w}) * (\lambda', \mathbf{w}') = (\lambda\lambda' - q(\mathbf{w}, \mathbf{w}'), \lambda\mathbf{w}' + \lambda'\mathbf{w})$$

induces a structure of Jordan algebra on $\mathbb{C} \times W$ with unity $\mathbf{e} = (1, \mathbf{0})$.

3. Let $\mathbb{A}$ be the complexification of one of the four Hurwitz's algebras $\mathbb{R}, \mathbb{C}, \mathbb{H}$ or $\mathbb{O}$ with the natural induced conjugation on $< \mathbf{e} >^\perp$ and denote by $\mathrm{Herm}_3(\mathbb{A})$ the algebra of Hermitian 3×3 matrices with coefficients in $\mathbb{A}$:

$$\mathrm{Herm}_3(\mathbb{A}) = \left\{ \begin{pmatrix} r_1 & \overline{x_3} & \overline{x_2} \\ x_3 & r_2 & \overline{x_1} \\ x_2 & x_1 & r_3 \end{pmatrix} \;\middle|\; \begin{array}{l} x_1, x_2, x_3 \in \mathbb{A} \\ r_1, r_2, r_3 \in \mathbb{C} \end{array} \right\}.$$

Then the symmetrized product $M \bullet N = \frac{1}{2}(MN + NM)$ induces on $\mathrm{Herm}_3(\mathbb{A})$ a structure of Jordan algebra. Albert proved in [4] that the Jordan algebra $\mathrm{Herm}_3(\mathbb{O})$ is not special.

4. If

$$W_n = \{A \in M_{n \times n}(\mathbb{C}) \; : \; A = A^t\} \subset M_{n \times n}(\mathbb{C}),$$

then W_n is a Jordan subalgebra of $M_{n \times n}(\mathbb{C})^+ = (M_{n \times n}(\mathbb{C}), \bullet, I_{n \times n})$, where $\bullet$ is the symmetrized product defined in 1) above. Indeed, one immediately verifies that $A = A^t$ and $B = B^t$ imply $A \bullet B = (A \bullet B)^t$.

Definition 6.5.9 (Operators $L_\mathbf{x}$ and $R_\mathbf{x}$) Let $\mathbf{A} = (V, *)$ be a $\mathbb{C}$–algebra. The *left multiplication by* $\mathbf{x} \in \mathbf{A}$ is the linear map $L_\mathbf{x} : V \to V$ defined by $L_\mathbf{x}(\mathbf{y}) = \mathbf{x} * \mathbf{y}$.

Analogously, the *right multiplication by* $\mathbf{x} \in \mathbf{A}$ is the linear map $R_\mathbf{x} : V \to V$ defined by $R_\mathbf{x}(\mathbf{y}) = \mathbf{y} * \mathbf{x}$.

This formalism allows the restatement of the previous definitions in terms of left and right multiplications. Indeed, $\mathbf{e} \in \mathbf{A}$ is the identity element if and only if $L_\mathbf{e}$ and $R_\mathbf{e}$ are the identity map on V.

The algebra $\mathbf{A} = (V, *)$ is commutative if and only if $L_\mathbf{x} = R_\mathbf{x}$ for every $\mathbf{x} \in \mathbf{A}$. The algebra $\mathbf{A}$ is associative if and only if $L_{\mathbf{x}*\mathbf{y}} = L_\mathbf{x} \circ L_\mathbf{y}$ for every $\mathbf{x}, \mathbf{y} \in \mathbf{A}$, where $\circ$ is the ordinary composition of functions on V. A commutative algebra with identity $\mathbf{A} = (V, *, \mathbf{e})$ is a Jordan algebra if and only if $L_\mathbf{x} \circ L_{\mathbf{x}^2} = L_{\mathbf{x}^2} \circ L_\mathbf{x}$ for every $\mathbf{x} \in \mathbf{A}$.

Given an element $\mathbf{x} \in \mathbf{A} = (V, *, \mathbf{e})$ we shall indicate by $\mathbb{C}[\mathbf{x}] \subset \mathbf{A}$ the subalgebra generated by $\mathbf{e}$ and by the powers of $\mathbf{x}$, $\{\mathbf{x}^n\}_{n \in \mathbb{N}}$. Let us remark that an algebra $A = (V, *)$ is power associative if and only if for every $\mathbf{x} \in \mathbf{A}$ the subalgebra generated by the powers of $\mathbf{x}$ is associative (and hence also commutative).

Definition 6.5.10 (Minimal Polynomial of an Element and Rank of a Power Associative Algebra) Let $\mathbf{A} = (V, *, \mathbf{e})$ be a power associative $\mathbb{C}$-algebra with unit element $\mathbf{e}$. For $\mathbf{x} \in \mathbf{A}$ we can define the *evaluation morphism at* $\mathbf{x}$

$$\phi_{\mathbf{x}} : \mathbb{C}[t] \to \mathbb{C}[\mathbf{x}]$$

by sending the polynomial $p(t) = \sum_{i=0}^{d} a_i t^i \in \mathbb{C}[t]$ to the element

$$\phi_{\mathbf{x}}(p(t)) = \sum_{i=0}^{d} a_i \mathbf{x}^i \in \mathbb{C}[\mathbf{x}] \subset A.$$

Since $\mathbf{A}$ is power associative the map $\phi_{\mathbf{x}}$ is a homomorphism of associative and commutative $\mathbb{C}$-algebras and

$$\ker(\phi_{\mathbf{x}}) = \{p(t) \in \mathbb{C}[t] \ : \ \phi_{\mathbf{x}}(p(t)) = \mathbf{0}_{\mathbf{A}}\} \subset \mathbb{C}[t]$$

is an ideal in $\mathbb{C}[t]$. Since we assume $\dim(V) = n$, $\ker(\phi_{\mathbf{x}}) \neq 0$ and we define *the minimal polynomial* $m_{\mathbf{x}}(t) \in \mathbb{C}[t]$ *of* $\mathbf{x}$ as the monic generator of the ideal $\ker(\phi_{\mathbf{x}})$.

Clearly $\deg(m_{\mathbf{x}}(t)) \leq \dim(V)$ and more precisely

$$\deg(m_{\mathbf{x}}(t)) = \min\{s > 0 \ : \ \mathbf{e}, \mathbf{x}, \ldots, \mathbf{x}^s \text{ are linearly dependent}\} = \dim_{\mathbb{C}}(\mathbb{C}[\mathbf{x}]).$$
$$(6.30)$$

Finally, we can define the *rank of* $\mathbf{A}$ as the integer

$$\mathrm{rk}(\mathbf{A}) = \max\{\deg(m_{\mathbf{x}}(t)), \ \mathbf{x} \in \mathbf{A}\} = \max\{\dim(\mathbb{C}[\mathbf{x}]), \ \mathbf{x} \in \mathbf{A}\}. \qquad (6.31)$$

An element $\mathbf{x} \in \mathbf{A}$ is called *regular* if $\dim(\mathbb{C}[\mathbf{x}]) = \mathrm{rk}(A)$.

Looking back at Example 6.5.8 we have $\mathrm{rk}(\mathbf{A}) = \mathrm{rk}(\mathbf{A}^+)$, that the Jordan algebra in part 2) has rank equal to two (see also Example 6.5.15 below) while the examples in 3) have rank three. The Cayley–Hamilton Theorem implies that $\mathrm{rk}(M_{n \times n}(\mathbb{C})) \leq n$. Since a diagonal matrix in $M_{n \times n}(\mathbb{C})$ with n different entries has minimal polynomial of degree n we conclude that $\mathrm{rk}(M_{n \times n}(\mathbb{C})) = n$ and that regular elements are precisely the matrices for which the minimal polynomial coincides with the characteristic polynomial. Moreover, for the Jordan subalgebra $W_n \subset M_{n \times n}(\mathbb{C})^+$ defined in Example 6.5.8 part 4) we have $\mathrm{rk}(W_n) = n = \mathrm{rk}(M_{n \times n}(\mathbb{C})^+)$.

The next result shows that in a power associative algebra there exists a *universal minimal polynomial* which evaluated in a regular element yields the minimal polynomial of the element. This generalizes the familiar situation occurring for square matrices with entries in a field, where one proves the same property for the *universal characteristic polynomial*.

Proposition 6.5.11 ([59, Proposition II.2.1]) *Let* $\mathbf{A} = (V, *, \mathbf{e})$ *be a power associative* $\mathbb{C}$*-algebra of finite dimension n and of rank* $r = \mathrm{rk}(\mathbf{A})$. *Then:*

1. *the set of regular elements is an open dense subset in the Zariski topology on V;*
2. *there exist unique homogeneous polynomials* $s_j(t_1, \dots, t_n) \in \mathbb{C}[t_1, \dots t_n]$ *of degree* j, $j = 1, \dots, r$, *such that for every regular element* $\mathbf{x} \in \mathbf{A}$ *we have*

$$m_{\mathbf{x}}(t) = t^r - s_1(\mathbf{x})t^{r-1} + s_2(\mathbf{x})t^{r-2} + \dots + (-1)^r s_r(\mathbf{x}). \qquad (6.32)$$

Proof Let $\mathbf{y} \in \mathbf{A}$ be a fixed regular element, which exists by definition of rank of $\mathbf{A}$. Then there exists $\mathbf{v}_{r+1}, \dots, \mathbf{v}_n \in V$ such that $\mathscr{A} = \{\mathbf{e}, \mathbf{y}, \dots, \mathbf{y}^{r-1}, \mathbf{v}_{r+1}, \dots, \mathbf{v}_n\}$ is a basis of V. For every $\mathbf{v} \in V$ let $[\mathbf{v}]_{\mathscr{A}}$ denote the column vector of the coordinates of $\mathbf{v}$ with respect to $\mathscr{A}$. Let $f(t_1, \dots, t_n) \in \mathbb{C}[t_1, \dots, t_n]$ be defined by the formula

$$f(\mathbf{x}) = \det([[\mathbf{e}]_{\mathscr{A}} | [\mathbf{x}]_{\mathscr{A}} | \dots | [\mathbf{x}^{r-1}]_{\mathscr{A}} | [\mathbf{v}_{r+1}]_{\mathscr{A}} | \dots | [\mathbf{v}_n]_{\mathscr{A}}]).$$

Since $f(\mathbf{y}) = 1, f(t_1, \dots, t_n) \neq 0$ and the set of regular elements of $\mathbf{A}$ is precisely $V \setminus \{\mathbf{x} \in V : f(\mathbf{x}) = 0\}$, which is an open non-empty subset in the Zariski topology on V.

Let $\mathbf{x} \in \mathbf{A}$ be a regular element. If

$$m_{\mathbf{x}}(t) = t^r - s_1(\mathbf{x})t^{r-1} + s_2(\mathbf{x})t^{r-2} + \dots + (-1)^r s_r(\mathbf{x}),$$

then

$$\mathbf{x}^r = s_1(\mathbf{x})\mathbf{x}^{r-1} - s_2(\mathbf{x})\mathbf{x}^{r-2} + \dots + (-1)^{r+1} s_r(\mathbf{x}). \qquad (6.33)$$

Let

$$u_j(\mathbf{x}) = \det([[\mathbf{e}]_{\mathscr{A}} | \dots | [\mathbf{x}^{j-1}]_{\mathscr{A}} | [\mathbf{x}^r]_{\mathscr{A}} | [\mathbf{x}^{j+1}]_{\mathscr{A}} | \dots | [\mathbf{v}_{r+1}]_{\mathscr{A}} | \dots | [\mathbf{v}_n]_{\mathscr{A}}]).$$

The usual rules for the calculation of the determinant together with (6.33) yield

$$s_j(\mathbf{x}) = (-1)^{j-1} \frac{u_{r-j}(\mathbf{x})}{f(\mathbf{x})}.$$

Therefore $s_j(\mathbf{x})$ is a rational function of $\mathbf{x}$. Let $c_{\mathbf{x}}(t)$ be the characteristic polynomial of the left multiplication by $\mathbf{x}$, $L_{\mathbf{x}} : V \to V$. By the Cayley–Hamilton Theorem $c_{\mathbf{x}}(L_{\mathbf{x}}) = \mathbf{0}_{\mathrm{End}(V)}$ so that

$$\phi_{\mathbf{x}}(c_{\mathbf{x}}(t)) = (c_{\mathbf{x}}(L_{\mathbf{x}}))(\mathbf{e}) = \mathbf{0}_{\mathrm{End}(V)}(\mathbf{e}) = \mathbf{0}_{\mathbf{A}}$$

and $c_{\mathbf{x}}(t) \in \ker(\phi_{\mathbf{x}})$. Thus by Gauss' Lemma the rational functions $s_j(\mathbf{x})$ are polynomials, which are clearly homogeneous of degree j. □

Let us remark that since the polynomial identity (6.33) holds on the dense open subset of regular elements of $\mathbf{A}$, it is satisfied by every element of $\mathbf{A}$. For non-regular elements the minimal polynomial divides the "universal minimal polynomial" of $\mathbf{A}$.

Definition 6.5.12 (Determinant and Cofactor in a Power Associative Algebra)
Let $\mathbf{A} = (V, *, \mathbf{e})$ be a power associative $\mathbb{C}$-algebra of finite dimension n and of rank $r = \mathrm{rk}(\mathbf{A})$ and let $x \in \mathbf{A}$ be an arbitrary element. Letting the notation be as above we can define *the cofactor of* $\mathbf{x}$:

$$\mathbf{x}^{\#} = (-1)^{r+1}[\mathbf{x}^{r-1} - s_1(\mathbf{x})\mathbf{x}^{r-2} + s_2(\mathbf{x})\mathbf{x}^{r-3} + \ldots + (-1)^{r-1}s_{r-1}(\mathbf{x})\mathbf{e}] \in \mathbf{A} \qquad (6.34)$$

and we can define the *determinant of* $\mathbf{x}$:

$$\det(\mathbf{x}) = s_r(\mathbf{x}) \in \mathbb{C}. \qquad (6.35)$$

To prove the analogue of (6.27) in this general setting we need the following properties of the determinant in a power associative algebra.

Lemma 6.5.13 *Let* $\mathbf{A} = (V, *, \mathbf{e})$ *be a power associative $\mathbb{C}$-algebra of finite dimension n and of rank $r = \mathrm{rk}(\mathbf{A})$. If $\mathbf{x} \in \mathbf{A}$ is a regular element and if $\mathbf{y} \in \mathbb{C}[\mathbf{x}]$ then:*

$$\det(\mathbf{x} * \mathbf{y}) = \det(\mathbf{x})\det(\mathbf{y}). \qquad (6.36)$$

Proof Let $\mathbf{x}$ be a fixed regular element and let $f_{\mathbf{x}} : \mathbb{C}[\mathbf{x}] \to \mathbb{C}[\mathbf{x}]$ be the restriction to $\mathbb{C}[\mathbf{x}]$ of $L_{\mathbf{x}}$. If $A_{\mathbf{x}}$ is the matrix of $f_{\mathbf{x}}$ with respect to the basis $\mathscr{B} = \{\mathbf{e}, \mathbf{x}, \ldots, \mathbf{x}^{r-1}\}$, then a direct computation, taking into account (6.33), immediately yields $\det(A_{\mathbf{x}}) = s_r(\mathbf{x})$.

If $\mathbf{y} \in \mathbb{C}[\mathbf{x}]$ is a general regular element, then $\mathbf{x} * \mathbf{y} \in \mathbb{C}[\mathbf{x}]$ is also regular by the continuity of $f_{\mathbf{x}}$ and by Proposition 6.5.11. Let $A_{\mathbf{y}}$ be the matrix of the restriction of $L_{\mathbf{y}}$ to $\mathbb{C}[\mathbf{x}]$ with respect to the basis $\mathscr{B}$ and let $A_{\mathbf{x}*\mathbf{y}}$ be the matrix of the restriction of $L_{\mathbf{x}*\mathbf{y}}$ to $\mathbb{C}[\mathbf{x}]$ with respect to the basis $\mathscr{B}$. Since $\mathbf{y}$ and $\mathbf{x} * \mathbf{y}$ are regular elements $\{\mathbf{e}, \mathbf{y}, \ldots, \mathbf{y}^{r-1}\}$ and $\{\mathbf{e}, \mathbf{x}*\mathbf{y}, \ldots, (\mathbf{x}*\mathbf{y})^{r-1}\}$ are also bases of $\mathbb{C}[\mathbf{x}] = \mathbb{C}[\mathbf{y}] = \mathbb{C}[\mathbf{x}*\mathbf{y}]$. Thus we can calculate the determinant with respect to an arbitrary basis obtaining $\det(A_{\mathbf{y}}) = s_r(\mathbf{y})$ and $\det(A_{\mathbf{x}*\mathbf{y}}) = s_r(\mathbf{x} * \mathbf{y})$. Since the algebra $\mathbb{C}[\mathbf{x}]$ is associative and commutative

$$L_{\mathbf{x}*\mathbf{y}} = L_{\mathbf{x}} \circ L_{\mathbf{y}}$$

holds so that the usual Binet formula yields

$$s_r(\mathbf{x} * \mathbf{y}) = \det(A_{\mathbf{x}*\mathbf{y}}) = \det(A_{\mathbf{x}})\det(A_{\mathbf{y}}) = s_r(\mathbf{x})s_r(\mathbf{y}), \qquad (6.37)$$

proving (6.36). Since (6.36) holds on an open non-empty subset of $\mathbb{C}[\mathbf{x}]$, it holds for every $\mathbf{y} \in \mathbb{C}[\mathbf{x}]$.   □

We have now developed all the tools to prove the Laplace formulas for power associative algebras with unity, whose proof can be achieved exactly as for the analogous formulas in $M_{n \times n}(\mathbb{K})$.

Corollary 6.5.14 (Laplace Formula for Power Associative Algebras) *Let* $\mathbf{A} = (V, *, \mathbf{e})$ *be a power associative* $\mathbb{C}$-*algebra of finite dimension* n *and of rank* $r = \mathrm{rk}(\mathbf{A})$. *Let* $\mathbf{x} \in \mathbf{A}$. *Then:*

$$\mathbf{x} * \mathbf{x}^{\#} = \det(\mathbf{x})\mathbf{e} = \mathbf{x}^{\#} * \mathbf{x}; \tag{6.38}$$

$$\det(\mathbf{x}^{\#}) = \det(\mathbf{x})^{r-1}; \tag{6.39}$$

$$(\mathbf{x}^{\#})^{\#} = \det(\mathbf{x})^{r-2}\mathbf{x}. \tag{6.40}$$

Let us see how to obtain these formulas in an explicit example.

Example 6.5.15 (Laplace Formulas for Quadratic Algebras of Rank 2) Let us recall the definition in Example 6.5.8 part 2). Let $q : W \to \mathbb{C}$ be a quadratic form on the vector space W. For $(\lambda, \mathbf{w}), (\lambda', \mathbf{w}') \in \mathbb{C} \times W$, the product

$$(\lambda, \mathbf{w}) * (\lambda', \mathbf{w}') = (\lambda\lambda' - q(\mathbf{w}, \mathbf{w}'), \lambda\mathbf{w}' + \lambda'\mathbf{w})$$

induces a structure of rank two Jordan algebra on $\mathbb{C} \times W$ with unity $\mathbf{e} = (1, \mathbf{0})$, indicated by $\mathbf{A} = (\mathbb{C} \times W, *, \mathbf{e})$. Indeed, if $\mathbf{x} = (\lambda, \mathbf{w})$, then

$$\mathbf{x}^2 - 2\lambda\mathbf{x} + (\lambda^2 + q(\mathbf{w}))\mathbf{e} = \mathbf{0}_{\mathbb{C} \times W},$$

yielding

$$(\lambda, \mathbf{w})^{\#} = (\lambda, -\mathbf{w}), \quad \det((\lambda, \mathbf{w})) = \lambda^2 + q(\mathbf{w}),$$

$$((\lambda, \mathbf{w})^{\#})^{\#} = (\lambda, \mathbf{w}).$$

Let $\widetilde{\mathbf{A}}$ be the power associative algebra $\mathbb{C} \times \mathbf{A}$ with unity $\widetilde{\mathbf{e}} = (1, 1, \mathbf{0})$. An element of $\widetilde{\mathbf{A}}$ has the form $\mathbf{y} = (\mu, \lambda, \mathbf{w})$ and we have:

$$(\mu, \lambda, \mathbf{w})^{\#} = (\lambda^2 + q(\mathbf{w}), \mu\lambda, -\mu\mathbf{w}), \quad \det((\mu, \lambda, \mathbf{w})) = \mu(\lambda^2 + q(\mathbf{w})),$$

$$((\lambda, \mathbf{w}))^{\#})^{\#} = \mu(\lambda^2 + q(\mathbf{w}))(\lambda, \mathbf{w}).$$

We now define invertible elements in a power associative algebra with unity.

Definition 6.5.16 (Invertible Elements in a Power Associative Algebra) Let $A = (V, *, e)$ be a power associative $\mathbb{C}$-algebra. An element $x \in A$ is said to be *invertible* in A if there exists a $y \in \mathbb{C}[x]$ such that

$$x * y = e = y * x. \tag{6.41}$$

Since $\mathbb{C}[x]$ is associative, if such a $y \in \mathbb{C}[x]$ exists, then it is unique and it will be denoted by x^{-1}.

In the non-associative setting, which also includes associative algebras, this is the unique possible definition since otherwise an element can admit infinitely many elements satisfying the identity (6.41), as shown in the next example.

Example 6.5.17 Let $W_2 \subset M_{2 \times 2}(\mathbb{C})^+$ be the Jordan subalgebra of symmetric matrices, see Example 6.5.8 part 4) for notation. Let $\alpha \in \mathbb{C}$ and let

$$\mathbf{a} = \begin{pmatrix} 1 & 0 \\ 0 & -1 \end{pmatrix}, \quad \mathbf{b}_\alpha = \begin{pmatrix} 1 & \alpha \\ \alpha & -1 \end{pmatrix}.$$

Then for every $\alpha \in \mathbb{C}$:

$$\mathbf{a} \bullet \mathbf{b}_\alpha = \begin{pmatrix} 1 & 0 \\ 0 & 1 \end{pmatrix} = \mathbf{b}_\alpha \bullet \mathbf{a}$$

and $\mathbf{b}_\alpha \in \mathbb{C}[\mathbf{a}]$ if and only if $\alpha = 0$. Hence there exist infinitely many $y \in A$ satisfying (6.41) with $x = \mathbf{a}$ fixed. In particular, $(W_2, \bullet, e)$ is not associative. The element $\mathbf{a}$ is invertible and $\mathbf{a}^{-1} = \mathbf{b}_0 = \mathbf{a}$.

The Laplace formulas provide an efficient way to determine the invertible elements of a power associative algebra and to calculate their inverse. The proof is now completely obvious since $\det(e) = 1$.

Corollary 6.5.18 *Let* $A = (V, *, e)$ *be a power associative $\mathbb{C}$-algebra of finite dimension n and of rank $r = \mathrm{rk}(A)$. Let $x \in A$. Then x is invertible if and only if $\det(x) \neq 0$.*

Moreover, in this case

$$x^{-1} = \frac{x^\#}{\det(x)}, \tag{6.42}$$

with $x^\#$ homogenous map of degree $r - 1$.

We can finally introduce the notion of birational involution associated to a power associative algebra.

Definition 6.5.19 (Birational Involution Associated to a Power Associative Algebra) Let $A = (V, *, e)$ be a power associative $\mathbb{C}$-algebra of finite dimension

$n + 1 \geq 3$ and of rank $r = \mathrm{rk}(\mathbf{A}) \geq 2$. Then the *birational involution associated to the power associative algebra with unity* $\mathbf{A}$ is the rational map

$$[\#_\mathbf{A}] : \mathbb{P}^n = \mathbb{P}(\mathbf{A}) \dashrightarrow \mathbb{P}(\mathbf{A}) = \mathbb{P}^n$$

sending $[\mathbf{x}]$ into $[\mathbf{x}^\#]$. Laplace's formula (6.40) implies that $[\#_\mathbf{A}]$ is birational involution of type $(r-1, r-1)$.

We now define another interesting class of birational involutions connected with the inversion map in (6.42).

Definition 6.5.20 (Birational Involutions of Spampinato Type) Let $\mathbf{A} = (V, *, \mathbf{e})$ be a power associative $\mathbb{C}$-algebra with unit of finite dimension $n + 1 \geq 3$ and of rank $r = \mathrm{rk}(\mathbf{A}) \geq 2$. The inverse map

$$\mathbf{x} \mapsto \mathbf{x}^{-1} = \frac{\mathbf{x}^\#}{\det(\mathbf{x})}$$

on $\mathbf{A}$ naturally induces a rational map

$$\widetilde{[\#_\mathbf{A}]} : \mathbb{P}(\mathbb{C} \times \mathbf{A}) \dashrightarrow \mathbb{P}(\mathbb{C} \times \mathbf{A}),$$

defined by

$$\widetilde{[\#_\mathbf{A}]}([(\mu, \mathbf{x})]) = [(\det(\mathbf{x}), \mu \mathbf{x}^\#)].$$

The formulas in Corollary 6.5.14 yield

$$\widetilde{[\#_\mathbf{A}]}^2([(\mu, \mathbf{x})]) = [(\mu \det(\mathbf{x}))^{r-1}(\mu, \mathbf{x})],$$

so that the map is a birational involution of bidegree (r, r)

Let $\widetilde{\mathbf{A}} = \mathbb{C} \times \mathbf{A}$ be the product of $\mathbb{C}$ and $\mathbf{A}$, which is a power associative algebra of dimension $n + 2$ with unity $\widetilde{\mathbf{e}} = (1, \mathbf{e})$. Then one has $(\mu, \mathbf{x})^\# = (\det(\mathbf{x}), \mu \mathbf{x}^\#)$ and $\det((\mu, \mathbf{x})) = \mu \det(\mathbf{x})$ so that the map $\widetilde{[\#_\mathbf{A}]}$ is associated to the power associative algebra $\widetilde{\mathbf{A}}$.

A Cremona transformation of bidegree (r, r) will be called *of Spampinato type* if it is linearly equivalent to the adjoint of a direct product $\mathbb{C} \times \mathbf{A}$ where $\mathbf{A}$ is a power associative algebra of rank r.

Birational involutions of Spampinato type were firstly investigated by N. Spampinato and C. Carbonaro Marletta, see [25, 26, 180], producing examples of interesting Cremona involutions in higher dimensional projective spaces. Elementary quadratic transformations introduced in Example 6.5.4 are linearly equivalent to a birational involution of Spampinato type, as we now verify.

Example 6.5.21 (Canonical Forms of Elementary Quadratic Transformations of $\mathbb{P}^n$*)*
Let the notation be as in Example 6.5.4 and in Example 6.5.8 part 2). If $Q \subset \mathbb{P}^{n+1}$
is an irreducible quadric hypersurface we can choose homogeneous projective
coordinates $(z_0 : \ldots : z_{n+1})$ on $\mathbb{P}^{n+1}$ such that $p_1 = (1 : 0 : \ldots : 0) \in Q$ with
$T_{p_1}Q$ the hyperplane of equation $z_{n+1} = 0$ and $p_2 = (0 : \ldots : 0 : 1) \in Q$ with $T_{p_2}Q$
the hyperplane of equation $z_0 = 0$.

Therefore the equation of Q has the form $z_0 z_{n+1} = \overline{q}(z_1, \ldots, z_n)$ with
$\overline{q}(z_1, \ldots, z_n)$ a quadratic form. Modulo another change of coordinates involving
only the variables $z_1, \ldots, z_n$, we can also suppose $\overline{q}(z_1, \ldots, z_n) = z_1^2 + q(z_2, \ldots, z_n)$,
$q(z_2, \ldots, z_n)$ a quadratic form, so that Q has equation

$$z_0 z_{n+1} = z_1^2 + q(z_2, \ldots, z_n).$$

Thus Q has an affine parametrization of the form $(1 : z_1 : \ldots : z_n : z_1^2 + q(z_2, \ldots, z_n))$. On the hyperplane H of equation $z_0 = 0$ we naturally have
homogeneous coordinates $(z_1 : \ldots : z_n : z_{n+1})$. Then H can be chosen as the target
of the projection from p_1, $\pi_{p_1} : Q \dashrightarrow \mathbb{P}^n$. Thus

$$\pi_{p_1}^{-1}(z_1 : \ldots : z_{n+1}) = (z_1^2 + q(z_2, \ldots, z_n) : -z_1 z_{n+1} : \ldots : -z_n z_{n+1} : -z_{n+1}^2) \in Q.$$

Finally, we get

$$\pi_{p_2} \circ \pi_{p_1}^{-1}(z_1 : \ldots : z_{n+1}) = (z_1^2 + q(z_2, \ldots, z_n) : -z_1 z_{n+1} : -z_2 z_{n+1} : \ldots : -z_n z_{n+1}),$$

where $\pi_{p_2} : Q \dashrightarrow \mathbb{P}^n$ is the projection onto the hyperplane of equation $z_{n+1} = 0$.
Therefore $\pi_{p_2} \circ \pi_{p_1}^{-1}$ is linearly equivalent to

$$\phi(z_1 : \ldots : z_{n+1}) = (z_1^2 + q(z_2, \ldots, z_n) : z_1 z_{n+1} : -z_2 z_{n+1} : - \ldots : -z_n z_{n+1}),$$

which is obviously linearly equivalent to the birational involution associated to the
algebra $\widetilde{\mathbf{A}}$ in Example 6.5.15. In particular, every elementary quadratic transformation is of Spampinato type.

The previous example has been worked out in detail in order to point out the
complexity of finding a canonical form of a birational involution, or better of
exhibiting inside the linear equivalence class of a birational involution a representa-
tive corresponding to inversion/cofactors in a power associative algebra.

Let us recall that a Cremona transformation $\phi : \mathbb{P}^n \dashrightarrow \mathbb{P}^n$ is called *semi-special*
if its base locus scheme is smooth and *special* if its base locus scheme is smooth and
irreducible. For the semi-special quadro-quadric Cremona transformation described
in Proposition 6.4.7 we have exhibited a corresponding power associative algebra
in its linear equivalence class: for 1) is the algebra defined in Example 6.5.21; for
cases 2),…, 5) the algebras defined in part 3) of Example 6.5.8. By a matter of fact
we remark that the last four examples are *simple* Jordan algebras while the first one
is *semi-simple* since in that case the corresponding q (and/or Q) has maximal rank.

It is not clear at all, even after this long introduction to the subject, that an arbitrary quadro-quadric Cremona transformation is linearly equivalent to a birational involution associated to a power associative algebra of rank three. This is indeed the case by the following recent result.

Theorem 6.5.22 ([156, Theorem 3.4]) *For every linear equivalence class of a quadro-quadric Cremona transformation* $\phi : \mathbb{P}^n \dashrightarrow \mathbb{P}^n$, *there exists a complex rank three Jordan algebra* $\mathbf{J}_\phi$ *of dimension* $n + 1$ *such that* $[\#_\mathbf{J}] = [\phi]$.

In particular, every quadro-quadric Cremona transformation is linearly equivalent to a birational involution associated to a rank three Jordan algebra.

For the proof we refer to the original paper which is now in press, available on the web and hopefully soon published.

Remark 6.5.23 A Jordan algebra of rank 1 is isomorphic to $\mathbb{C}$ (with the standard multiplicative product). It is a classical result that any rank 2 Jordan algebra is isomorphic to an algebra as in Example 6.5.8.(2). Jordan algebras of rank 3 have not yet been classified in arbitrary dimension.

Apparently we have restricted the class of power associative algebras needed to represent quadro-quadric birational Cremona transformations to the subclass of Jordan algebras. Since a commutative power associative algebra of rank 3 with unity is necessarily a Jordan algebra of the same rank by a recent result in [54, Corollary 13] Theorem 6.5.22 does not effectively restrict the class of algebras involved. In any case the theory of Jordan algebras is richer and more developed than that of power associative algebras and at various steps we have used different tools of Jordan theory to prove Theorem 6.5.22, see e.g. [131].

In the opposite direction, by applying Theorem 6.5.22 to the birational involution given by the adjoint of a commutative power-associative algebra of rank three with unity, one could deduce a new proof of [54, Corollary 13] mentioned above.

It is worth noting that the adjoint identity (6.40) can be used to characterize Jordan algebras of rank three under some non-degenerateness hypothesis of $\det(\mathbf{x})$, see [133] for details and precise statements/hypotheses.

Besides giving a different proof of Corollary 6.4.6, Theorem 6.5.22 is a very powerful instrument for classifying quadro-quadric Cremona transformations because rank three Jordan algebras of low dimension can be classified with methods coming from linear algebra (idempotent elements, projections and so on). This instrument has been successfully applied in [157] to obtain the classification of quadro-quadric Cremona transformations in $\mathbb{P}^n$ for $n = 3, 4, 5$. The case $n = 3$ was classically known to Enriques and Conforto, see [38], and it has been recently revisited in [149] with the usual methods of Algebraic Geometry. Our approach via the classification of Jordan algebras is extraordinary efficient for $n = 3$ and not so difficult for $n = 4, 5$. In the last two cases the complete classification was previously unknown, see [157] for a taste of the methods, for details and for related references.

6.6 The *XJC*-Correspondence

In Sect. 6.4 we proved that there is a one-to-one correspondence between the set of quadro-quadric Cremona transformations of $\mathbb{P}^{n-1}$, modulo linear equivalence, and the set of $\overline{X}^n(3,3) \subset \mathbb{P}^{2n+1}$, different from rational normal scrolls, modulo projective equivalence. We shall refer to the first set as the *C*-set, or *C*-world in the terminology of [156], and to the second one as the *X*-set, or *X*-world in the terminology of [156]. Theorem 6.5.22 proved that there is a one-to-one correspondence between the *C*-world and the *J*-world, the set of complex Jordan algebras of dimension n modulo *isotopy*, see [156] for the definition.

To each complex Jordan algebra **J** of rank three and dimension n we can associate a $\overline{X}^n(3,3) \subset \mathbb{P}^{2n+1}$, called the *twisted cubic over the cubic Jordan algebra* **J** and indicated by $X_{\mathbf{J}}$, which is nothing but $X_{[\#_{\mathbf{J}}]}$ in the notation of Sect. 6.4. Thus

$$X_{\mathbf{J}} \subset \mathbb{P}(\mathbb{C} \oplus \mathbf{J} \oplus \mathbf{J} \oplus \mathbb{C})$$

has parametrization

$$(1 : \mathbf{x}) \to (1 : \mathbf{x} : \mathbf{x}^{\#} : \det(\mathbf{x})).$$

We have thus constructed two maps: one from the *J*-set to the *X*-set and the other one from the *J*-set to the *C*-set. One could ask if these maps are one-to-one, establishing the so-called *XJC-correspondence* which essentially identifies the three different worlds. This fact is a consequence of Theorem 6.5.22 and of Theorem 6.4.5 which together imply the following.

Theorem 6.6.1 ([156, Theorem 3.4]) *Let* $X = \overline{X}^n(3,3) \subset \mathbb{P}^{2n+1}$, *not a rational normal scroll. Then there exists a complex rank three Jordan algebra* $\mathbb{J}$ *of dimension* n *such that* X *is projectively equivalent to* $X_{\mathbf{J}}$.

In conclusion the classes of extremal $X(3,3)$ and of quadro-quadric Cremona transformations are also in bijection with the isotopy classes of rank three complex Jordan algebras. This correspondence has been dubbed the *XJC*-correspondence in [156]. For example, it shows that smoothness of X corresponds to semi-simple Jordan algebras in the *J*-world and to semi-special quadro-quadric Cremona transformations in the *C*-world. The *XJC*-correspondence has many applications and allowed us to introduce the notion of simplicity and semi-simplicity for quadro-quadric Cremona transformations and for $X^{r+1}(3,3)$, see [155].

The reader may wish to consult [155–157] for more details on these and other applications.

Chapter 7
Hypersurfaces with Vanishing Hessian

Hypersurfaces with vanishing hessian were studied systematically for the first time in the fundamental paper [78], where Gordan and Noether analyze Hesse's claims in [90, 91] according to which these hypersurfaces should be necessarily cones.

If $X = V(f) \subset \mathbb{P}^N$ is a reduced complex hypersurface, the hessian of f (or by abusing the terminology a little, the hessian of X), indicated by hess_X, is the determinant of the matrix of the second derivatives of the form f, that is the determinant of the so-called hessian matrix of f, see Sect. 7.1 for precise definitions and notation.

Of course, cones have vanishing hessian and Hesse claimed twice in [90] and in [91] that a hypersurface $X \subset \mathbb{P}^N$ is a cone if $\mathrm{hess}_X = 0$. Clearly the claim is true if $\deg(f) = 2$ so that the first relevant case for the problem is that of cubic hypersurfaces. One immediately sees that $V(x_0 x_3^2 + x_1 x_3 x_4 + x_2 x_4^2) \subset \mathbb{P}^4$ is a cubic hypersurface with vanishing hessian but not a cone (for example, because the first partial derivatives of the equation are linearly independent).

Actually the question is quite subtle because, as was first pointed out in [78], the claim is true for $N \leq 3$ and in general false for $N \geq 4$. The cases $N = 1, 2$ are easily handled but beginning from $N = 3$ the question is related to non-trivial characterizations of cones among developable hypersurfaces or, from a differential point of view, to a characterization of algebraic cones (or of algebraic cylinders in the affine setting) among hypersurfaces with zero gaussian curvature at every regular point, see Sect. 7.2.2.

Gordan and Noether's approach to the problem and their proofs for the classification for $N \leq 4$ have been revisited recently in modern terms in [127] for $N = 3$ and in [72] for $N = 4$, see also [152, 153]. We shall present here a complete and detailed account of Gordan–Noether Theory as well its wide range of applications in different areas of mathematics including differential geometry, commutative algebra (see Sect. 7.2.4) and the theory of PDE.

© Springer International Publishing Switzerland 2016
F. Russo, *On the Geometry of Some Special Projective Varieties*,
Lecture Notes of the Unione Matematica Italiana 18,
DOI 10.1007/978-3-319-26765-4_7

In [78] is constructed a series of projective hypersurfaces in $\mathbb{P}^N$ for every $N \geq 4$, which generically are not cones and to which the explicit example recalled above belongs, see also [152, 153] and above all [37, Sect. 2]. Moreover, Gordan and Noether also classified all hypersurfaces with vanishing hessian for $N \leq 4$, proving that they are either cones or $N = 4$ and the hypersurfaces belong to their series of examples, see *loc. cit.*, [62, 72]. The work of Perazzo considered the classification of cubic hypersurfaces with vanishing hessian for $N \leq 6$, see [151]. Perazzo's ideas and techniques are very interesting and deeply inspired the paper [77], which filled in some gaps contained in [151] and which contains the classification of cubic hypersurfaces in $\mathbb{P}^N$ for $N \leq 6$, see Sect. 7.6.

As far as we know, no explicit classification result is known for $N \geq 5$ and for degree greater than three, leaving room for further research on the subject.

Hypersurfaces with vanishing hessian have been forgotten for a long time despite their appearance in various problems of a different nature. For example, the cubic hypersurface recalled above, very well known to classical algebraic geometers, is celebrated in the modern algebraic-differential geometry literature as the *Bourgain–Sacksteder Hypersurface* (see [2, 3, 60]). Moreover, the regular points of a hypersurface with vanishing hessian are all *parabolic* and represent a natural generalization of the flex points of plane curves (see [31, 37]). The divisibility properties of the hessian with respect to the original form have interesting geometric consequences, see [31, 37, 170] and Sect. 7.2.2 here; for instance *developable* hypersurfaces are those for which $\text{hess}_X = 0 \pmod{f}$, see [3, 60, 170] and Sect. 7.2.2.

Many classes of hypersurfaces with vanishing hessian, not cones, are somehow surprising since they are ruled by a family of linear spaces along which the hypersurface is not developable, so that this ruling is different from the one given by the fibers of the Gauss map. These examples and their generalizations are known in differential geometry as *twisted planes*, see for example [60]. Despite the huge number of papers dedicated to this subject by differential geometers, very few classification or structure results have been obtained. In our opinion the global point of view provided by polarity, which has been overlooked until now by differential geometers, is a stronger tool to treat these objects, see Sect. 7.2.1.

In conclusion, several open problems spanning from the classification of hypersurfaces with vanishing hessian in $\mathbb{P}^5$ to the construction of new series of examples seem to deserve some attention, also for their connections with other areas and/or for the recent applications in other fields, see e.g. Sect. 7.2.4.

7.1 Preliminaries, Definitions, Statement of the Problem and of the Classical Results

Let $f(x_0, \ldots, x_N) \in \mathbb{K}[x_0, \ldots, x_N]_d$, $d \geq 1$, $\mathbb{K}$ an algebraically closed field of characteristic zero and f without multiple irreducible factors. Let $X = V(f) \subset \mathbb{P}^N$ be the associated degree d projective hypersurface. Let

$$\text{Hess}_X = \text{H}(f) = \left[\frac{\partial^2 f}{\partial x_i \partial x_j} \right]_{0 \le i,j \le N},$$

be the *hessian matrix of X* (or of f, in this case indicated by $H(f)$).

Clearly $\text{H}(f) = \mathbf{0}_{(N+1)\times(N+1)}$ if and only if $d = 1$. Thus, from now on, we shall suppose $d \ge 2$. Let

$$\text{hess}_X = \det(\text{Hess}_X)$$

be the **hessian (determinant)** of X (or of f, in this case indicated by $h(f)$).

There are two possibilities:

1. either $\text{hess}_X = 0$ or
2. $\text{hess}_X \in \mathbb{C}[x_0, \dots, x_N]_{(N+1)(d-2)}$.

We shall be interested in case (1), that is in *hypersurfaces with vanishing hessian (determinant)*.

Remark 7.1.1 Let $\widehat{A} \in GL_{N+1}(\mathbb{C})$, let $\widehat{\mathbf{x}} = \widehat{A}\mathbf{x}$ and let $\widehat{f}(\mathbf{x}) = f(\widehat{A}\mathbf{x})$. We shall say that $\widehat{f}$ is *linearly equivalent to* f, indicated by $\widehat{f} \sim f$.

Let us state some immediate consequences of the definitions, whose proof is left to the reader.

1. If $\frac{\partial f}{\partial x_i} = 0$ for some i, then $h(f) = 0$.
2. $h(\widehat{f}) = 0$ if and only if $h(f) = 0$.
3. If $\frac{\partial f}{\partial x_0}, \dots, \frac{\partial f}{\partial x_N}$ are linearly dependent, then $h(f) = 0$.

The next result is also well known.

Proposition 7.1.2 (Characterizations of Cones) *Let* $X = V(f) \subset \mathbb{P}^N$ *be a hypersurface,* $d = \deg(X) \ge 2$. *Then the following conditions are equivalent:*

(i) X *is a cone;*
(ii) *there exists a point* $p \in X$ *of multiplicity* d;
(iii) *the partial derivatives* $\frac{\partial f}{\partial x_0}, \frac{\partial f}{\partial x_1}, \dots, \frac{\partial f}{\partial x_N}$ *of* f *are linearly dependent;*
(iv) *up to a projective transformation,* f *depends on at most N variables;*
(v) *the dual variety of* X, $X^* \subset (\mathbb{P}^N)^*$, *is degenerate, i.e.* $< X^* > \subsetneq \mathbb{P}^{N*}$.

We are now in position to state Hesse's Claim.

Claim (Hesse's Claim, 1851 and 1859, [90, 91])

$$\text{hess}_X = 0 \implies X = V(f) \subset \mathbb{P}^N \text{ is a cone.}$$

Or equivalently

$$h(f) = 0 \implies \frac{\partial f}{\partial x_0}, \dots, \frac{\partial f}{\partial x_N} \text{ are linearly dependent.}$$

Remark 7.1.3 We now collect some other easy consequences of the definitions and point out some well-known cases where Hesse's Claim is true.

- For $N = 1$ first year calculus yields Hesse's Claim.
- Suppose $N \geq 2$, $d = 2$ and let $B \in M_{(N+1)\times(N+1)}(\mathbb{K})$, $B = B^t$. Then

$$f(\mathbf{x}) = \mathbf{x}^t \cdot B \cdot \mathbf{x} \sim x_0^2 + \ldots + x_r^2$$

with $r + 1 = \mathrm{rk}(B)$ and with $h(f) = \det(B)$. Thus for $d = 2$, $X = V(f) \subset \mathbb{P}^N$ is a cone if and only if $\mathrm{hess}_X = 0$.

Thus from now on we shall suppose $N \geq 2$ and $d \geq 3$. We now state the main contributions of Gordan and Noether in [78]. We shall prove each of these statements in the following sections as different applications of the Gordan–Noether Identity, the main tool developed in [78] to derive significant geometrical consequences from the vanishing of the hessian determinant.

Theorem 7.1.4 ([78]) *Let $X = V(f) \subset \mathbb{P}^N$ be a hypersurface with vanishing hessian with $d = \deg(X) \geq 3$. Then:*

1. *If $N \leq 3$, Hesse's Claim is true.*
2. *For every $N \geq 4$ and for every $d \geq 3$ there exist counterexamples to Hesse's Claim.*
3. *For $N = 4$ all examples, not cones, are classified and belong to a unique class.*

Example 7.1.5 ("Un Esempio Semplicissimo", Perazzo [151]) Let $N \geq 4$ and let

$$f(x_0, x_1, x_2, x_3, x_4, \ldots, x_N) = x_0 x_3^2 + x_1 x_3 x_4 + x_2 x_4^2 + x_5^3 + \ldots + x_N^3.$$

This example was considered by Perazzo around 1900 in [151] and later, for $N = 4$, by differential geometers, who usually call it the *Bourgain (Sacksteder) twisted plane*, see for example [2, 3, 60].
We have

$$\frac{\partial f}{\partial x_0} = x_3^2, \ \frac{\partial f}{\partial x_1} = x_3 x_4, \ \frac{\partial f}{\partial x_2} = x_4^2, \ \frac{\partial f}{\partial x_3} = 2x_0 x_3 + x_1 x_4,$$

$$\frac{\partial f}{\partial x_4} = x_1 x_3 + 2x_2 x_4, \ \frac{\partial f}{\partial x_i} = 3x_i^2 \ \forall i = 5, \ldots, N.$$

Thus the partial derivatives of f are linearly independent but algebraically dependent. Indeed,

$$\frac{\partial f}{\partial x_0} \frac{\partial f}{\partial x_2} - \left(\frac{\partial f}{\partial x_1}\right)^2 = x_3^2 x_4^2 - (x_3 x_4)^2 = 0$$

and this last property implies $h(f) = 0$, a fact proved below in Sect. 7.2.1.

Although Hesse's Claim is false, Gordan–Noether pointed out that it is *birationally true* notwithstanding the condition $\text{hess}_X = 0$ is not invariant under birational transformations, as we shall see in Example 7.3.13. Indeed, Gordan–Noether proved the following notable result as an application of their theory. We shall reprise and prove it later in Theorem 7.3.11.

Theorem 7.1.6 ([78]) *Let $X = V(f) \subset \mathbb{P}^N$ be a hypersurface of degree $d \geq 2$ and with $\text{hess}_X = 0$. Then there exists a Cremona transformation*

$$\Phi : \mathbb{P}^N \dashrightarrow \mathbb{P}^N$$

such that $\Phi(X)$ is a cone.

Moreover, the Cremona transformation can be explicitly constructed from f in such a way that the equation of $\phi(X)$ depends on at most N variables.

For example, the Perazzo hypersurface

$$X = V(x_0 x_3^2 + x_1 x_3 x_4 + x_2 x_4^2) \subset \mathbb{P}^4$$

is Cremona equivalent to

$$V(x_1 x_3 + x_2 x_4) \subset \mathbb{P}^4,$$

which is clearly a cone with vertex the point $(1 : 0 : 0 : 0 : 0)$, see Example 7.3.12, and also Cremona equivalent to

$$V(x_0 x_3^2 + x_2 x_4^2) \subset \mathbb{P}^4,$$

a cone with vertex the point $(0 : 1 : 0 : 0 : 0)$.

Before developing Gordan–Noether Theory we shall survey the appearance of Hesse's Claim and of the Gordan–Noether Theorem in several different settings.

7.2 Instances and Relevance of Hesse's Claim in Geometry and in Commutative Algebra

7.2.1 The Polar Map

Let

$$\nabla_f = \nabla_X : \mathbb{P}^N \dashrightarrow \mathbb{P}^{N*}$$

be *the polar (or gradient) map of $X = V(f) \subset \mathbb{P}^N$*, defined by

$$\nabla_f(p) = (\frac{\partial f}{\partial x_0}(p) : \ldots : \frac{\partial f}{\partial x_N}(p)).$$

The base locus scheme of ∇_X is the scheme

$$\mathrm{Sing}(X) := V(\frac{\partial f}{\partial x_0}, \dots, \frac{\partial f}{\partial x_N}) \subset \mathbb{P}^N.$$

Let

$$Z := \overline{\nabla_f(\mathbb{P}^N)} \subseteq \mathbb{P}^{N*}$$

be *the polar image of* $\mathbb{P}^N$

The restriction of ∇_f to X is the Gauss map of X:

$$\mathcal{G}_X = \nabla_{f|X} : \quad X \quad \dashrightarrow \quad \mathbb{P}^{N*}$$
$$p \in X_{\mathrm{reg}} \;\rightarrow\; \mathcal{G}_X(p) = [T_p X]$$

so that

$$X^* := \overline{\mathcal{G}_X(X)} \subset Z$$

is the dual variety of X.

We can consider the rational map ∇_f as the quotient by the natural $\mathbb{K}^*$-action of the affine morphism

$$\nabla_f : \mathbb{K}^{N+1} \to \mathbb{K}^{N+1}$$

defined in the same way. From this perspective we have the following Key Formula:

$$\mathrm{Hess}(f) = \mathrm{Jac}(\nabla_f : \mathbb{K}^{N+1} \to \mathbb{K}^{N+1}), \tag{7.1}$$

that is, the Hessian matrix of f is the Jacobian matrix of the affine morphism ∇_f.

Moreover, by generic smoothness applied to the rational map ∇_f we get from (7.1)

$$T_{\nabla_f(p)} Z = \mathbb{P}(\mathrm{Im}(\mathrm{H}(f)(p))) \subseteq \mathbb{P}^{N*} \quad \forall p \in \mathbb{P}^N \text{ general,} \tag{7.2}$$

yielding

$$\dim Z = \mathrm{rk}(\mathrm{H}(f)) - 1. \tag{7.3}$$

Therefore $\mathrm{h}(f) = 0$ if and only if $Z \subsetneq \mathbb{P}^{N*}$. Thus $\mathrm{h}(f) = 0$ if and only if

$$\exists g \in \mathbb{K}[y_0, \dots, y_N] \setminus 0 \;:\; g(\frac{\partial f}{\partial x_0}, \dots, \frac{\partial f}{\partial x_N}) = 0,$$

which in turn happens if and only if $\frac{\partial f}{\partial x_0}, \ldots, \frac{\partial f}{\partial x_N}$ are algebraically dependent. We have thus reinterpreted Hesse's Claim in the following way.

Claim (Hesse's Claim Revisited)

$\frac{\partial f}{\partial x_0}, \ldots, \frac{\partial f}{\partial x_N}$ algebraically dependent $\implies \frac{\partial f}{\partial x_0}, \ldots, \frac{\partial f}{\partial x_N}$ linearly dependent.

Put in this way one begins to believe that the Claim might also be false in low dimension. Indeed, in this case the validity of Hesse's Claim has very interesting interpretations which we shall now derive. Before this we recall a lemma relating the polar map of a hypersurface with the polar map of a general hyperplane section, whose proof is left to the reader.

Lemma 7.2.1 ([37, Lemma 3.10]) *Let* $X = V(f) \subset \mathbb{P}^N$ *be a hypersurface. Let* $H = \mathbb{P}^{N-1}$ *be a hyperplane not contained in* X, *let* $h = H^*$ *be the corresponding point in* $\mathbb{P}^{N*}$ *and let* π_h *denote the projection from the point* h. *Then:*

$$\nabla_{X \cap H} = \pi_h \circ (\nabla_{X|H}).$$

In particular, $\overline{\nabla_{X \cap H}(H)} \subseteq \pi_h(Z)$.

7.2.2 Curvature and h(f)

The next result is well known and is a straightforward consequence of Euler's Formula.

Lemma 7.2.2 *Let* $f \in \mathbb{K}[x_0, \ldots, x_N]_d$. *Then*

$$x_0^2 \, h(f) = (d-1)^2 \begin{vmatrix} \frac{d}{d-1}f & \frac{\partial f}{\partial x_1} & \cdots & \frac{\partial f}{\partial x_N} \\ \frac{\partial f}{\partial x_1} & \frac{\partial^2 f}{\partial x_1^2} & \cdots & \frac{\partial^2 f}{\partial x_1 \partial x_N} \\ \vdots & \vdots & \ddots & \vdots \\ \frac{\partial f}{\partial x_N} & \frac{\partial^2 f}{\partial x_1 \partial x_N} & \cdots & \frac{\partial^2 f}{\partial x_N^2} \end{vmatrix} \tag{7.4}$$

When dealing with Gaussian curvature we shall implicitly assume $\mathbb{K} = \mathbb{R}$ (or sometimes $\mathbb{K} = \mathbb{C}$.) The next formula was first stated for $N = 3$ by Gauss, who calculated it explicitly without expressing it in the determinant form (for obvious reasons).

Proposition 7.2.3 (Gauss Curvature Formula, [73, Sect. 9] and [171, p. 13]) *Let* $X = V(f) \subset \mathbb{P}^N$ *be a hypersurface of degree* $d \geq 1$, *let* $p \in X \setminus \mathrm{Sing}(X)$, $p \in$

$\mathbb{P}^N \setminus V(x_0) = \mathbb{A}_0^N$, and let $K(p)$ be the Gaussian curvature of X. Then

$$K(p) = -\frac{\begin{vmatrix} 0 & \frac{\partial f}{\partial x_1}(p) & \cdots & \frac{\partial f}{\partial x_N}(p) \\ \frac{\partial f}{\partial x_1}(p) & \frac{\partial^2 f}{\partial x_1^2}(p) & \cdots & \frac{\partial^2 f}{\partial x_1 \partial x_N}(p) \\ \vdots & \vdots & \ddots & \vdots \\ \frac{\partial f}{\partial x_N}(p) & \frac{\partial^2 f}{\partial x_1 \partial x_N}(p) & \cdots & \frac{\partial^2 f}{\partial x_N^2}(p) \end{vmatrix}}{\left((\frac{\partial f}{\partial x_1}(p))^2 + \ldots + (\frac{\partial f}{\partial x_N}(p))^2\right)^{\frac{N+1}{2}}}. \tag{7.5}$$

In particular, $K(p) = 0$ for every p as in the statement if and only if $h(f) = 0$ (mod. f), *that is, if and only if f divides* $h(f)$.

Example 7.2.4 We shall reinterpret the previous results for small N.
Let $N = 2$, let $C = V(f) \subset \mathbb{P}^2$ be a reducible curve of degree d and let

$$H(C) := V(h(f)) \subset \mathbb{P}^2$$

be the *Hessian curve of C*. Let us observe that $H(C) = \mathbb{P}^2$ if and only if $h(f) = 0$.

Then f divides $h(f)$ if and only if $C \subseteq H(C)$, that is, if and only if every $p \in C \setminus \mathrm{Sing}(C)$ is a flex, which happens if and only if C is a union of lines.

Thus also in the simplest case $N = 2$ a curve C for which f divides $h(f)$ is not necessarily a cone since the irreducible components of C might not pass through a fixed point. Here is the simplest possible example. For $f(x_0, x_1, x_2) = x_0 x_1 x_2$, we have $h(f) = 2x_0 x_1 x_2 \neq 0$ and $h(f) = 0 \pmod{f}$.

Theorem 7.1.4 indeed assures us that C is a cone if and only if $h(f) = 0$, which is a stronger condition.

To appreciate the depth of the Gordan–Noether Theorem 7.1.4 for $N = 3$ we recall a more or less well-known result in the theory of differentiable surfaces in $\mathbb{R}^3$, see for example [60, Sects. 0.4 and 0.5].

Theorem 7.2.5 (Characterization of Developable Algebraic Surfaces) *Let $S = V(f) \subset \mathbb{P}^3$ be an irreducible surface. Then the following conditions are equivalent:*

1. $K(p) = 0 \ \forall p \in S_{\mathrm{reg}}$, *that is, f divides $h(f)$;*
2. *S is a cone or S is the developable surface of tangent lines to an irreducible curve $C \subset \mathbb{P}^3$*

In differential geometry surfaces as in (2) are usually called *developable*, which simply means that $\overline{\mathscr{G}_S(S)}$ is a curve (recall that $\mathscr{G}_S(S)$ is a point if and only if S is a plane). Theorem 7.1.4 is an interesting and stronger refinement, which as far as we know does not appear in any modern text of differential (projective) geometry. We restate Theorem 7.1.4 in this particular case.

Theorem 7.2.6 (The Gordan–Noether Theorem in $\mathbb{P}^3$) *Let $S = V(f) \subset \mathbb{P}^3$ be an irreducible surface. Then S is a cone if and only if $\mathrm{h}(f) = 0$.*

The proof is postponed until Sect. 7.3.1 after the development of Gordan–Noether Theory in the following subsections.

7.2.3 What Does the Condition f Divides h(f) Measure?

Let

$$\mathscr{G}_X : X \dashrightarrow X^* \subseteq Z \subseteq \mathbb{P}^{N*}$$

be the Gauss map of the reducible hypersurface $X = V(f) \subset \mathbb{P}^N$.

If A is a matrix with entries in $K[x_0, \ldots, x_N]$ and if $f \in K[x_0, \ldots, x_N]$, then $\mathrm{rk}_{(f)} A$ denotes the *rank of A modulo (f)*, that is, the maximal order of a minor not belonging to the ideal generated by f. With this notation, obviously $\mathrm{rk}\, A = \mathrm{rk}_{(0)}\, A$.

Lemma 7.2.7 ([170, Theorem 2]) *Let $X = V(f) = X_1 \cup \ldots \cup X_r \subset \mathbb{P}^N$, with $X_i = V(f_i), f = f_1 \cdots f_r$ and $p_i \in X_i$ general. Then*

$$\mathrm{rk}(\mathrm{d}\,\mathscr{G}_X)_{p_i} = \mathrm{rk}_{(f_i)} H(f) - 2. \tag{7.6}$$

In particular,

$$\dim(X_i^*) = \mathrm{rk}(\mathrm{d}\,\mathscr{G}_X)_{p_i} = \mathrm{rk}_{(f_i)} H(f) - 2 \le \mathrm{rk}(H(f)) - 2 \le \dim(Z) - 1.$$

Proof See also [170, Theorem 2] and [60, Sect. 0.3]. The simple argument presented below is due to J. Kollár. Let $\widehat{X} = V(f) \subset \mathbb{C}^{N+1}$ be the affine cone over X and let $[\mathbf{v}_i] = p_i$. Then $\mathbf{w} \in T_{\mathbf{v}_i}\widehat{X}$ if and only if $\nabla_f(\mathbf{v}_i)^t \cdot \mathbf{w} = 0$. From Euler's Formula we get

$$\mathbf{v}_i^t \cdot (H(f)(\mathbf{v}_i)) = (d-1)\nabla_f(\mathbf{v}_i)$$

so that

$$\mathbf{v}_i^t \cdot (H(f)(\mathbf{v}_i)) \cdot \mathbf{w} = (d-1)\nabla_f(\mathbf{v}_i)^t \cdot \mathbf{w} = 0.$$

Let

$$(\mathrm{d}\,\mathscr{G}_X)_{\mathbf{v}_i} : T_{\mathbf{v}_i}\widehat{X} \to \mathbb{C}^{N+1}.$$

Then

$$(d\mathscr{G}_X)_{\mathbf{v}_i}(T_{\mathbf{v}_i}\widehat{X}) \subseteq \nabla_f(\mathbf{v}_i)^t \cdot \mathbf{x} = 0$$

and the conclusion follows from the following elementary fact: if $L : V \to W$ is a linear map, if $V_1 \subset V$ and if $W_1 \subset W$ are linear subspaces with $\mathrm{codim}(V_1) = \mathrm{codim}(W_1) = 1$, such that $L(V) \not\subset W_1$ and such that $L(V_1) \subseteq W_1$, then $\mathrm{rk}(L_{|V_1}) = \mathrm{rk}(L) - 1$. $\square$

We point out an immediate consequence for further reference.

Corollary 7.2.8 *Let $X = V(f) \subset \mathbb{P}^N$ be a hypersurface with vanishing hessian. Then*

$$X^* \subsetneq Z \subsetneq (\mathbb{P}^N)^*.$$

7.2.4   Weak and Strong Lefschetz Properties for Standard Artinian Gorenstein Graded Algebras

We now present an unexpected appearance of hypersurfaces with vanishing hessian as exceptions to the ordinary multiplication law in algebras having a structure similar to the even part of the cohomology ring of a smooth complex projective manifold.

Let $X \subset \mathbb{P}^N_{\mathbb{C}}$ be a smooth irreducible complex projective variety of dimension $n = \dim(X)$ endowed with the euclidean topology.

The cohomology groups with coefficients in the field $\mathbb{C}$ will be indicated by

$$H^i(X) := H^i(X; \mathbb{C}).$$

As is well known, $\dim_{\mathbb{C}}(H^i(X)) < \infty$ for every $i \geq 0$ and $H^i(X) = \mathbf{0}$ for $i > 2n$.

Let $[H] \in H^2(X)$ be the class of a hyperplane section of X. For every integer $k \geq 1$ the cap product defines a $\mathbb{C}$-linear map:

$$\bullet[H]^k : H^i(X) \to H^{i+2k}(X)$$
$$[Y] \to [H]^k \cap [Y]$$

We recall the following fundamental result of S. Lefschetz.

Theorem 7.2.9 (Hard Lefschetz Theorem) *Let $X \subset \mathbb{P}^N_{\mathbb{C}}$ be a smooth irreducible complex projective variety of dimension $n \geq 1$. Then $\forall q = 1, \ldots, n$*

$$\bullet[H]^q : H^{n-q}(X) \to H^{n+q}(X) \tag{7.7}$$

is an isomorphism.

The following consequence of the Hard Lefschetz Theorem inspired the algebraic notions we shall introduce in a moment.

Corollary 7.2.10 *Let the notation be as above. Then:*

$$\bullet[H]^k : H^i(X) \to H^{i+2k}(X) \tag{7.8}$$

is injective for $i \leq n - k$ and surjective for $i \geq n - k$.

Proof Suppose $i \leq n - k$. The composition

$$H^i(X) \xrightarrow{\bullet[H]^k} H^{i+2k}(X) \xrightarrow{\bullet[H]^{n-k-i}} H^{2n-i}(X)$$

is an isomorphism by Theorem 7.2.9, so that the first map is injective.

Analogously if $i \geq n - k$, the composition

$$H^{2n-i-2k}(X) \xrightarrow{\bullet[H]^{i-n+k}} H^i(X) \xrightarrow{\bullet[H]^k} H^{i+2k}(X)$$

is an isomorphism by Theorem 7.2.9, so that the second map is surjective. $\square$

The Poincaré Duality Theorem assures us that

$$H^i(X) \simeq (H^{2n-i}(X))^*$$

for every $i = 0, \ldots, n$, the isomorphism being induced by the cap product (recall that $H^{2n}(X) \simeq \mathbb{C}$). Some notable classes of finite-dimensional algebras are characterized by this property, which we now formalize.

Definition 7.2.11 Let $\mathbb{K}$ be a field and let

$$A = \bigoplus_{i=0}^{d} A_i$$

be an Artinian associative and commutative graded $\mathbb{K}$-algebra with $A_0 = \mathbb{K}$ and $A_d \neq 0$. The integer d is usually called the *socle degree of A*. Let

$$\bullet : A_i \times A_{d-i} \to A_d$$
$$(\alpha, \beta) \to \alpha \bullet \beta$$

be the restriction of the multiplication in A.

We say that *A satisfies the Poincaré Duality Property* if:

(i) $\dim_{\mathbb{K}}(A_d) = 1$;
(ii) $\bullet : A_i \times A_{d-i} \to A_d \simeq \mathbb{K}$ is non-degenerate for every $i = 0, \ldots, [\frac{d}{2}]$.

The algebra A is said to be *standard* if

$$A \simeq \frac{\mathbb{K}[x_0, \dots, x_N]}{I},$$

as graded algebras, with $I \subset \mathbb{K}[x_0, \dots, x_N]$ a homogeneous ideal. This condition implies $\sqrt{I} = (x_0, \dots, x_N)$ because $(x_0, \dots, x_N)^r \subseteq I$ for $r \geq d + 1$.

To each Artinian graded $\mathbb{K}$ algebra $A = \oplus_{i=0}^d A_i$ as above, letting $a_i = \dim_{\mathbb{K}} A_i$, we can associate its **h**-*vector* (or *Hilbert vector*)

$$\mathbf{h} = (1, a_1, \dots, a_d).$$

For algebras satisfying the Poincaré Duality Property we have $a_d = 1$ and $a_{d-i} = a_i$ for every $i = 1, \dots, [\frac{d}{2}]$.

The **h**-vector of A is said to be *unimodal* if there exists an integer $t \geq 1$ such that

$$1 \leq a_1 \leq \dots \leq a_t \geq a_{t+1} \geq \dots \geq a_{d-1} \geq 1.$$

The notion of the Poincaré Duality Property was inspired by the observation that the even part of the cohomology ring with coefficient in a characteristic zero field of a compact orientable manifold of even real dimension satisfies the previous property due to Poincaré Duality Theorem.

We now introduce the definition of a Gorenstein ring which, despite its apparent abstractness, is a property shared by the even part of the cohomology ring of compact orientable manifolds.

Definition 7.2.12 Let $(R, m, \mathbb{K})$ be a local ring. Then R is called a *local Gorenstein ring* if it has finite injective dimension as an R-module. A commutative ring R is called *Gorenstein* if the localization at each prime ideal is a local Gorenstein ring.

The following characterization of Artinian graded Gorenstein algebras shows the connection between the algebraic notions of Gorestein ring and Poincaré Duality.

Proposition 7.2.13 ([74], [128, Proposition 2.1]) *Let A be a graded Artinian $\mathbb{K}$-algebra. Then A satisfies the Poincaré Duality Property if and only if it is Gorenstein.*

Example 7.2.14 Let

$$Q = \mathbb{K}[\frac{\partial}{\partial x_0}, \dots, \frac{\partial}{\partial x_N}]$$

and let $F(\mathbf{x}) \in \mathbb{K}[x_0, \dots, x_N]_d$ be a homogeneous polynomial of degree $d \geq 1$. Then for every $G \in Q$ we shall indicate by $G(F) \in \mathbb{K}[x_0, \dots, x_N]$ the polynomial obtained by applying the differential operator G to the polynomial F. Define

$$\text{Ann}_Q(F) = \{G \in Q \ : \ G(F) = 0\} \subset Q.$$

Then $\mathrm{Ann}_Q(F) \subset Q$ is a homogenous ideal and

$$A = \frac{Q}{\mathrm{Ann}_Q(F)}$$

is a standard Artinian Gorenstein graded $\mathbb{K}$-algebra with $A_i = 0$ for $i > d$ and $A_d \neq 0$.

The Theory of Inverse Systems developed by Macaulay yields a nice characterization of standard Artinian Gorenstein graded $\mathbb{K}$-algebras, which is surely well known to the experts in the field. A short proof of a slightly more general result can be found in [128, Theorem 2.1].

Theorem 7.2.15 *Let*

$$A = \bigoplus_{i=0}^{d} A_i \simeq \frac{\mathbb{K}[x_0, \ldots, x_N]}{I}$$

be an Artinian standard graded $\mathbb{K}$-algebra. Then A is Gorenstein if and only if there exists an $F \in \mathbb{K}[x_0, \ldots, x_N]_d$ such that $A \simeq Q / \mathrm{Ann}_Q(F)$.

We now come to the definition of Lefschetz Properties, originally developed by R. Stanley in [183], see also [83] for an expanded treatment. In the sequel we shall strictly follow the presentation in [192] and in [128].

Definition 7.2.16 Let $\mathbb{K}$ be a field and let

$$A = \bigoplus_{i=0}^{d} A_i$$

be an Artinian associative and commutative graded $\mathbb{K}$-algebra with $A_d \neq 0$.

The algebra A is said to have *the Strong Lefschetz Property*, briefly *SLP*, if there exists an element $L \in A_1$ such that the multiplication map

$$\bullet L^k : A_i \to A_{i+k}$$

is of maximal rank, that is injective or surjective, $\forall\, 0 \leq i \leq d$ and $\forall\, 0 \leq k \leq d - i$.

An element $L \in A_1$ satisfying the previous property will be called *a strong Lefschetz element of A*.

The algebra A is said to have *the Weak Lefschetz Property*, briefly *WLP*, if there exists an element $L \in A_1$ such that the multiplication map

$$\bullet L : A_i \to A_{i+1}$$

is of maximal rank, that is injective or surjective, $\forall\, 0 \leq i \leq d - 1$.

An element $L \in A_1$ satisfying the last property will be called *a Lefschetz element of A*.

A is said to have *the Strong Lefschetz Property in the narrow sense* if there exists an element $L \in A_1$ such that the multiplication map

$$\bullet L^{d-2i} : A_i \to A_{d-i}$$

is an isomorphism $\forall\ i = 0, \ldots, [\frac{d}{2}]$.

Remark 7.2.17 Since we shall always deal with infinite fields (more precisely algebraically closed fields of characteristic zero), if there exists a Lefschetz element or a strong Lefschetz element, then the general element of A_1, in the sense of the Zariski topology, shares the same property.

If A satisfies the *WLP*, then the **h**-vector of A is unimodal. The converse is not true as shown by simple examples, see [128]. Moreover, there are Artinian Gorenstein algebras whose **h**-vector is not unimodal.

If a graded Artinian $\mathbb{K}$-algebra A satisfies the *SLP* in the narrow sense, then the **h**-vector of A is unimodal and symmetric, that is $a_i = a_{d-i}$ (see the proof of Corollary 7.2.10).

Conversely, for a graded Artinian $\mathbb{K}$-algebra having a symmetric **h**-vector, the notion of *SLP* and *SLP in the narrow sense* coincide. This is the case for Gorenstein Artinian graded $\mathbb{K}$-algebras on which we shall mainly focus.

In a moment we shall see examples of Artinian Gorenstein graded algebras having unimodal **h**-vector but not satisfying the *SLP* (or the *WLP*), whose construction relies on the existence of a homogeneous polynomial with vanishing hessian determinant depending on all the variables modulo linear changes of coordinates.

Definition 7.2.18 Let

$$A = \frac{Q}{\mathrm{Ann}_Q(F)}$$

be a standard Artinian Gorenstein graded $\mathbb{K}$-algebra with $A_i = 0$ for $i > d$ and $A_d \neq 0$, $F \in \mathbb{K}[x_0, \ldots, x_N]_d$.

Without loss of generality we can assume $(\mathrm{Ann}_Q(F))_1 = 0$, that is, that there does not exist a linear change of coordinates such that F does not depend on all the variables; equivalently the partial derivatives of F are linearly independent. Under this hypothesis, which we shall assume from now on, $\{x_0, \ldots, x_N\}$ is a basis of A_1, that is, $a_1 = N + 1 = a_{d-1}$ (recall that $a_i = \dim(A_i)$). Let us define

$$\mathrm{hess}^{(1)} F = \det(H(F)),$$

that is, $\mathrm{hess}^{(1)} F = \mathrm{h}(F)$.

Let $2 \leq i \leq [d/2]$ and let $\mathbf{B}_i = \{\alpha_1^{(i)}, \ldots, \alpha_{a_i}^{(i)}\}$ be a basis of A_i. Let us define the matrix $\mathrm{Hess}_{\mathbf{B}_i}^{(i)} F$ as the $a_i \times a_i$ matrix whose elements are given by

$$(\mathrm{Hess}_{\mathbf{B}_i}^{(i)} F)_{m,n} = (\alpha_m^{(i)}(\frac{\partial}{\partial \mathbf{x}}) \bullet \alpha_n^{(i)}(\frac{\partial}{\partial \mathbf{x}}))(F) \in \mathbb{K}[x_0, \ldots, x_N]_{d-2i}.$$

Finally, define

$$\text{hess}_{\mathbf{B}_i}^{(i)} F = \det(\text{Hess}_{\mathbf{B}_i}^{(i)} F) \in \mathbb{K}[x_0, \ldots, x_N]_{a_i(d-2i)}.$$

The definition depends on the choice of the basis $\mathbf{B}_i$'s. Choosing different bases the value of the (i)-hessian of F is altered by the multiplication by a non-zero element of $\mathbb{K}$. Since we are mainly interested in the vanishing of these polynomials we could omit the reference to the basis and simply write $\text{hess}^{(i)} F$.

We shall need the following elementary result, known as the **differential Euler Identity**, whose proof is left to the reader.

Lemma 7.2.19 *Let* $G \in \mathbb{K}[x_0, \ldots, x_N]_e$ *and let* $L = a_0 \frac{\partial}{\partial x_0} + \ldots + a_N \frac{\partial}{\partial x_N} \in Q_1$. *Then*

$$L^e(G) = e! \cdot G(a_0, \ldots, a_N).$$

The connection between the huge amount of algebraic definitions introduced so far and the contents of this chapter is finally made clear by the next result.

Theorem 7.2.20 ([128, 192]) *Let the notation be as above. An element* $L = a_0 x_0 + \ldots + a_N x_N \in A_1$ *is a strong Lefschetz element of* $A = Q/\text{Ann}_Q(F)$ *if and only if*

(i) $F(a_0, \ldots, a_N) \neq 0$ *and*
(ii) $\text{hess}^{(i)} F(a_0, \ldots, a_N) \neq 0$ *for all* $i = 1, \ldots, [d/2]$.

Proof The identification $A_d \simeq \mathbb{K}$ is obtained by letting $G \in A_d$ act on F as a differential operator. Since $\deg(G) = d = \deg(F)$ we get $G(F) \in \mathbb{K}$.

The Poincaré Duality Property holds for A so that to define a linear map $f : A_i \to A_{d-i}$ is the same as giving a bilinear map $\phi : A_i \times A_i \to A_d \simeq \mathbb{K}$ via the identification between A_i and A_{d-i} given by multiplication in A.

Consider the multiplication linear map $\bullet L^{d-2i} : A_i \to A_{d-i}$ with L as in the statement. The associated bilinear map $\phi_i : A_i \times A_i \to A_d \simeq \mathbb{K}$ is defined by

$$\phi_i(\xi, \eta) = [(L^{d-2i} \bullet \xi) \bullet \eta](F)$$

and it is symmetric by the commutativity of the product in A. Moreover, $\bullet L^{d-2i}$ is an isomorphism if and only if ϕ_i is non-degenerate.

Choose a basis $\mathbf{B}_i = \{\alpha_1^{(i)}, \ldots, \alpha_{a_i}^{(i)}\}$ of A_i. Then the symmetric matrix H^i associated to ϕ_i with respect to the basis $\mathscr{B}_i$ has elements

$$H_{m,n}^i = [(L^{d-2i} \bullet \alpha_m^{(i)}) \bullet \alpha_m^{(i)}](F)$$

$$= L^{d-2i}([\alpha_m^{(i)} \bullet \alpha_m^{(i)}](F)) \overset{\text{Lemma 7.2.19}}{=} (d-2i)! \cdot \{([\alpha_m^{(i)} \bullet \alpha_m^{(i)}](F))(a_0, \ldots, a_N)\}$$

$$= (d-2i)! \cdot [(\text{Hess}_{\mathscr{B}_i}^{(i)} F)_{m,n}(a_0, \ldots, a_N)].$$

In conclusion, $\bullet L^{d-2i}$ is an isomorphism for every $i = 1, \ldots, [d/2]$ if and only if (i) and (ii) hold. □

Corollary 7.2.21 ([128, 192]) *Let the notation be as above. Then:*

1. $A \simeq \frac{Q}{\mathrm{Ann}_Q(F)}$, $F \in \mathbb{K}[x_0, \ldots, x_N]_d$ *such that* $(\mathrm{Ann}_Q(F))_1 = 0$, *satisfies the SLP if and only if* $\mathrm{hess}^{(i)}(F) \neq 0$ *for every* $i = 1, \ldots, [d/2]$.
2. *Let the hypotheses and notation be as in (1). If* $d \leq 4$, *then* A *satisfies the SLP if and only if* $\mathrm{h}(F) \neq 0$. *In particular, for* $N \leq 3$, *every such* A *satisfies the SLP due to Theorem 7.1.4 (see Corollary 7.3.9 below for the proof of this statement).*
3. *For every* $N \geq 4$ *and for* $d = 3, 4$ *a polynomial* $F \in \mathbb{K}[x_0, \ldots, x_N]_d$ *with vanishing hessian and with* $(\mathrm{Ann}_Q(F))_1 = 0$ *produces an example of a graded Artinian Gorenstein algebra* $A = \oplus_{i=0}^d A_i$ *not satisfying the SLP.*

R. Gondim applied Theorem 7.2.20 to prove two wider generalizations of Corollary 7.2.21. One can consult [76] to see the examples leading to these applications and inspired by the constructions in [72, 78, 151]. The first result deals with *SLP*.

Corollary 7.2.22 ([76]) *For each pair* $(N, d) \notin \{(3, 3), (3, 4)\}$ *with* $N \geq 3$ *and with* $d \geq 3$, *there exist infinitely many standard graded Artinian Gorenstein algebras* $A = \oplus_{i=0}^d A_i$ *of codimension* $\dim A_1 = N + 1 \geq 4$ *and socle degree* d *that do not satisfy the Strong Lefschetz Property.*

Furthermore, for each $L = \sum_{i=0}^N a_i X_i \in A_1$ *we can choose arbitrarily the level* k *where the map*

$$\bullet L^{d-2k} : A_k \to A_{d-k}$$

is not an isomorphism.

The second application concerns unimodality and the *WLP*.

Corollary 7.2.23 ([76]) *For each pair* $(N, d) \neq (3, 3)$ *with* $N \geq 3$ *and odd* $d = 2q+1 \geq 3$, *there exist infinitely many standard graded Artinian Gorenstein algebras* $A = \oplus_{i=0}^d A_i$ *with* $\dim A_1 = N + 1$ *and socle degree* d *with unimodal Hilbert vector and that do not satisfy the Weak Lefschetz Property.*

7.3   The Gordan–Noether Identity

Let us introduce some definitions.

Definition 7.3.1 (Map $\psi_g : \mathbb{P}^N \dashrightarrow \mathbb{P}^N$) Let $f \in \mathbb{K}[x_0, \ldots, x_n]_d$, $f = f_{\mathrm{red}}$, and let $X = V(f) \subset \mathbb{P}^N$ be the associated degree d hypersurface. Suppose $h(f) = 0$.
   We deduce from Corollary 7.2.8 that

$$X^* \subsetneq Z \subsetneq \mathbb{P}^{N*}.$$

In particular, there exists an irreducible $g \in \mathbb{K}[y_0, \ldots, y_N]_e$, $g \neq 0$, such that

$$g(\nabla_f(\mathbf{x})) = g(\frac{\partial f}{\partial x_0}(\mathbf{x}), \ldots, \frac{\partial f}{\partial x_N}(\mathbf{x})) = 0.$$

By definition $Z \subseteq T = V(g) \subset \mathbb{P}^N$ with equality holding if codim$(Z) = 1$. We shall make the following equivalent hypothesis on g:

(i) there exists an $i \in \{0, \ldots, N\}$ such that $\frac{\partial g}{\partial y_i}(\nabla_f(\mathbf{x})) \neq 0$;

(ii) $Z \nsubseteq \text{Sing}(T)$.

The generators of minimal degree in the homogeneous ideal $I(Z) \subset \mathbb{K}[y_0, \ldots, y_N]$ clearly satisfy condition (i).

Under the previous equivalent hypothesis the *Gordan–Noether map associated to g*:

$$\psi_g = \nabla_g \circ \nabla_f : \mathbb{P}^N \dashrightarrow \mathbb{P}^N \tag{7.9}$$

is well defined.

Definition 7.3.2 (Functions h_i) Let $g \in \mathbb{K}[x_0, \ldots, x_N]_e$, $e \geq 1$, be as in Definition 7.3.1. Let us remark that $e = 1$ implies that X is a cone and conversely for a cone there exists a g satisfying the previous conditions and having $e = 1$.

Since $g(\nabla_f(\mathbf{x})) = 0$, Euler's Formula implies

$$0 = e \cdot g(\nabla_f(\mathbf{x})) = \sum_{i=0}^{N} \frac{\partial f}{\partial x_i}(\mathbf{x}) \frac{\partial g}{\partial y_i}(\nabla_f(\mathbf{x})). \tag{7.10}$$

Let

$$\rho(\mathbf{x}) = \text{g.c.d.}(\frac{\partial g}{\partial y_0}(\nabla_f(\mathbf{x})), \ldots, \frac{\partial g}{\partial y_N}(\nabla_f(\mathbf{x})))$$

so that

$$\rho(\mathbf{x}) \cdot h_i(\mathbf{x}) = \frac{\partial g}{\partial y_i}(\nabla_f(\mathbf{x})). \tag{7.11}$$

Therefore g.c.d.$(h_0, \ldots, h_N) = 1$ and

$$\psi_g = (h_0 : \ldots : h_N) : \mathbb{P}^N \dashrightarrow \mathbb{P}^N.$$

We collect some identities which are immediate consequences of $g(\nabla_f(\mathbf{x})) = 0$.

Proposition 7.3.3 *Let the notation be as above. Then:*

$$\sum_{i=0}^{N} \frac{\partial f}{\partial x_i}(\mathbf{x}) h_i(\mathbf{x}) = 0. \tag{7.12}$$

- *For every $j = 0, \ldots, N$*

$$\sum_{i=0}^{N} \frac{\partial^2 f}{\partial x_j \partial x_i}(\mathbf{x}) h_i(\mathbf{x}) = 0. \tag{7.13}$$

- *Letting $g_j = \frac{\partial g}{\partial y_j}(\nabla_f(\mathbf{x})), j = 0, \ldots, N$, we have:*

$$\sum_{i=0}^{N} \frac{\partial g_j}{\partial x_i}(\mathbf{x}) h_i(\mathbf{x}) = 0. \tag{7.14}$$

Proof From (7.10) we get

$$0 = \sum_{i=0}^{N} \frac{\partial f}{\partial x_i}(\mathbf{x}) \frac{\partial g}{\partial y_i}(\mathbf{x})(\nabla_f(\mathbf{x})) = \rho(\mathbf{x})[\sum_{i=0}^{N} \frac{\partial f}{\partial x_i}(\mathbf{x}) h_i(\mathbf{x})],$$

proving (7.12).

Since $g(\nabla_f(\mathbf{x})) = 0$, we deduce

$$\frac{\partial}{\partial x_j} g(\nabla_f) = 0 \ \ \forall \, i = 0, \ldots, N$$

so that the Chain Rule yields

$$\sum_{i=0}^{N} \frac{\partial^2 f}{\partial x_j \partial x_i}(\mathbf{x}) h_i(\mathbf{x}) = 0.$$

We have

$$\sum_{i=0}^{N} \frac{\partial g_j}{\partial x_i} h_i = \sum_{i=0}^{N}(\sum_{k=0}^{N} \frac{\partial g_j}{\partial(\frac{\partial f}{\partial x_k})} \frac{\partial^2 f}{\partial x_k \partial x_i}) h_i = \sum_{k=0}^{N} \frac{\partial g_j}{\partial(\frac{\partial f}{\partial x_k})}(\sum_{i=0}^{N} \frac{\partial^2 f}{\partial x_k \partial x_i} h_i) \stackrel{(7.13)}{=} 0.$$

□

The importance of the previous identities is well illustrated by the main tool developed in [78]. We shall follow the approach rewritten in [127], see also [193].

Theorem 7.3.4 (Gordan–Noether Identity) *Let $F \in \mathbb{K}[x_0, \ldots, x_N]_m$, let $\mathbb{K}' \supseteq \mathbb{K}$ be a field extension, let ψ_g be the map defined in (7.9) (see also Definition 7.3.1) and let the notation be as above. Then:*

$$\sum_{i=0}^{N} \frac{\partial F}{\partial x_i}(\mathbf{x}) h_i(\mathbf{x}) = 0 \iff F(\mathbf{x}) = F(\mathbf{x} + \lambda \psi_g(\mathbf{x})) \ \forall \, \lambda \in \mathbb{K}', \ \ \forall \, \mathbf{x} \in \mathbb{K}^{N+1}.$$

Proof Let

$$F(\mathbf{x} + \lambda \psi_g(\mathbf{x})) - F(\mathbf{x}) = \sum_{k=1}^{m} \Phi_k(\mathbf{x})\lambda^k,$$

with $\Phi_k(\mathbf{x}) \in \mathbb{K}[\mathbf{x}]$ and with

$$\Phi_1(\mathbf{x}) = \sum_{i=0}^{N} \frac{\partial F}{\partial x_i}(\mathbf{x})h_i(\mathbf{x}).$$

If $F(\mathbf{x} + \lambda\psi_g(\mathbf{x})) = F(\mathbf{x})$, then $\Phi_1 = 0$.

Suppose now $\Phi_1 = 0$. We shall prove that for every $k = 1, \ldots, m - 1$, $\Phi_k = 0$ implies $\Phi_{k+1} = 0$. From (7.14) and from Leibnitz's Rule we deduce

$$(\#) := \sum_{j=0}^{N} h_j \cdot \frac{\partial}{\partial x_j}(g_{i_1} \cdots g_{i_k}) = 0.$$

Therefore

$$(\#\#) := \sum_{j=0}^{N} h_j \cdot \sum_{i_1,\ldots,i_k} \frac{\partial^k F}{\partial x_{i_1} \ldots \partial x_{i_k}} \cdot \frac{\partial}{\partial x_j}\left(\frac{g_{i_1} \cdots g_{i_k}}{k!}\right) = 0.$$

Thus if

$$\Phi_k = \sum_{i_1,\ldots,i_k} \frac{\partial^k F}{\partial x_{i_1} \ldots \partial x_{i_k}} \cdot \frac{h_{i_1} \cdots h_{i_k}}{k!} = 0,$$

then

$$\Phi_{k+1} = \frac{1}{(k+1)\rho^k}\sum_{j=0}^{N} h_j\left[\sum_{i_1,\ldots,i_k} \frac{\partial}{\partial x_j}\left(\frac{\partial^k F}{\partial x_{i_1} \ldots \partial x_{i_k}}\right) \cdot \frac{g_{i_1} \cdots g_{i_k}}{k!}\right] + \frac{1}{(k+1)\rho^k} \cdot (\#\#)$$

$$= \frac{1}{(k+1)\rho^k}\sum_{j=0}^{N} h_j\left[\sum_{i_1,\ldots,i_k} \frac{\partial}{\partial x_j}\left(\frac{\partial^k F}{\partial x_{i_1} \ldots \partial x_{i_k}}\right) \cdot \frac{g_{i_1} \cdots g_{i_k}}{k!} + \sum_{i_1,\ldots,i_k} \frac{\partial^k F}{\partial x_{i_1} \ldots \partial x_{i_k}} \cdot \frac{\partial}{\partial x_j}\left(\frac{g_{i_1} \cdots g_{i_k}}{k!}\right)\right]$$

$$= \frac{1}{(k+1)\rho^k}\sum_{j=0}^{N} h_j \frac{\partial(\Phi_k \cdot \rho^k)}{\partial x_j} = 0.$$

$\square$

To prove an interesting consequence of the Gordan–Noether Identity we need a preliminary step.

Lemma 7.3.5 *Let $F = F_1 \cdot F_2 \in \mathbb{K}[x_0, \ldots, x_N]_m$ with $F_i \in \mathbb{K}[x_0, \ldots, x_N]_{m_i}$, $i = 1, 2$ and let $\mathbb{K}' \supseteq \mathbb{K}$ be an extension of fields. Then $\forall\ \lambda \in \mathbb{K}'$, $\forall\ \mathbf{x} \in \mathbb{K}^{N+1}$ we have*

$$F(\mathbf{x}) = F(\mathbf{x} + \lambda \psi_g(\mathbf{x})) \Longleftrightarrow F_i(\mathbf{x}) = F_i(\mathbf{x} + \lambda \psi_g(\mathbf{x}))\ \forall\ i = 1, 2.$$

Proof One implication is obvious so we shall prove only the relevant one. Let

$$F_i(\mathbf{x} + \lambda \psi_g(\mathbf{x})) = \sum_{j=0}^{m_i} A_j^i(\mathbf{x}) \lambda^j \in (\mathbb{K}[\mathbf{x}])[\lambda],$$

with $A_j^i(\mathbf{x}) \in \mathbb{K}[\mathbf{x}]$.
From

$$F_1(\mathbf{x} + \lambda \psi_g(\mathbf{x})) \cdot F_2(\mathbf{x} + \lambda \psi_g(\mathbf{x})) = F(\mathbf{x} + \lambda \psi_g(\mathbf{x})) = F(\mathbf{x}) \in \mathbb{K}[\mathbf{x}]$$

we deduce $m_1 = m_2 = 0$ since $(\mathbb{K}[\mathbf{x}])[\lambda]$ is an integral domain. In conclusion

$$F_i(\mathbf{x} + \lambda \psi_g(\mathbf{x})) = A_0^i(\mathbf{x}) = F_i(\mathbf{x})$$

for $i = 1, 2$. □

The previous lemma and the definition of the h_i's yield the following key formula.

Corollary 7.3.6 *For every $i = 0, \ldots, N$ we have*

$$\sum_{j=0}^{N} h_j \cdot \frac{\partial}{\partial x_j} h_i = 0, \tag{7.15}$$

so that the h_i's satisfy the Gordan–Noether Identity.

7.3.1 Hesse's Claim for N = 2, 3

We list some consequences of the Gordan–Noether Identity, either for later reference or for immediate applications to the cases $N = 2, 3$.

Lemma 7.3.7 *Let $X = V(f) \subset \mathbb{P}^N$ be a hypersurface with vanishing hessian and let g be as in Definition 7.3.1. Then*

$$\psi_g(\mathbf{x}) = \psi_g(\mathbf{x} + \lambda \psi_g(\mathbf{x})), \tag{7.16}$$

$\forall\ \lambda \in \mathbb{K}$ and $\forall\ \mathbf{x} \in \mathbb{K}^{N+1}$.

In particular,

$$\psi_g(\mathbb{P}^N) \subseteq V(h_0, \ldots, h_N) = \text{Bs}(\psi_g) \subset \mathbb{P}^N.$$

Proof Let us remark that (7.15) and Theorem 7.3.4 yield

$$h_j(\mathbf{x}) = h_j(\mathbf{x} + \lambda \psi_g(\mathbf{x})) \qquad (7.17)$$

$\forall \lambda \in \mathbb{K}, \forall \mathbf{x} \in \mathbb{K}^{N+1}$ and $\forall j = 0, \ldots, N$. Thus

$$\psi_g(\mathbf{x}) = \psi_g(\mathbf{x} + \lambda \psi_g(\mathbf{x}))$$

$\forall \lambda \in \mathbb{K}$ and $\forall \mathbf{x} \in \mathbb{K}^{N+1}$, proving (7.16).

In particular, from (7.17) we deduce $h_i(\psi_g(\mathbf{x})) = 0, \forall i = 0, \ldots, N$ and $\forall \mathbf{x} \in \mathbb{K}^{N+1}$, yielding $\psi_g(\mathbb{P}^N) \subseteq V(h_0, \ldots, h_N) = \text{Bs}(\psi_g) \subset \mathbb{P}^N$.   □

Corollary 7.3.8 ([78]) *Let $X = V(f) \subset \mathbb{P}^2$ be a reduced curve of degree $d \geq 3$ with vanishing hessian. Then X is a cone, that is, the union of d lines passing through the vertex of X.*

Proof By Lemma 7.3.7 we deduce that $\psi_g(\mathbb{P}^N) = q \in \text{Bs}(\psi_g) \subset \mathbb{P}^2$. Then for every $p \in X \setminus (\text{Bs}(\psi_g) \cap X)$ and for every $\lambda \in \mathbb{K}$ we have

$$0 = f(p) = f(p + \lambda \psi_g(p)) = f(p + \lambda q),$$

yielding $< p, q > \subseteq X$.   □

Corollary 7.3.9 ([78]) *Let $X = V(f) \subset \mathbb{P}^3$ be a reduced surface of degree $d \geq 3$ with vanishing hessian. Then X is a cone, that is X is either the union of d planes passing through a line or it is a cone with vertex a point over the union of irreducible plane curves.*

Proof Corollary 7.2.8 implies $X^* \subsetneq Z \subsetneq \mathbb{P}^{3^*}$ and moreover by Lemma 7.3.7 $\psi_g(\mathbb{P}^3) \subseteq V(h_0, h_1, h_2, h_3) \subset \mathbb{P}^3$.

If $\dim(Z) = 1$, then Lemma 7.2.1 assures us that a general hyperplane section of X is a reduced plane curve of degree d with vanishing hessian and hence the union of d lines passing through a point. This immediately implies that X is a union of d planes passing through a line.

From now on we can suppose that $Z = V(g) \subset \mathbb{P}^3$ is a surface and that $Z^* = \psi_g(\mathbb{P}^3)$. There are only two possibilities: either $\dim(Z^*) = 0$ or $\dim(Z^*) = 1$ and Z^* is an irreducible component of $\text{Bs}(\psi_g)$. In the first case, arguing as in the proof of Corollary 7.3.8, we deduce that X is a cone of vertex the point Z^*. We now exclude the second case.

The Gordan–Noether Identity implies

$$\psi_g(p) = \psi_g(p + \lambda \psi_g(p)) \quad \forall \lambda \in \mathbb{K} \ \forall p \in \mathbb{P}^3.$$

Let $q_1, q_2 \in Z^*$ be general points. Then $\psi_g^{-1}(q_i)$ is a union of cones whose vertex contains q_i. Thus $\overline{\psi_g^{-1}(q_i)}$ is also a union of cones whose vertex contains q_i. Clearly

$$\overline{\psi_g^{-1}(q_1)} \cap \overline{\psi_g^{-1}(q_2)} \subseteq V(h_0, h_1, h_2, h_3) = \mathrm{Bs}(\psi_g)$$

and this intersection is a union of curves. Let $r \in \overline{\psi_g^{-1}(q_1)} \cap \overline{\psi_g^{-1}(q_2)}$, $r \neq q_i$, $i = 1, 2$. Then we claim that

$$\langle r, q_i \rangle \subseteq \overline{\psi_g^{-1}(q_i)} \cap \mathrm{Bs}(\psi_g),$$

for $i = 1, 2$. Indeed, if there existed $q \in < r, q_i > \cap \psi_g^{-1}(q_i)$ then we would have

$$\psi_g(q) = \psi_g(q + \lambda q_i),$$

yielding

$$\psi_g(r) = \psi_g(q) \neq 0$$

and $r \notin \mathrm{Bs}(\psi_g)$, contrary to our hypothesis.

Therefore from the generality of the points $q_1, q_2 \in Z^*$ we deduce

$$< q_1, q_2 >= Z^* \subseteq \mathrm{Bs}(\psi_g).$$

This is impossible because the dual of an irreducible surface cannot be a line. □

Corollary 7.3.10 *Let $X = V(f) \subset \mathbb{P}^N$ be a hypersurface with vanishing hessian. If $\dim(\nabla_f(\mathbb{P}^N)) \leq 2$, then X is a cone with positive dimensional vertex.*

Proof By Lemma 7.2.1, Corollaries 7.3.8 and 7.3.9 we deduce that a general hyperplane section of X is a cone so that X is also a cone with positive-dimensional vertex. □

7.3.2 Cremona Equivalence with a Cone

In this subsection we provide another interesting application of the Gordan–Noether Identity, see also [193].

Theorem 7.3.11 ([78]) *Let $X = V(f) \subset \mathbb{P}^N$ be a hypersurface of degree $d \geq 3$ with vanishing hessian. Then there exists a Cremona transformation*

$$\Phi : \mathbb{P}^N \dashrightarrow \mathbb{P}^N$$

such that $\Phi(X)$ is a cone.

Moreover, the Cremona transformation can be explicitly constructed from f in such a way that the equation of $\Phi(X)$ depends on at most N variables.

Proof Let the notation be as above and assume that X is not a cone. By definition of the h_i's, see (7.11),

$$(h_0(\mathbf{x}), \ldots, h_N(\mathbf{x})) \neq \mathbf{0}$$

and, without loss of generality, we can suppose $h_N(\mathbf{x}) \neq 0$ and set

$$\lambda = -\frac{x_N}{h_N(\mathbf{x})} \in \mathbb{K}(x_0, \ldots, x_N) \supseteq \mathbb{K}.$$

Let us remark that for every $i \in \{0, \ldots, N\}$ such that $h_i \neq 0$ we can apply the same construction. By the Gordan–Noether Identity for the h_i's we deduce

$$h_i(\mathbf{x}) = h_i\left(\mathbf{x} - \frac{x_N}{h_N(\mathbf{x})}\psi_g(\mathbf{x})\right) = h_i\left(x_0 - \frac{x_N}{h_N}h_0, \ldots, x_{N-1} - \frac{x_N}{h_N}h_{N-1}, 0\right). \quad (7.18)$$

Define the rational map $\Phi : \mathbb{P}^N \dashrightarrow \mathbb{P}^N$ by

$$\begin{cases} x_0' &= x_0 - \frac{h_0(\mathbf{x})}{h_N(\mathbf{x})}x_N \\ &\vdots \\ x_i' &= x_i - \frac{h_i(\mathbf{x})}{h_N(\mathbf{x})}x_N \\ &\vdots \\ x_{N-1}' &= x_{N-1} - \frac{h_{N-1}(\mathbf{x})}{h_N(\mathbf{x})}x_N \\ x_N' &= x_N \end{cases}$$

The map Φ is birational since by (7.18) we have

$$\begin{cases} x_0 &= x_0' + \frac{h_0(\mathbf{x})}{h_N(\mathbf{x})}x_N' = x_0' + \frac{h_0(x_0', \ldots, x_{N-1}', 0)}{h_N(x_0', \ldots, x_{N-1}', 0)}x_N' \\ &\vdots \\ x_i &= x_i' + \frac{h_i(\mathbf{x})}{h_N(\mathbf{x})}x_N' = x_i' + \frac{h_i(x_0', \ldots, x_{N-1}', 0)}{h_N(x_0', \ldots, x_{N-1}', 0)}x_N' \\ &\vdots \\ x_{N-1} &= x_{N-1}' + \frac{h_{N-1}(\mathbf{x})}{h_N(\mathbf{x})}x_N' = x_{N-1}' + \frac{h_{N-1}(x_0', \ldots, x_{N-1}', 0)}{h_N(x_0', \ldots, x_{N-1}', 0)}x_N' \\ x_N &= x_N' \end{cases}$$

Then by the Gordan–Noether Identity applied to f we get

$$f(\mathbf{x}) = f\left(\mathbf{x} - \frac{x_N}{h_N(\mathbf{x})}\psi_g(\mathbf{x})\right) = f(x_0', \ldots, x_{N-1}', 0)$$

and the transformed equation does not depend on x_N. To compute the equation of $\Phi(X)$ one might have to cancel some irreducible factors of the transformed equation as we shall illustrate in the next example.   □

Example 7.3.12 (Two Different Cremona Equivalences of the Perazzo Hypersurface in $\mathbb{P}^4$) Let

$$f(\mathbf{x}) = x_0 x_3^2 + x_1 x_3 x_4 + x_2 x_4^2.$$

We saw in Example 7.1.5 that we can take $g(\mathbf{y}) = y_0 y_2 - y_1^2$. Thus by definition

$$(h_0, \ldots, h_4) = (x_4^2, -2x_3 x_4, x_3^2, 0, 0).$$

Since $h_0 = x_4^2 \neq 0$, we can define the Cremona transformation $\Phi : \mathbb{P}^4 \dashrightarrow \mathbb{P}^4$ given by

$$
\begin{cases}
x_0' = x_0 \\
x_1' = x_1 + \frac{2x_3}{x_4} x_0 \\
x_2' = x_2 - \frac{x_3^2}{x_4^2} x_0 \\
x_3' = x_3 \\
x_4' = x_4
\end{cases}
\qquad
\begin{cases}
x_0 = x_0' \\
x_1 = x_1' - \frac{2x_3'}{x_4'} x_0' \\
x_2 = x_2' + \frac{x_3'^2}{x_4'^2} x_0' \\
x_3 = x_3' \\
x_4 = x_4'
\end{cases}
$$

Thus

$$f(0, x_1', x_2', x_3', x_4') = x_4'(x_1' x_3' + x_2' x_4')$$

and

$$\Phi(V(x_0 x_3^2 + x_1 x_3 x_4 + x_2 x_4^2)) = V(x_1 x_3 + x_2 x_4) \subset \mathbb{P}^4.$$

Since $h_1 = -2x_3 x_4 \neq 0$, we can also define $\varphi : \mathbb{P}^4 \dashrightarrow \mathbb{P}^4$ given by

$$
\begin{cases}
x_0' = x_0 + \frac{x_4^2}{2x_3 x_4} x_1 \\
x_1' = x_1 \\
x_2' = x_2 + \frac{x_3^2}{2x_3 x_4} x_1 \\
x_3' = x_3 \\
x_4' = x_4
\end{cases}
\qquad
\begin{cases}
x_0 = x_0' - \frac{x_4'^2}{2x_3' x_4'} x_1' \\
x_1 = x_1' \\
x_2 = x_2' - \frac{x_3'^2}{2x_3' x_4'} x_1' \\
x_3 = x_3' \\
x_4 = x_4'
\end{cases}
$$

In this case

$$f(x_0', 0, x_2', x_3', x_4') = x_0' x_3'^2 + x_2' x_4'^2$$

so that

$$\varphi(V(x_0 x_3^2 + x_1 x_3 x_4 + x_2 x_4^2)) = V(x_0 x_3^2 + x_2 x_4^2) \subset \mathbb{P}^4.$$

The condition of vanishing hessian is sufficient to be birational to a cone but it is not necessary, as shown by the next example.

Example 7.3.13 (Cremona Linearization of the Veronese Surface, [35]) Let

$$f(\mathbf{x}) = \begin{vmatrix} x_{0,0} & x_{0,1} & x_{0,2} \\ x_{0,1} & x_{1,1} & x_{1,2} \\ x_{0,2} & x_{1,2} & x_{2,2} \end{vmatrix}.$$

A direct computation yields $h(f) = -16f^2 \neq 0$.

The hypersurface $X = V(f) \subset \mathbb{P}^5$ is the secant variety to the Veronese surface $S = v_2(\mathbb{P}^2) \subset \mathbb{P}^5$.

Let us define $\Phi : \mathbb{P}^5 \dashrightarrow \mathbb{P}^5$ and $\Phi^{-1} : \mathbb{P}^5 \dashrightarrow \mathbb{P}^5$ by

$$\begin{cases} x'_{0,0} = x_{0,0} \\ x'_{0,1} = x_{0,1} \\ x'_{0,2} = x_{0,2} \\ x'_{1,1} = x_{1,1} - x_{0,1}^2 \\ x'_{1,2} = x_{1,2} - x_{0,1}x_{0,2} \\ x'_{2,2} = x_{2,2} - x_{0,2}^2 \end{cases} \qquad \begin{cases} x_{0,0} = x'_{0,0} \\ x_{0,1} = x'_{0,1} \\ x_{0,2} = x'_{0,2} \\ x_{1,1} = x'_{1,1} + x'^2_{0,1} \\ x_{1,2} = x'_{1,2} + x'_{0,1}x'_{0,2} \\ x_{2,2} = x'_{2,2} + x'^2_{0,2} \end{cases}$$

Then one verifies that

$$\Phi(S) = V(x_{1,1}, x_{1,2}, x_{2,2}) \subset \mathbb{P}^5$$

and that

$$\Phi(V(f)) = V(x_{1,1}x_{2,2} - x_{1,2}^2) \subset \mathbb{P}^5.$$

Therefore $\Phi(V(f))$ is a cone with vertex the plane $\Phi(S)$ but $h(f) \neq 0$.

7.3.3 Applications of the Gordan–Noether Identity to the Polar Map

We shall always identify $\mathbb{P}^{N^{**}} = (\mathbb{P}^{N^*})^*$ with the original $\mathbb{P}^N$ so that by definition and with the previous identifications we have

$$Z^* = \overline{\bigcup_{z \in Z_{reg}} (T_z Z)^*} \subset \mathbb{P}^N. \tag{7.19}$$

Let $X = V(f) \subset \mathbb{P}^N$ be a reduced hypersurface such that $h(f) = 0$, let $T = V(g) \subset \mathbb{P}^{N^*}$ be a hypersurface such that $Z \subseteq T$ and such that

$$g_i := \frac{\partial g}{\partial y_i}(\frac{\partial f}{\partial x_0}, \ldots, \frac{\partial f}{\partial x_N}) \neq 0 \quad \text{for at least one} \quad i \in \{0, \ldots, N\}. \qquad (*)$$

Set $T_Z^* := \overline{\psi_g(\mathbb{P}^N)}$ and note that, by definition of ψ_g and of Z^*, $T_Z^* \subset Z^*$. Therefore by taking $\alpha + 1 = \mathrm{codim}(Z)$ polynomials $g^0, \ldots g^\alpha$ defining Z locally around a point $z \in Z_{\mathrm{reg}}$ we deduce that

$$(T_z Z)^* = \langle (T_z V(g^0))^*, \ldots, (T_z V(g^\alpha))^* \rangle \subset Z^*.$$

Thus for a general point $r \in (T_{\nabla_X(p)} Z)^*$, $p \in \mathbb{P}^N$ such that $\nabla_X(p) \in Z_{\mathrm{reg}}$, there exist $a_0, \ldots, a_\alpha \in \mathbb{K}$ such that, letting $\underline{a} = (a_0, \ldots, a_\alpha) \in \mathbb{K}^{\alpha+1}$ and $g_{\underline{a}} = \sum_{i=0}^\alpha a_i g^i$,

$$r = \sum_{i=0}^\alpha a_i \psi_{g^i}(p) = \psi_{g_{\underline{a}}}(p). \qquad (7.20)$$

Moreover, $g_{\underline{a}}$ also satisfies $(*)$.

The following result is another immediate consequence of the Gordan–Noether Identity and of the previous remarks.

Corollary 7.3.14 *Let $X = V(f) \subset \mathbb{P}^N$ be a hypersurface with vanishing hessian and let the notation be as above. Then*

i) *for every $p \in \mathbb{P}^N \setminus \mathrm{Sing}\, X$ such that $\nabla_X(p) \in Z_{\mathrm{reg}}$ we have*

$$< p, (T_{\nabla_X(p)} Z)^* > \subseteq \nabla_X^{-1}(\nabla_X(p));$$

ii) *for $p \in \mathbb{P}^N$ general, the irreducible component of $\overline{\nabla_X^{-1}(\nabla_X(p))}$ passing through p is $< p, (T_{\nabla_X(p)} Z)^* >$. In particular, for $p \in \mathbb{P}^N$ general $\overline{\nabla_X^{-1}(\nabla_X(p))}$ is a union of linear spaces of dimension equal to $\mathrm{codim}(Z)$ passing through $(T_{\nabla_X(p)} Z)^*$.*

iii)

$$Z^* \subseteq \mathrm{Sing}\, X. \qquad (7.21)$$

Proof Let the notation be as above. By (7.13) and Theorem 7.3.4 we get

$$\frac{\partial f}{\partial x_i}(\mathbf{x}) = \frac{\partial f}{\partial x_i}(\mathbf{x} + \lambda \psi_g(\mathbf{x})) \quad \text{for every} \quad j = 0, \ldots, N, \qquad (7.22)$$

for every $\lambda \in \mathbb{K}$ and for every $g \in \mathbb{K}[y_0, \ldots, y_N]_e$ such that $g(\frac{\partial f}{\partial x_0}, \ldots, \frac{\partial f}{\partial x_N}) = 0$ and such that $(*)$ holds.

Let $p \in \mathbb{P}^N \setminus \text{Sing}\,X$ such that $\nabla_X(p) \in Z_{\text{reg}}$. Let $r \in (T_{\nabla_X(p)}Z)^*$ be a general point. By (7.20) we can suppose $r = \psi_g(p)$ so that (7.22) yields that the line $< p, r >$ is contracted to the point $\nabla_X(p)$ and that $r =< p, r > \cap \text{Sing}\,X$. The generality of r implies that the linear space $\mathbb{P}^{\text{codim}(Z)} =< p, (T_{\nabla_X(p)}Z)^* >$ is contained in $\nabla_X^{-1}(\nabla_X(p))$, proving i) and also that $(T_{\nabla_X(p)}Z)^* \subset \text{Sing}\,X$. Moreover, since, by definition, Z^* is ruled by the linear spaces $(T_{\nabla_X(p)}Z)^*$ part iii) immediately follows.

If $p \in \mathbb{P}^N$ is general, then, by generic smoothness, the irreducible component of $\nabla_X^{-1}(\nabla_X(p))$ passing through p has dimension $\text{codim}(Z)$ and it is smooth at p so that it coincides with $< p, (T_{\nabla_X(p)}Z)^* >$, proving ii). $\qquad\square$

7.4 The Gordan–Noether–Franchetta Classification in $\mathbb{P}^4$ and Examples in Arbitrary Dimension

7.4.1 Gordan–Noether, Franchetta, Permutti and Perazzo Examples

We now describe some series of examples of hypersurfaces in $\mathbb{P}^N$, $N \geq 4$, with vanishing Hessian and which are not cones, following [62, 78] and above all [37, Sect. 2.3].

Definition 7.4.1 (Gordan–Noether Hypersurfaces, [37, 78, 127]) Assume $N \geq 4$ and fix integers $t \geq m + 1$ such that $2 \leq t \leq N - 2$ and $1 \leq m \leq N - t - 1$.

Consider forms $h_i(y_0, \ldots, y_m) \in \mathbb{K}[y_0, \ldots, y_m]$, $i = 0, \ldots, t$, of the same degree, and also forms $\psi_j(x_{t+1}, \ldots, x_N) \in \mathbb{K}[x_{t+1}, \ldots, x_N]$, $j = 0, \ldots, m$, of the same degree. For $\ell = 1, \ldots, t - m$, let $a_{u,v}^{(\ell)} \in k$ for $u = 1, \ldots, t - m - 1$, $v = 0, \ldots, t$. Introduce the following homogeneous polynomials, all of the same degree:

$$Q_\ell(x_0, \ldots, x_N) := \det \begin{pmatrix} x_0 & \cdots & x_t \\ \frac{\partial h_0}{\partial \psi_0} & \cdots & \frac{\partial h_t}{\partial \psi_0} \\ \cdots & \cdots & \cdots \\ \frac{\partial h_0}{\partial \psi_m} & \cdots & \frac{\partial h_t}{\partial \psi_m} \\ a_{1,0}^{(\ell)} & \cdots & a_{1,t}^{(\ell)} \\ \cdots & \cdots & \cdots \\ a_{t-m-1,0}^{(\ell)} & \cdots & a_{t-m-1,t}^{(\ell)} \end{pmatrix}$$

where $\ell = 1, \ldots, t - m$ and $\frac{\partial h_i}{\partial \psi_j}$ stands for the derivative $\frac{\partial h_i}{\partial y_j}$ computed at $y_j = \psi_j(x_{t+1}, \ldots, x_N)$ for $i = 0, \ldots, t$ and $j = 0, \ldots, m$. Let s denote the common degree of the polynomials Q_ℓ. Taking the Laplace expansion along the first row, one has an expression of the form:

$$Q_\ell = M_{\ell,0}x_0 + \ldots + M_{\ell,t}x_t,$$

where $M_{\ell,i}$, $\ell = 1, \ldots, t - m$, $i = 0, \ldots, t$ are homogeneous polynomials of degree $s - 1$ in $x_{t+1}, \ldots, x_N$.

Fix an integer $d > s$ and set $\mu = [\frac{d}{s}]$. Fix biforms $P_k(z_1, \ldots, z_{t-m}; x_{t+1}, \ldots, x_N)$ of bidegree $k, d - ks$, $k = 0, \ldots, \mu$. Finally, set

$$f(x_0, \ldots, x_N) := \sum_{k=0}^{\mu} P_k(Q_1, \ldots, Q_{t-m}, x_{t+1}, \ldots, x_N), \tag{7.23}$$

which is a form of degree d in $x_0, \ldots, x_N$.

The polynomial f is called a **Gordan–Noether polynomial** (or a *GN-polynomial*) of type (N, t, m, s), and so is any polynomial which can be obtained from it by a projective change of coordinates. Accordingly, a **Gordan–Noether hypersurface** (or a *GN-hypersurface*) of type (N, t, m, s) is the hypersurface $V(f) \subset \mathbb{P}^N$, where f is a GN-polynomial of type (N, t, m, s).

Let f be a GN-polynomial of type (N, t, m, s). The **core of** $V(f)$ is the t-dimensional subspace $\Pi \subset V(f)$ defined by the equations $x_{t+1} = \ldots = x_N = 0$.

The main point of the Gordan–Noether construction is that a GN-polynomial has vanishing Hessian. For a proof, see [37, Proposition 2.9]. Another proof closer to Gordan–Noether's original is contained in [127].

Following [37, 152] we give a geometric description of a GN-hypersurface of type (N, t, m, s) as follows.

Proposition 7.4.2 ([152] and [37, Proposition 2.11]) *Let $X = V(f) \subset \mathbb{P}^N$ be a GN-hypersurface of type (N, t, m, s) and degree d. Set $\mu = [\frac{d}{s}]$. Then*

1. *$X = V(f)$ has multiplicity $d - \mu$ at the general point of its core Π.*
2. *The general $(t+1)$-dimensional subspace $\Pi_\xi \subset \mathbb{P}^N$ through Π cuts out on $V(f)$, off Π, a cone of degree μ whose vertex is an m-dimensional subspace $\Gamma_\xi \subset \Pi$.*
3. *As Π_ξ varies the corresponding subspace Γ_ξ describes the family of tangent spaces to an m-dimensional unirational subvariety $S(f)$ of Π.*
4. *If $V(f)$ is general and $\mu > N - t - 2$ then $V(f)$ is not a cone.*
5. *The general GN-hypersurface is irreducible and has vanishing hessian.*
6. *Every GN-hypersurface of type (N, t, m, s) has vanishing hessian.*

Definition 7.4.3 (Franchetta Hypersurface, [62]) A reduced hypersurface $X = V(f) \subset \mathbb{P}^4$ of degree d is said to be a *Franchetta hypersurface* if it is swept out by a one-dimensional family Σ of planes such that:

- all the planes of the family Σ are tangent to a plane rational curve C (of degree $p > 1$) lying on X;
- the family Σ and the curve C are such that for a general hyperplane $H = \mathbb{P}^3 \subset \mathbb{P}^4$ passing through C, the intersection $H \cap X$, off the linear span of C, is the union of planes of Σ all tangent to the curve C in the same point p_H.

Remark 7.4.4 Note that by Proposition 7.4.2 a GN-hypersurface $X = V(f) \subset \mathbb{P}^4$ of type $(4, 2, 1, s)$ is a Franchetta hypersurface with core the linear span of the curve

C. Conversely, Permutti proved in [152] that a Franchetta hypersurface $V(f) \subset \mathbb{P}^4$ is a GN-hypersurface of type $(4, 2, 1, s)$. In particular, a Franchetta hypersurface has vanishing Hessian. This fact can be proved directly, as in [152] or [37, Proposition 2.18].

We first survey the construction of the generalization of the Gordan–Noether–Permutti hypersurfaces considered in [153] and then the further extensions of [37, Sect. 2]. For details and proofs we shall refer directly to the above cited papers.

Definition 7.4.5 (Permutti Hypersurfaces, [153]) This class of examples is an extension of the Gordan–Noether construction to the case $t = m + 1$.

Fix integers N, t such that $N \geq 2$, $1 \leq t \leq N - 2$. Fix $t + 1$ homogeneous polynomials $M_0(x_{t+1}, \ldots, x_N), \ldots, M_t(x_{t+1}, \ldots, x_N)$ of the same degree $n - 1$ in the variables $x_{t+1}, \ldots, x_N$ and assume that they are algebraically dependent over K— which will be automatic if $N \leq 2t$ because then the number $N - t$ of variables is smaller than the number $t + 1$ of polynomials.

Set $Q = M_0 x_0 + \ldots + M_t x_t$, a form of degree n. Fix an integer $d > n$ and set $\mu = [\frac{d}{n}]$. Further fix forms $P_k(x_{t+1}, \ldots, x_N)$ of degree $d - kn$ in $x_{t+1}, \ldots, x_N$, $k = 0, \ldots, \mu$. The form of degree d

$$f(x_0, \ldots, x_N) = \sum_{k=0}^{\mu} Q^k P_k(x_{t+1}, \ldots, x_N),$$

or any form obtained thereof by a linear change of variables, will be called a **Permutti polynomial**, or a P-*polynomial* of type (N, t, n). Accordingly, the corresponding hypersurface $V(f) \subset \mathbb{P}^N$ will be called a **Permutti hypersurface** or P-*hypersurface* of type (N, t, n), with *core* the t-dimensional subspace Π with equations $x_{t+1} = \cdots = x_N = 0$. It is immediate that a GN-polynomial of type $(N, t, t - 1, n)$ is a P-polynomial of type (N, t, n).

One can easily prove the analogue of Proposition 7.4.2 in Permutti's setup.

Proposition 7.4.6 ([37, Proposition 2.13]) *Let* $X = V(f) \subset \mathbb{P}^N$ *be a general P-hypersurface of type* (N, t, n) *and degree* d. *Set* $\mu = [\frac{d}{n}]$. *Then*

1. *$V(f)$ has multiplicity $d - \mu$ at the general point of its core Π.*
2. *The general $(t + 1)$-dimensional subspace Π' through Π cuts out on $V(f)$, off Π, a cone of degree at most μ, consisting of μ subspaces of dimension t which all pass through a subspace Γ of Π' of dimension $t - 1$.*
3. *As Π' varies the corresponding Γ describes a unirational family of dimension $\chi \leq \min\{t - 1, r - t - 1\}$.*
4. *If $\mu > r - t - 2$, then $V(f)$ is a cone if and only if the forms $M_0, \ldots, M_t$ are linearly dependent over k. This in turn happens as soon as either $t = 1$, or $n = 1, 2$.*
5. *$V(f)$ is irreducible.*
6. *Every P-hypersurface of type (N, t, n) has vanishing hessian.*

For P-hypersurfaces $X = V(f) \subset \mathbb{P}^N$ one can describe the dual variety $X^* \subset \mathbb{P}^{N*}$. Note that, as ξ varies in the subspace $\bar{\Pi}$ with equations $x_0 = \cdots = x_t = 0$, then the subspace $\Gamma_\xi^\perp$ of dimension $N - t$ varies describing a cone $W \subset \mathbb{P}^{N*}$, of dimension $N - t - 1$ with vertex $\Pi^\perp$ which contains the subspace $\Pi_\xi^\perp$ of dimension $N - t - 2$. Indeed, one proves that

Proposition 7.4.7 ([37, Proposition 2.14]) *Let $X = V(f) \subset \mathbb{P}^N$ be a general P-hypersurface of type (N, t, n) and degree d. Let $\mu = [\frac{d}{n}]$ and let the notation be as above. Then:*

(i) *$X^* \subset W$, where $W \subset \mathbb{P}^{N*}$ is a cone over a unirational variety of dimension $\chi \leq \min\{t - 1, N - t - 1\}$ whose vertex is the orthogonal of the core Π of X.*

(ii) *The general ruling of the cone $W \subset \mathbb{P}^{N*}$ is an $(N - t)$-dimensional subspace through $\Pi^\perp$ which cuts X^*, off $\Pi^\perp$, in μ subspaces of dimension $N - t - 1$ all passing through the same subspace of $\Pi^\perp$ of dimension $N - t - 2$. Hence $\dim(X^*) = \min\{N - 2, 2(N - t - 1)\}$.*

Conversely, if $X \subset \mathbb{P}^N$ is the dual of such a variety, then $X \subset \mathbb{P}^N$ is a P-hypersurface.

From this we also see that a general P-hypersurface is not a cone. In addition, one has:

Proposition 7.4.8 ([37, Proposition 2.15]) *Let $X = V(f) \subset \mathbb{P}^N$ be a general P-hypersurface of type (N, t, n). Then $Z = W \subset \mathbb{P}^{N*}$, and therefore $\dim(Z^*) = \min\{N - 1, 2(N - t) - 1\}$.*

Remark 7.4.9 As pointed out in [37, Remark 2.16], the case $t = N - 2$ deserves special attention. Indeed, under these hypotheses X^* is a scroll surface with a line directrix $L = \Pi^\perp$ of multiplicity $e \geq \mu$, where μ is the number of lines of the ruling of X^*, different from L, and passing through a general point of L.

It is a subvariety of the three-dimensional rational cone W over a curve with vertex L, and the general plane ruling of the cone cuts X^* along μ lines of X^*, all passing through the same point of L. In particular, for $\mu = 1$, the dual X^* is a rational scroll. According to Proposition 7.4.8, we have $Z = W$, hence $\dim(Z) = 3$. This should be compared with Corollary 7.3.10.

If $t = 2$ the two constructs of GN-hypersurfaces and P-hypersurfaces coincide. For $N = 4$ this is the only value of t which leads to hypersurfaces which are not cones. The case $N = 4$ is well understood due to the results of Gordan–Noether in [78] and of Franchetta in [62]. In the next section we shall provide another proof following strictly the approach in [72], which is a combination of Gordan–Noether Theory with the geometrical approach of Franchetta.

The case of cubic hypersurfaces has been studied in some detail by U. Perazzo (see [151]) and more recently in [77]. We will partly generalize Perazzo's results. Inspired by the construction of P-hypersurfaces and by Perazzo's results, we will give new examples of hypersurfaces with vanishing Hessian, which are extensions of some P-hypersurfaces.

Definition 7.4.10 (Perazzo Examples, [151]) Let $X = V(f) \subset \mathbb{P}^N$ be a hypersurface which contains a subspace Π of dimension t such that the general subspace Π_ξ of dimension $t+1$ through Π cuts out on $V(f)$ a cone with a vertex Γ_ξ of dimension s. Assume that $s \geq N - t - 1$. By extended analogy, we will call Π the *core* of $V(f)$ and call $V(f)$ an H-*hypersurface* of type (N, t, s). Notice that a P-hypersurface of type (N, t, n) with $N \leq 2t$ is also an H-hypersurface of type $(N, t, t-1)$.

As for P-hypersurfaces, we can introduce the cone $W \subset \mathbb{P}^{N*}$ with vertex $\Pi^\perp$, which is swept out by the $(N-s-1)$-dimensional subspaces $\Gamma_\xi^\perp$ as Π_ξ varies among all subspaces of dimension $t + 1$ containing Π.

A special case of an H-hypersurface is that of a hypersurface $X = V(f) \subset \mathbb{P}^N$ of degree d containing a subspace Π of dimension t whose general point has multiplicity $d - \mu > 0$ for $V(f)$, such that the general subspace Π_ξ of dimension $t+1$ through Π cuts out on $V(f)$, off Π, a union of μ subspaces of dimension t, with $\mu \leq 2t - N + 1$. In this situation, we will call $X = V(f) \subset \mathbb{P}^N$ an R-*hypersurface* of type (N, t, μ).

With these definitions we can recall the following result.

Proposition 7.4.11 *An* H-*hypersurface* $X = V(f) \subset \mathbb{P}^N$ *of type* (N, t, s) *has vanishing Hessian. Moreover,* $Z = W \subset \mathbb{P}^{N*}$.

Remark 7.4.12 As described in [37, Remark 2.19], the duals of R-hypersurfaces of degree d and type $(N, N - 2, \mu)$ are very interesting and particular. Indeed, $X = V(f) \subset \mathbb{P}^N$ is an R-hypersurface as above, its dual $X^* \subset \mathbb{P}^{N*}$ is a scroll surface with a line directrix L of multiplicity $e \geq \mu$, where $\mu \leq N - 3$ is the number of lines of the ruling of X^*, different from L, and passing through a general point of L.

An R-hypersurface with $\mu = 1$ is a hypersurface of degree d with a core Π of dimension t whose general point has multiplicity $d - 1$ for the hypersurface, and moreover $2t \geq r$. This is the case considered by Perazzo in [151, p. 343], where he proves that these hypersurfaces have vanishing Hessian.

7.4.2 A Geometrical Proof of the Gordan–Noether and Franchetta Classification of Hypersurfaces in $\mathbb{P}^4$ with Vanishing Hessian

In this subsection we shall prove in a geometrical way but using Gordan–Noether Identities that hypersurfaces in $\mathbb{P}^4$ with vanishing Hessian are either cones or Franchetta hypersurfaces. We shall deeply rely on the treatment developed by Garbagnati and Repetto in [72].

We begin with a preliminary result.

Lemma 7.4.13 *Let* $X = V(f) \subset \mathbb{P}^4$ *be a hypersurface of degree* $d \geq 3$, *not a cone. If* $X = V(f)$ *has vanishing Hessian, then* $Z^* \subset \mathbb{P}^4$ *is an irreducible plane rational* *curve. Equivalently Z is a cone with vertex a line over an irreducible plane rational curve.*

Proof By Corollary 7.3.10, we can suppose $\dim(Z) = 3$ and let $Z = V(g) \subset \mathbb{P}^{4*}$. Since X is not a cone $1 \leq \dim(Z^*) \leq 2$, where the last equality follows from $Z^* \subseteq Bs(\psi_g)$.

Assume first $\dim(Z^*) = 2$ and let $r_1, r_2 \in Z^*$ be general points. Then there exists a

$$t \in \overline{\psi_g^{-1}(r_1)} \cap \overline{\psi_g^{-1}(r_2)}.$$

Arguing as in the proof of Corollary 7.3.9 we deduce $< r_i, t > \subseteq Bs(\psi_g)$ and hence $< r_i, t > \subseteq Z^*$. Thus Z^* is a cone since it is a surface ruled by lines with the property that two general lines of the ruling intersect. This would imply that $Z \subset \mathbb{P}^{4*}$ is a plane and hence that $X \subset \mathbb{P}^4$ is a cone, contrary to our hypothesis.

Thus $\dim(Z^*) = 1$ and $\dim(Bs(\psi_g)) = 2$. Let $r_1, r_2 \in Z^*$ be general points. Then

$$\dim(\overline{\psi_g^{-1}(r_1)} \cap \overline{\psi_g^{-1}(r_2)}) = 2$$

and

$$\overline{\psi_g^{-1}(r_1)} \cap \overline{\psi_g^{-1}(r_2)} \subseteq Bs(\psi_g)$$

imply that there exists an irreducible surface P, irreducible component of $Bs(\psi_g)$, such that $Z^* \subseteq P$ and such that $P \subseteq \overline{\psi_g^{-1}(r_1)} \cap \overline{\psi_g^{-1}(r_2)}$ for all $r_1, r_2 \in Z^*$ general. Since $< r_i, p > \subseteq P$ for all $p \in P$ general by the Gordan–Noether Identity for Ψ_g, we deduce $Z^* \subseteq Vert(P)$, yielding $P = \mathbb{P}^2 = \langle Z^* \rangle$. $\qquad\square$

The description given in Lemma 7.4.13 is the main ingredient to prove that a projective hypersurface $X = V(f) \subset \mathbb{P}^4$ with vanishing Hessian, not a cone, is a Franchetta hypersurface.

Theorem 7.4.14 (Gordan–Noether–Franchetta Classification, [62, 78]) *Let $X = V(f) \subset \mathbb{P}^4$ be an irreducible hypersurface of degree $d \geq 3$, not a cone. The following conditions are equivalent:*

1. *X has vanishing Hessian.*
2. *X is a Franchetta hypersurface.*
3. *X^* is a scroll surface of degree d, having a line directrix L of multiplicity e, sitting in a three-dimensional rational cone $W(f)$ with vertex L, and the general plane ruling of the cone cuts X^* off L along $\mu \leq e$ lines of the scroll, all passing through the same point of L.*
4. *X is a general GN-hypersurface of type $(4, 2, 1, s)$, with $\mu = [\frac{d}{s}]$, which has a plane of multiplicity $d - \mu$.*

In particular, $X^* \subset \mathbb{P}^{N^*}$ is smooth if and only if $d = 3$, X^* is a rational normal scroll of degree 3 and X contains a plane, the orthogonal of the line directrix of X^*, with multiplicity 2; equivalently X is the projection of $\mathbb{P}^1 \times \mathbb{P}^2 \subset \mathbb{P}^5$ from an external point.

Proof Conditions (2) and (3) are easily seen to be equivalent (the directrix line L of X^* is the dual of the plane which is the linear span of the curve C of the Franchetta hypersurface).

By Remark 7.4.4, the equivalence of (2) and (4) is clear. Condition (4) implies (1) by Proposition 7.4.2. Thus to finish the proof it is sufficient to prove that a hypersurface $X = V(f) \subset \mathbb{P}^4$ with vanishing Hessian, not a cone, is a Franchetta hypersurface, or equivalently that X is as in case (3).

By Lemma 7.4.13, we know that $Z^* \subset \mathrm{Sing}(X) \subset X = V(f)$ is an irreducible plane rational curve, whose linear span $\varPi = \mathbb{P}^2$ is an irreducible component of $\mathrm{Bs}(\psi_g)$ and of $\mathrm{Sing}(X)$. Therefore $Z \subset \mathbb{P}^{4^*}$ is a cone of vertex the line $L = \varPi^* = \mathbb{P}^1$ over an irreducible plane curve $\varGamma$, the dual of Z^* as a plane curve.

Consider now a general hyperplane $H \subset \mathbb{P}^4$ passing through the plane $\varPi$. The intersection $X \cap H$ is a hypersurface of degree d in $H = \mathbb{P}^3$ containing the plane $\varPi$ with a certain multiplicity $m > 0$. Note also that the point $h = H^* \in L = \varPi^*$ (because $\varPi \subset H$), whence $\pi_h(Z) \subset \mathbb{P}^3$ is a non-degenerate surface naturally embedded in the dual space of H. More precisely, $\pi_h(Z)$ is a cone with vertex the point $p_L = \pi_h(L)$ over the plane curve $\hat{\varGamma} = \pi_h(\varGamma) \simeq \varGamma$.

By Lemma 7.2.1 we infer that $\nabla_{X \cap H} \subset \pi_h(Z) \subset \mathbb{P}^{3^*}$, so that $X \cap H \subset H = \mathbb{P}^3$ has vanishing Hessian. By Corollary 7.3.8 it follows that either $X \cap H$ is a cone over a plane curve with vertex a point or $X \cap H$ consists of $d - m$ distinct planes, eventually counted with multiplicity, passing through a line. In the first case $\nabla_{X \cap H}$ would be a plane in $\mathbb{P}^{3^*}$, which is impossible because $\pi_h(Z)$ is a non-degenerate cone with vertex a point.

Therefore the closure of $\nabla_{X \cap H}$ is a line l_H in $H^* = \mathbb{P}^3$. The line l_H is contained in $\overline{\nabla_X(H)}$ and, by Lemma 7.2.1, $\overline{\nabla_X(H)} \subseteq \varPi_H := \langle h, l_H \rangle$. We now prove that $\overline{\nabla_X(H)} = \varPi_H$. Indeed, for a general point q, $\nabla_X^{-1}(\nabla_X(q))$ is a line, we call it L_q. The closure of the fiber of $\nabla_{X|H}$ passing through q is either the point q or the line L_q. If it were the point q, $\dim(\overline{\nabla_X(H)}) = 3$, which is impossible, because $\overline{\nabla_X(H)} \subseteq \varPi_H$. Hence it is the line L_q and $\dim(\overline{\nabla_X(H)}) = 2$, i.e. $\overline{\nabla_X(H)} = \varPi_H$.

Therefore $X \cap H$, off $\varPi$, is a union of $d - m$ planes passing through the line T. Moreover,

$$\nabla_{X \cap H}(X \cap H) = \{p_1, \ldots, p_{d-m}\} \subset l_H$$

and

$$\overline{\nabla_X(X \cap H)} = \langle h, p_1 \rangle \cup \ldots \cup \langle h, p_{d-m} \rangle \subset \varPi_H = \nabla_X(H).$$

Varying $H \supset \Pi$ we deduce that X^* is a scroll surface, having as line directrix L and such that the general plane ruling of the cone Z cuts X^* off L along $d - m = \mu$ lines of the scroll, all passing through the same point $h \in L$. The scroll surface $X^* \subset \mathbb{P}^{4*}$ is a non-developable surface, in fact $(X^*)^* = X$ is a hypersurface in $\mathbb{P}^4$ (cf. [37, Sect. 1.2]). Moreover, $\deg(X^*) = \deg((X^*)^*) = \deg(X) = d$ (cf. [37, Sect. 1.2]) and $X \subset \mathbb{P}^4$ is a hypersurface as in (3).

This geometrical description also assures us that a general H through Π cuts X along $d - m$ distinct planes. For such a general H, let $z = \psi_g(H) \in Z^*$ (Z^* is the plane curve dual to Z with respect to the plane Π). Let Π_H^* be the dual of the plane with respect to the ambient space $\mathbb{P}^4$. Since $\Pi_H^* = T_z(Z^*) = T$, the line of intersection of the planes in $X \cap H$ is the tangent line to the plane curve Z^* at the point z. In conclusion, $X \subset \mathbb{P}^4$ is a Franchetta hypersurface, where we can take as the one-dimensional family Σ of planes contained in X exactly the intersection of a general $\mathbb{P}^3$ through Π with X and we consider as the curve C (cf. Definition 7.4.3) the curve Z^*.   □

7.5   The Perazzo Map of Hypersurfaces with Vanishing Hessian

Let us introduce the so-called *Reciprocity Law of Polarity* to be used later on in the analysis of the geometry of Z^*. We define first the notion of *degree s polar hypersurface of $X = V(f) \subset \mathbb{P}^N$*.

Definition 7.5.1 (s^{th}-**Polar Hypersurface to** $X = V(f) \subset \mathbb{P}^N$) For $s = 1, \ldots, d-1$ and for every $p \in \mathbb{P}^N$ the *degree s polar of X with respect to p* is the hypersurface

$$H_p^s(f) := V\left(\sum_{i_0 + \ldots + i_N = s} \frac{\partial^s f}{\partial x_0^{i_0} \ldots \partial x_N^{i_N}}(p) x_0^{i_0} \ldots x_N^{i_N} \right) \subset \mathbb{P}^N.$$

By definition $\deg(H_p^s(f)) = s$ if the polynomial on the right in the above expression is not identically zero. Otherwise we naturally put $H_p^s(f) = \mathbb{P}^N$. For $s = 1$ the hyperplane $H_p^1(f)$ is the polar hyperplane of X with respect to p, which will be indicated simply by H_p. For $s = 2$ the hypersurface $H_p^2(f)$ is a quadric hypersurface whose associated symmetric matrix is $\mathrm{Hess}(f)(p)$ and we shall put, by abusing notation, $Q_p = H_p^2(f)$ if the reference to f is well understood.

We need a classical result used repeatedly in the sequel. For a proof and other applications one can consult the first chapter of [48].

Proposition 7.5.2 (Reciprocity Law of Polarity) *Let $X = V(f) \subset \mathbb{P}^N$ be a degree d hypersurface. Then for every $s = 1, \ldots, d-1$ and for every pair of distinct points $p, q \in \mathbb{P}^N$ we have*

$$p \in H_q^s(f) \iff q \in H_p^{d-s}(f).$$

In particular,

$$\{p \in X \; : \; \text{mult}_p(X) \geq s\} = \bigcap_{q \in \mathbb{P}^N} H_q^{d-s+1}. \tag{7.24}$$

Perazzo introduced in [151] the notion of (Perazzo) rank of a cubic hypersurface with vanishing hessian, which we now extend to the general case. Although he does not explicitly define the rational map described below, its use was implicit in his analysis, see *loc. cit.*

Definition 7.5.3 (Perazzo Map, [151]) Let $X = V(f) \subset \mathbb{P}^N$ be a reduced hypersurface with vanishing hessian, let $\nabla_X : \mathbb{P}^N \dashrightarrow \mathbb{P}^{N*}$ be its polar map and let $Z = \overline{\nabla_X(\mathbb{P}^N)} \subsetneq \mathbb{P}^{N*}$ be its polar image. *The Perazzo map of X is the rational map:*

$$\mathbf{P}_X : \mathbb{P}^N \dashrightarrow \mathbb{G}(\text{codim}(Z) - 1, N)$$
$$p \mapsto (T_{\nabla_X(p)}Z)^*$$

which is defined at least in the open set $\nabla_X^{-1}(Z_{\text{reg}})$, where Z_{reg} is the locus of smooth points of Z.

With this notation we have that for a general point $p \in \mathbb{P}^N$, $\text{Sing}\, Q_p = (T_{\nabla_X(p)}Z)^*$. Moreover, by definition $\mathbf{P}_X$ is the composition of ∇_X with the dual Gauss map of Z.

The image of the Perazzo map will be denoted by

$$W_X = \overline{\mathbf{P}_X(\mathbb{P}^N)} \subset \mathbb{G}(\text{codim}(Z) - 1, N),$$

or simply by W, and $\mu = \dim W$ is called the *Perazzo rank of X*.

Remark 7.5.4 If $\mu = 0$, then $Z \subset \mathbb{P}^{N*}$ is a linear space and X is a cone such that $\text{Vert}(X) = Z^* = \mathbb{P}^{\text{codim}(Z)-1} \subseteq X$.

Therefore if $X = V(f) \subset \mathbb{P}^N$ is a hypersurface with vanishing hessian, not a cone, necessarily $\mu \geq 1$.

If $\text{codim}(Z) = 1$ and if $Z = V(g)$, then $\mathbf{P}_X = \psi_g$, where ψ_g is the Gordan–Noether map defined by g, see Definition 7.3.1.

The following result will be useful when determining the structure of particular classes of cubic hypersurfaces with vanishing hessian defined in the sequel. First, let us remark that for a cubic hypersurface and for $r, s \in \mathbb{P}^N$ we have: $r \in \text{Sing}\, Q_s$ if and only if $s \in \text{Sing}\, Q_r$.

Theorem 7.5.5 ([151], [77, Theorem 2.5]) *Let $X = V(f) \subset \mathbb{P}^N$ be a cubic hypersurface with vanishing hessian. Let $w = [(T_{\nabla_X(p)}Z)^*] \in W_X \subset \mathbb{G}(\text{codim}(Z) - 1, N)$ be a general point and let $r \in (T_{\nabla_X(p)}Z)^*$ be a general point with $p \in \mathbb{P}^N$ general. Then:*

$$\overline{\mathbf{P}_X^{-1}(w)} = \bigcap_{r \in (T_{\nabla_X(p)}Z)^* = \text{Sing}\, Q_p} \text{Sing}\, Q_r = \mathbb{P}_w^{N-\mu}. \tag{7.25}$$

Proof By definition $\overline{\mathbf{P}_X^{-1}(w)}$ is the closure of the set of all (general) $p' \in \mathbb{P}^N$ such that $\mathrm{Sing}(Q_{p'}) = \mathrm{Sing}(Q_p)$. This happens if and only if $r \in \mathrm{Sing}(Q_p)$ implies $r \in \mathrm{Sing}(Q_{p'})$ (or vice versa by symmetry and by the generality of $p, p' \in \mathbb{P}^N$), which in turn happens if and only if $p' \in \mathrm{Sing}(Q_r)$ for every $r \in \mathrm{Sing}\, Q_p$, yielding the first equality in (7.25) and concluding the proof. $\square$

Remark 7.5.6 Formula (7.25) is a particular case of general results on linear systems of quadrics *with tangential defect*, see [42, 119] and particularly [8, Corollary 1] where a proof of (7.25) is deduced from these general facts. As shown by the examples in the last pages of [8], the fibers of the Perazzo map are not necessarily linear for hypersurfaces of degree greater than three. The general structure of the fibers of the Perazzo map for arbitrary hypersurfaces is, to the best of our knowledge, unknown.

The linearity of the fibers has strong consequences for the geometry of a cubic hypersurface with vanishing hessian. Let us recall some easy and well-known facts on *congruences of order one of linear spaces*, that is irreducible families $\Theta \subset \mathbb{G}(\beta, N)$ of linear spaces of dimension $\beta > 0$ such that through a general point of $\mathbb{P}^N$ there passes a unique member of the family. Clearly the previous condition forces $\dim(\Theta) = N - \beta$ and that the tautological map $p : \mathscr{U} \to \mathbb{P}^N$ from the universal family $\pi : \mathscr{U} \to \Theta$ is birational onto $\mathbb{P}^N$.

Let the notation be as above and let

$$V = \{q \in \mathbb{P}^N \; ; \; \#(p^{-1}(q)) \geq 2\} = \{q \in \mathbb{P}^N \; ; \; \dim(p^{-1}(q)) > 0\} \subset \mathbb{P}^N$$

be the so-called *jump (or branch) locus of* Θ.

The easiest example of congruences of linear spaces of dimension β is given by the family of linear spaces of dimension $\beta + 1$ passing through a fixed linear space $L = \mathbb{P}^\beta \subset \mathbb{P}^N$.

This motivates the following definition.

Definition 7.5.7 An irreducible cubic hypersurface $X \subset \mathbb{P}^N$ with vanishing hessian, not a cone, will be called a *Special Perazzo Cubic Hypersurface* if the general fibers of its Perazzo map form a congruence of linear spaces of dimension $N - \mu$ passing through a fixed $\mathbb{P}^{N-\mu-1}$.

Special Perazzo Cubic Hypersurfaces will be described from the geometric and algebraic point of view in Sect. 7.6.1 following very closely the original treatment of Perazzo, providing canonical forms and also an interesting projective characterization in any dimension. Let us remark that for Special Perazzo Cubic Hypersurfaces the linear span of two general fibers is a $\mathbb{P}^{N-\mu+1}$ which is strictly contained in $\mathbb{P}^N$ if $\mu > 1$. For $\mu = 1$ the general fibers determine a pencil of hyperplanes and this case can be treated in arbitrary dimension as we shall see below.

7.6 Cubic Hypersurfaces with Vanishing Hessian and Their Classification for $N \leq 6$

We now survey the classification of cubic hypersurfaces in low dimension according to [77]. First we define an interesting class of cubic hypersurfaces with vanishing hessian which a posteriori will be seen to be equivalent to that of Special Cubic Perazzo Hypersurfaces.

7.6.1 Classes of Cubic Hypersurfaces with Vanishing Hessian According to Perazzo and Canonical Forms of Special Perazzo Cubic Hypersurfaces

The examples of hypersurfaces with vanishing hessian, not cones, considered in [78] and in Perazzo [151] are singular along linear spaces, see also [37].

We introduce some terminology in order to obtain a projective characterization of Special Perazzo Cubic Hypersurfaces.

This approach follows strictly the original work of Perazzo, who used the next results to provide simplified canonical forms for Special Perazzo Cubic Hypersurfaces, see [151, Sects. 14–18, pp. 344–350], and it is based on [77, Sect. 3].

The first definition is a generalization of the classical notion of *generatrix* of a cone with vertex a point, that is a line contained in the cone and passing through the vertex.

Definition 7.6.1 Let $Q \subset \mathbb{P}^N$ be a quadric of rank $\text{rk}(Q) = \beta \geq 3$ and let $\text{Vert}(Q) = \text{Sing } Q = \mathbb{P}^{N-\beta}$ be its vertex. A linear space $M = \mathbb{P}^\tau, \tau \geq 1$, is a *generator of Q* if

$$\text{Vert}(Q) \subsetneq M \subset Q.$$

The proof of the next result is left to the reader.

Lemma 7.6.2 *Let $Q \subset \mathbb{P}^N$ be a quadric of rank $\text{rk}(Q) = \beta$ and let $M = \mathbb{P}^\tau$ be a generator of Q. Then*

$$\bigcap_{m \in M \backslash \text{Vert}(Q)} T_m Q = \mathbb{P}(\nabla_Q(M))^* = \mathbb{P}^{2N-\beta-\tau}.$$

Definition 7.6.3 Let $Q \subset \mathbb{P}^N$ be a quadric, let M be a generator of Q and let L be a linear space containing M. The quadric is *tangent to the linear space L along the generator M* if $L \subseteq T_m Q$ for all $m \in M$.

Obviously a necessary condition for L being tangent to Q along $M = \mathbb{P}^\tau$ is

$$L \subseteq \bigcap_{m \in M \setminus \mathrm{Vert}(Q)} T_m Q = \mathbb{P}^{2N - \mathrm{rk}(Q) - \tau},$$

yielding $\dim(M) = \tau \leq \dim(L) \leq 2N - \mathrm{rk}(Q) - \tau$.

The next result is the core of the work of Perazzo on canonical forms. We shall assume $\mathrm{codim}(Z) = 1$ or equivalently $\mathrm{rk}(Q_p) = N$ for $p \in \mathbb{P}^N$ general.

We shall consider $\mathrm{Sing}\, Q_p \subset M \subset Q_p$ with $M = \mathbb{P}^\tau$ a common generator having equations: $x_{\tau+1} = \ldots = x_N = 0$. Furthermore, we shall suppose

$$L = \bigcap_{m \in M \setminus \mathrm{Vert}(Q_p)} T_m Q_p = \mathbb{P}^{N-\tau},$$

that is, L will always be the maximal linear subspace containing M along which the quadric Q_p can be tangent along the generator M. Moreover, we shall assume that L has equations: $x_{N-\tau+1} = \ldots = x_N = 0$. Under these hypothesis $\dim(L) = N - \tau \geq \tau = \dim(M)$ with equality if and only if $L = M$.

Theorem 7.6.4 ([151], [77, Theorem 3.6]) *Let $X = V(f) \subset \mathbb{P}^N$ be a cubic hypersurface having vanishing hessian, not a cone, and with $\mathrm{codim}(Z) = 1$. Suppose that the polar quadric of a general point of $\mathbb{P}^N$ is tangent to a fixed linear space $L = \mathbb{P}^{N-\tau}$ along a common generator M with $M = V(x_{\tau+1}, \ldots, x_N)$ and $L = V(x_{N-\tau+1}, \ldots, x_N)$. Then:*

$$f = \sum_{i=0}^{\tau} x_i C^i(x_{N-\tau+1}, \ldots, x_N) + D(x_{\tau+1}, \ldots, x_N), \tag{7.26}$$

where the C_i's are linearly independent quadratic forms depending only on the variables $x_{N-\tau+1}, \ldots, x_N$ and D is a cubic in the variables $x_{\tau+1}, \ldots, x_N$.

Conversely, for every cubic hypersurface $X = V(f) \subset \mathbb{P}^N$ with f as in (7.26) the polar quadric of a general point of $\mathbb{P}^N$ is tangent to $V(x_{N-\tau+1}, \ldots, x_N)$ along the common generator $V(x_{\tau+1}, \ldots, x_N)$. Moreover, such a cubic hypersurface has vanishing hessian and $\mathrm{codim}(Z) = 1$.

The next result is crucial to obtain a simplified canonical form for Special Perazzo Cubic Hypersurfaces and reveals an interesting projective characterization of these hypersurfaces. Our proof was completely inspired by the calculations made by Perazzo in [151, Sects. 1.14–1.16 pp. 344–349], where as always we shall suppose $\mathrm{codim}(Z) = 1$ (that is, $p = 0$ in Perazzo's notation).

Theorem 7.6.5 ([151], [77, Theorem 3.7]) *Let $X = V(f) \subset \mathbb{P}^N$ be an irreducible cubic hypersurface with vanishing Hessian, not a cone, and such that $\mathrm{codim}(Z) = 1$.*

Then the following conditions are equivalent:

i) $X \subset \mathbb{P}^N$ *is a Special Perazzo Cubic Hypersurface with* $\dim(Z^*) + 1 = \sigma$ *and with* $L = \bigcap_{w \in W \text{general}} \mathbb{P}_w^{N - \dim(Z^*)} = \mathbb{P}^{N-\sigma}$;

ii) $\dim(Z^*) = \sigma - 1$ *and the polar quadric* Q_p *of a general point* $p \in \mathbb{P}^N$ *is tangent to the linear space* $L = \mathbb{P}^{N-\sigma}$ *along a common generator* $M = \mathbb{P}^\sigma$ *of* Q_p *with* $\bigcap_{m \in M \backslash \text{Vert}(Q_p)} T_m Q_p = L$.

Moreover, $X \subset \mathbb{P}^N$ *is projectively equivalent to*

$$V(\sum_{i=0}^{\sigma} x_i C^i(x_{N-\sigma+1}, \dots, x_N) + D(x_{\sigma+1}, \dots, x_N)) \subset \mathbb{P}^N. \tag{7.27}$$

7.6.2   Cubics with Vanishing Hessian in $\mathbb{P}^N$ with $N \leq 6$

The geometric structure and the canonical forms of cubic hypersurfaces with vanishing hessian and with $\dim(Z^*) = 1$ are completely described in the next result.

Theorem 7.6.6 ([77, 151]) *Let* $X \subset \mathbb{P}^N$ *be a cubic hypersurface with vanishing hessian, not a cone, with* $\dim(Z^*) = 1$. *Then* $N \geq 4$, X *is a Special Perazzo Cubic Hypersurface and we have:*

1) $\text{codim}(Z) = 1$;
2) $SZ^* = \mathbb{P}^2 \subseteq \text{Sing} X$;
3) Z^* *is a conic;*
4) *If* $N = 4$, *then* X *is projectively equivalent to* $S(1, 2)^*$;
5) X *is projectively equivalent to*

$$V(x_0 x_{N-1}^2 + 2 x_1 x_{N-1} x_N + x_2 x_N^2 + D(x_3, x_4, \dots, x_N)).$$

For $N \leq 3$ a cubic hypersurface with vanishing hessian is easily seen to be a cone. We shall now consider the cases $N = 4, 5, 6$ with more emphasis on the case $N = 4$.

The next theorem is essentially a combination of classical results, more or less known nowadays, see [37, 72, 151] and Theorem 7.4.14 here.

Theorem 7.6.7 *Let* $X \subset \mathbb{P}^4$ *be an irreducible cubic hypersurface with vanishing hessian, not a cone. Then*

i) $(\text{Sing} X)_{\text{red}} = \mathbb{P}^2$;
ii) $X^* \simeq S(1, 2) \subset \mathbb{P}^4$;
iii) X *is projectively equivalent to a linear external projection of* $\text{Seg}(1, 2) \subset \mathbb{P}^5$;
iv) X *is projectively equivalent to* $V(x_0 x_3^2 + 2 x_1 x_3 x_4 + x_2 x_4^2) \subset \mathbb{P}^4$.

Proof First of all let us remark that codim(Z) = 1. On the contrary by cutting X with a general hyperplane H and by projecting Z from the general point [H] we would obtain a cubic hypersurface in $\mathbb{P}^3$ with vanishing hessian, which would be a cone. This would imply that X is a cone. Thus codim(Z) = 1 and reasoning as in Lemma 7.4.13 we get dim(Z^*) = 1. All conclusions now follow from Theorem 7.6.6, except for iv). Theorem 7.6.6 implies that X is projectively equivalent to $Y = V(x_0 x_3^2 + 2x_1 x_3 x_4 + x_2 x_4^2 + D(x_3, x_4)) \subset \mathbb{P}^4$ with $D(x_3, x_4) = (ax_3 + bx_4)x_3^2 + (cx_3 + dx_4)x_4^2$. The projective transformation $x_0' = x_0 + ax_3 + bx_4$, $x_1' = x_1$, $x_2' = x_2 + cx_3 + dx_4$, $x_3' = x_3$ and $x_4' = x_4$ sends Y into $V(x_0 x_3^2 + 2x_1 x_3 x_4 + x_2 x_4^2)$, as claimed.   □

The classification of cubic hypersurfaces in $\mathbb{P}^5$ and $\mathbb{P}^6$ can be summarized in the next theorem.

Theorem 7.6.8 ([151], [77, Sect. 4]) *Let $X = V(f) \subset \mathbb{P}^N$, $N = 5, 6$, be a cubic hypersurface with vanishing hessian, not a cone. Then* codim(Z) = 1 *and X is a Special Perazzo Cubic Hypersurface such that Z^* is either a conic or a surface in $\mathbb{P}^3$ of degree at most three, the last case occurring only for $N = 6$.*

Thus for $N = 4, 5, 6$ all cubic hypersurfaces with vanishing hessian are classified and we have a complete description from different points of view: geometric, algebraic or projective. For $N \geq 7$ the situation is more complicated and not all examples are of Special Perazzo type.

7.6.3 Examples in Higher Dimension

The classification of cubic hypersurfaces with vanishing hessian $X \subset \mathbb{P}^N$ for $N \leq 6$ showed that they all have codim(Z) = 1 and that they are Special Perazzo Cubic Hypersurfaces.

In [37, Sect. 2] two series of examples of hypersurfaces with vanishing hessian $X \subset \mathbb{P}^N$, $N \geq 5$, of sufficiently high degree and such that codim(Z) > 1 have been constructed, see [37, Proposition 2.15, Remarks 2.16 and 2.19] and Propositions 7.4.8 and 7.4.11 here.

Based on [77, Sect. 5] we now present some examples of cubic hypersurfaces in $\mathbb{P}^N$, $N \geq 7$, with codim(Z) sufficiently large. Later on we will also provide examples of cubic hypersurfaces with vanishing hessian not a cone in $\mathbb{P}^N$ with $N = 7, 13, 25$ which are not Special Perazzo Cubic Hypersurfaces showing that the result in Theorem 7.6.8 is sharp.

First we give two different constructions of cubic hypersurfaces with vanishing hessian, not cones, from examples in lower dimension dubbed *concatenation* and *juxtaposition*. By repeated use of these methods one can prove that for a fixed $\alpha \in \mathbb{N}$ there exists $N_0 = N_0(\alpha)$ such that for every $N \geq N_0$ there exists a cubic hypersurface $X \subset \mathbb{P}^N$ with vanishing hessian, not a cone, and such that codim(Z) = α.

Example 7.6.9 ([77, Sect. 5]) Given $g = x_0 x_3^2 + x_1 x_3 x_4 + x_2 x_4^2$ let $\tilde{g} = x_2 x_4^2 + x_5 x_4 x_7 + x_6 x_7^2$. We define the concatenation of g and $\tilde{g}$ as the form f obtained by summing up the monomials without repetition, that is

$$f = x_0 x_3^2 + x_1 x_3 x_4 + x_2 x_4^2 + x_5 x_4 x_7 + x_6 x_7^2.$$

Then $V(f) \subset \mathbb{P}^7$ is a cubic hypersurface with vanishing hessian, not a cone. Indeed, letting $f_i = \frac{\partial f}{\partial x_i}$ one easily verifies that the f_i's are linearly independent so that X is not a cone and Z is non-degenerate. Moreover,

$$f_0 f_2 = f_1^2 \quad \text{and} \quad f_2 f_6 = f_5^2,$$

yielding $\mathrm{codim}(Z) \geq 2$ and $\deg(Z) \geq 3$. Thus $\mathrm{codim}(Z) = 2$ since the polar image of every cubic hypersurface with vanishing hessian in $\mathbb{P}^5$, not a cone, is a quadric hypersurface.

Example 7.6.10 ([77, Sect. 5]) Let $g = x_0 x_3^2 + x_1 x_3 x_4 + x_2 x_4^2$ and let

$$f = x_0 x_3^2 + x_1 x_3 x_4 + x_2 x_4^2 + x_5 x_8^2 + x_6 x_8 x_9 + x_7 x_9^2$$

be the cubic form obtained by the juxtaposition of g with itself, that is, f is the sum of g and $\tilde{g} = x_5 x_8^2 + x_6 x_8 x_9 + x_7 x_9^2$ obtained by shifting the indexes of g.

As above it is possible to verify immediately that $X = V(f) \subset \mathbb{P}^9$ is not a cone and that the following algebraic relations hold $f_0 f_2 = f_1^2, f_5 f_7 = f_6^2$.

Example 7.6.11 ([77, Sect. 5]) Let $g = x_0 x_3^2 + x_1 x_3 x_4 + x_2 x_4^2$ and let

$$f = x_0 x_3^2 + x_1 x_3 x_4 + x_2 x_4^2 + x_5 x_8^2 + x_6 x_8 x_9 + x_7 x_9^2 + x_{10} x_{13}^2 + x_{11} x_{13} x_{14} + x_{12} x_{14}^2$$

be the cubic form obtained by the juxtaposition of g with itself two times.

Then $X = V(f) \subset \mathbb{P}^{14}$ is not a cone and the following algebraic relations hold: $f_0 f_2 = f_1^2, f_5 f_7 = f_6^2, f_{10} f_{12} = f_{11}^2$.

We conclude with an example of a cubic hypersurface $X \subset \mathbb{P}^7$ with vanishing hessian, not a cone, which is not a Special Perazzo Cubic Hypersurface. This example and its generalizations are included also for their relations with Severi varieties of dimension 4, 8 and 16. They were first considered in [77].

Example 7.6.12 ([77, Sect. 5]) Let

$$g = \det \begin{pmatrix} x_0 & x_1 & x_2 \\ x_3 & x_4 & x_5 \\ x_6 & x_7 & x_8 \end{pmatrix}. \tag{7.28}$$

Then $R = V(g) \subset \mathbb{P}^8$ is a cubic hypersurface. If we identify $\mathbb{P}^8$ with $\mathbb{P}(M_{3\times 3}(\mathbb{K}))$, then R is the locus of matrices of rank at most two and it is naturally identified with the secant variety of $W = \mathbb{P}^2 \times \mathbb{P}^2 \subset \mathbb{P}^8$, the locus of matrices of rank 1. Moreover, $\operatorname{Sing} R = W$ as schemes.

The polar map $\nabla_R : \mathbb{P}^8 \dashrightarrow \mathbb{P}^8$ is a birational involution sending a matrix $p = [A]$ to its cofactor matrix. Since the cofactor matrix of a rank two matrix has rank one and since $\nabla_{R|R} = \mathscr{G}_R$, we deduce that $\overline{\nabla_R(R)} = \mathscr{G}_R(R) = R^*$ is naturally identified with W. By the previous description ∇_R is an isomorphism on $\mathbb{P}^8 \setminus R$ and the closure of every positive dimensional fiber of ∇_R (and hence of every fiber of $\mathscr{G}_R$) is a $\mathbb{P}^3$. Indeed, by homogeneity it is sufficient to verify this for $\overline{\nabla_R^{-1}(q)}$ with $q = (0 : 0 : 0 : 0 : 0 : 0 : 0 : 0 : 1)$. The 3×3 matrices mapped by ∇_R to q are precisely the rank two matrices $X = [x_{i,j}]$ having $x_2 = x_3 = x_5 = x_6 = x_7 = x_8 = 0$, i.e. $\overline{\nabla_R^{-1}(q)} = V(x_2, x_5, x_6, x_7, x_8)$ is the closure of the orbit of

$$p = \begin{pmatrix} 1 & 0 & 0 \\ 0 & 1 & 0 \\ 0 & 0 & 0 \end{pmatrix} . \tag{7.29}$$

under the natural action.

Let

$$f = \det \begin{pmatrix} x_0 & x_1 & x_2 \\ x_3 & x_4 & x_5 \\ x_6 & x_7 & 0 \end{pmatrix} .$$

Then $x_8 = 0$ is the equation of $T_p R$ with $p \in R$ the point defined in (7.29).

Let

$$X = R \cap T_p R = V(f) \subset \mathbb{P}^7 = V(x_8) \subset \mathbb{P}^8.$$

Then $X \subset \mathbb{P}^7$ is a cubic hypersurface with vanishing hessian, not a cone. Indeed, the partial derivatives f_i's are linearly independent and we have

$$f_0 f_4 = x_2 x_5 x_6 x_7 = f_1 f_3.$$

More precisely, $Z = \overline{\nabla_X(\mathbb{P}^7)} = V(y_0 y_4 - y_1 y_3) \subset (\mathbb{P}^7)^*$ is a rank four quadric with vertex $V = V(y_0, y_1, y_3, y_4)$. Thus

$$Z^* = V(x_0 x_4 - x_1 x_3, x_2, x_5, x_6, x_7) \subset\!< Z^* >= V(x_2, x_5, x_6, x_7) = \mathbb{P}^3.$$

Hence $\langle Z^* \rangle = \overline{\nabla_R^{-1}(q)}$ is the fiber of the Gauss map of R passing through p, that is, the contact locus on R of the hyperplane $T_p R$, yielding $\langle Z^* \rangle \subseteq \operatorname{Sing}(X)$ (a fact which can also be verified directly). The variety Z^* is thus the locus of secant and tangent lines to W passing through $p \in R = SW$.

The hypersurface X is also singular along $\operatorname{Sing} R \cap T_p R = W \cap T_p R$. We claim that $(\operatorname{Sing} X)_{\mathrm{red}} = \langle Z^* \rangle \cup Y_1 \cup Y_2$, where each $Y_i \subset\!< Y_i >= \mathbb{P}^5$ is a Segre threefold

$\mathbb{P}^1 \times \mathbb{P}^2$ and where $\langle Y_1 \rangle \cap \langle Y_2 \rangle = \langle Z^* \rangle$. Indeed, $T_pR \cap W$ is a hyperplane section of W so that it has degree 6. Moreover, $T_pR \cap W$ contains the following two Segre threefolds:

$$\text{rk} \begin{pmatrix} x_0 & x_1 & x_2 \\ x_3 & x_4 & x_5 \\ 0 & 0 & 0 \end{pmatrix} = 1$$

lying in $V(x_6, x_7) = \mathbb{P}^5 \subset \mathbb{P}^7$ and

$$\text{rk} \begin{pmatrix} x_0 & x_1 & 0 \\ x_3 & x_4 & 0 \\ x_6 & x_7 & 0 \end{pmatrix} = 1$$

lying in $V(x_2, x_5) = \mathbb{P}^5 \subset \mathbb{P}^7$.

Let $\mathscr{P}_X : \mathbb{P}^7 \dashrightarrow Z^*$ be the Perazzo map of X. Thus for $w \in W = Z^*$ general we have $\overline{\mathscr{P}_X^{-1}(w)} = \mathbb{P}_w^5$. More precisely, if $w \in Z^*$, $w = (a_0 : a_1 : 0 : a_3 : a_4 : 0 : 0 : 0)$ with $a_0 a_4 - a_1 a_3 = 0$, then:

$$Q_w = \begin{pmatrix} 0 & 0 & 0 & 0 & 0 & 0 & 0 & 0 \\ 0 & 0 & 0 & 0 & 0 & 0 & 0 & 0 \\ 0 & 0 & 0 & 0 & 0 & 0 & -a_4 & a_3 \\ 0 & 0 & 0 & 0 & 0 & 0 & 0 & 0 \\ 0 & 0 & 0 & 0 & 0 & 0 & 0 & 0 \\ 0 & 0 & 0 & 0 & 0 & 0 & a_1 & -a_0 \\ 0 & 0 & -a_4 & 0 & 0 & a_1 & 0 & 0 \\ 0 & 0 & a_3 & 0 & 0 & -a_0 & 0 & 0 \end{pmatrix}$$

This matrix has rank two due to the relation $a_0 a_4 - a_1 a_3 = 0$ and

$$\text{Sing } Q_w = V(-a_4 x_6 + a_3 x_7, -a_4 x_2 + a_1 x_5)$$

for w general. Therefore two general fibers of the Perazzo map intersect in a $\mathbb{P}^3$ and $X = V(f) \subset \mathbb{P}^7$ is not a Special Perazzo Cubic Hypersurface.

Let us remark that $X^* \subset Z \subset (\mathbb{P}^7)^*$ is a fourfold which by duality is the projection of R^* from the point $\mathscr{G}_R(p) = q = (0 : 0 : \ldots : 0 : 1)$. One can alternatively deduce $\dim(X^*) = 4$ by first observing that T_pR cuts a general fiber of the Gauss map of R in a $\mathbb{P}^2$, which becomes the general fiber of the Gauss map of X so that $\dim(X^*) = \dim(X) - 2 = 4$.

To see the vanishing of the hessian of X geometrically one remarks that $\nabla_R(T_pR) = Q \subset \mathbb{P}^8$ is a quadric singular at $\nabla_R(p)$. Indeed, the restriction of ∇_R to T_pR is birational onto the image, which is a quadric hypersurface since ∇_R is given by quadratic equations and it is an involution; moreover the general positive dimensional fiber of the restriction of ∇_R to $T_p R$ is two dimensional while the fiber of the Gauss map through p has dimension 3. Thus projecting Q from $\nabla_R(p)$ one obtains a quadric hypersurface $Q \subset \mathbb{P}^7$ containing $\nabla_X(\mathbb{P}^7)$. Thus $X \subset \mathbb{P}^7$ has vanishing hessian and it is not difficult to prove that $Z_X = Q$, also by direct computations as seen above.

Remark 7.6.13 One can construct in a similar way examples of cubic hypersurfaces with vanishing hessian, not cones, such that codim$(Z) = 1$ and codim$(X^*, Z) > 1$. Indeed, by taking as R the secant variety to one of the two Severi varieties $W = \mathbb{G}(1, 5) \subset \mathbb{P}^{14}$, respectively $W = E_6 \subset \mathbb{P}^{26}$, and by considering their section by a tangent hyperplane one gets examples of cubic hypersurfaces $X \subset \mathbb{P}^N$ with $N = 13$, respectively $N = 25$, such that Z_X is a quadric hypersurface while X^* has dimension 8, respectively 16. Thus in the first case codim$(X^*, Z) = 4$ while in the second case codim$(X^*, Z) = 8$.

Let us recall that dim$(Z) = \mathrm{rk}\,\mathrm{Hess}_X - 1$ while dim$(X^*) = \mathrm{rk}_f\,\mathrm{Hess}_X - 2$ by Lemma 7.2.7, where $\mathrm{rk}_f\,\mathrm{Hess}_X$ denotes the rank of the matrix Hess_X modulo the ideal (f).

For the cubic hypersurfaces $X = V(f) = SW \subset \mathbb{P}^{\frac{3n}{2}+1}$, $n = 4, 8, 16$, deduced from the corresponding Severi varieties $W^n \subset \mathbb{P}^{\frac{3n}{2}+2}$, we have $\mathrm{rk}\,\mathrm{Hess}_X = \frac{3n}{2} + 1$ and $\mathrm{rk}_f\,\mathrm{Hess}_X = n + 2$ since codim$(X^*, Z) = \frac{n}{2}$. Thus for $n = 4, 8, 16$ we have $\mathrm{rk}\,\mathrm{Hess}_X > \mathrm{rk}_f\,\mathrm{Hess}_X$, something which is somehow unexpected and which can occur only for non-Special Cubic Perazzo Hypersurfaces.

In general it seems quite difficult to construct examples of (cubic) hypersurfaces with vanishing hessian, not cones, such that codim(X^*, Z) can be arbitrarily large. The question is related to the phenomena first studied by B. Segre in [172], see also [61]. We plan to come back to this intriguing problem elsewhere.

References

1. B. Ådlandsvik, Joins and higher secant varieties. Math. Scand. **61**, 213–222 (1987)
2. M.A. Akivis, V.V. Goldberg, Smooth lines on projective planes over two-dimensional algebras and submanifolds with degenerate Gauss maps, Beiträge Algebra Geom. **44**, 165–178 (2003)
3. M.A. Akivis, V.V. Goldberg, *Differential Geometry of Varieties with Degenerate Gauss Maps.* CMS Books in Mathematics, vol. 18 (Springer, New York, 2004)
4. A.A. Albert, On a certain algebra of quantum mechanics. Ann. Math. **35**, 65–73 (1934)
5. A.A. Albert, On power-associativity of rings. Summa Brasil. Math. **2**, 21–32 (1948)
6. A.A. Albert, On power associative rings. Trans. Am. Math. Soc. **64**, 552–593 (1948)
7. A.A. Albert, A theory of power associative commutative algebras. Trans. Am. Math. Soc. **69**, 503–527 (1950)
8. A. Alzati, Special linear systems and syzygies. Collect. Math. **59**, 239–254 (2008)
9. A. Alzati, J.C. Sierra, Quadro-quadric special birational transformations of projective spaces. Int. Math. Res. Not. **2015–1**, 55–77 (2015)
10. M.F. Atiyah, Complex fibre bundles and ruled surfaces. Proc. Lond. Math. Soc. **5**, 407–434 (1955)
11. L. Bădescu, Special chapters of projective geometry. Rend. Sem. Matem. Fisico Milano **69**, 239–326 (1999–2000)
12. L. Bădescu, *Projective Geometry and Formal Geometry.* Monografie Matematyczne, vol. 65 (Birkäuser Verlag, Basel, 2004)
13. L. Bădescu, M. Schneider, A criterion for extending meromorphic functions. Math. Ann. **305**, 393–402 (1996)
14. W. Barth, Submanifolds of low codimension in projective space, in *Proceedings of the International Congress of Mathematicians (Vancouver, 1974)*, Canadian Mathematical Congress, 1975, pp. 409–413
15. W. Barth, M.E. Larsen, On the homotopy groups of complex projective algebraic manifolds. Math. Scand. **30**, 88–94 (1972)
16. M.C. Beltrametti, P. Ionescu, On manifolds swept out by high dimensional quadrics. Math. Z. **260**, 229–234 (2008)
17. M.C. Beltrametti, A.J. Sommese, J. Wiśniewski, Results on varieties with many lines and their applications to adjunction theory, in *Complex Algebraic Varieties (Bayreuth, 1990)*. Lecture Notes in Mathematics, vol. 1507 (Springer, Berlin, Heidelberg, 1992), pp. 16–38
18. E. Bertini, *Introduzione alla Geometria Proiettiva degli Iperspazi*, seconda edizione (Casa Editrice Giuseppe Principato, Messina, 1923)

© Springer International Publishing Switzerland 2016
F. Russo, *On the Geometry of Some Special Projective Varieties*,
Lecture Notes of the Unione Matematica Italiana 18,
DOI 10.1007/978-3-319-26765-4

19. A. Bertram, L. Ein, R. Lazarsfeld, Vanishing theorems, a theorem of Severi and the equations defining projective varieties. J. Am. Math. Soc. **4**, 587–602 (1991)
20. E. Bompiani, Proprietà differenziali caratteristiche di enti algebrici. Rom. Acc. L. Mem. **26**, 452–474 (1921)
21. A. Brigaglia, La teoria generale delle algebre in Italia dal 1919 al 1937. Riv. Stor. Sci. **1**, 199–237 (1984)
22. J. Bronowski, The sum of powers as canonical expressions. Proc. Camb. Philos. Soc. **29**, 69–82 (1933)
23. J. Bronowski, Surfaces whose prime sections are hyperelliptic. J. Lond. Math. Soc. **8**, 308–312 (1933)
24. G. Castelnuovo, Massima dimensione dei sistemi lineari di curve piane di dato genere. Annali di Mat. **18**, 119–128 (1890)
25. C. Carbonaro Marletta, I sistemi omaloidici di ipersuperficie dell' S_4, legati alle algebre complesse di ordine 4, dotate di modulo. Rend. Accad. Sci. Fis. Mat. Napoli **15**, 168–201 (1949)
26. C. Carbonaro Marletta, La funzione inversa $y = x^{-1}$ in una algebra complessa semi-semplice. Boll. Accad. Gioenia Sci. Nat. Catania **2**, 195–201 (1953)
27. P.E. Chaput, Severi varieties. Math. Z. **240**, 451–459 (2002)
28. L. Chiantini, C. Ciliberto, Weakly defective varieties. Trans. Am. Math. Soc. **354**, 151–178 (2001)
29. L. Chiantini, C. Ciliberto, On the concept of k-secant order of a variety. J. Lond. Math. Soc. **73**, 436–454 (2006)
30. L. Chiantini, C. Ciliberto, F. Russo, On secant defective varieties, work in progress
31. C. Ciliberto, Ipersuperficie algebriche a punti parabolici e relative hessiane. Rend. Acc. Naz. Scienze **98**, 25–42 (1979–1980)
32. C. Ciliberto, On a property of Castelnuovo varieties. Trans. Am. Math. Soc. **303**, 201–210 (1987)
33. C. Ciliberto, F. Russo, Varieties with minimal secant degree and linear systems of maximal dimension on surfaces. Adv. Math. **200**, 1–50 (2006)
34. C. Ciliberto, F. Russo, On the classification of *OADP* varieties. SCIENCE CHINA Math. **54**, 1561–1575 (2011). Special issue dedicated to Fabrizio Catanese on the occasion of his 60th birthday
35. C. Ciliberto, M.A. Cueto, M. Mella, K. Ranestad, P. Zwiernik, Cremona linearizations of some classical varieties, in *Proceedings of the Conference "Homage to Corrado Segre"* . arXiv: 1403.1814, to appear
36. C. Ciliberto, M. Mella, F. Russo, Varieties with one apparent double point. J. Algebraic Geom. **13**, 475–512 (2004)
37. C. Ciliberto, F. Russo, A. Simis, Homaloidal hypersurfaces and hypersurfaces with vanishing Hessian. Adv. Math. **218**, 1759–1805 (2008)
38. F. Conforto, *Le superficie razionali* (Zanichelli, Bologna, 1939)
39. C.W. Curtis, *Linear Algebra—An Introductory Approach*. Undergraduate Texts in Mathematics, corrected 6th edn. (Springer, New York, 1997)
40. O. Debarre, *Higher-Dimensional Algebraic Geometry*. Universitext (Springer, New York, 2001).
41. O. Debarre, *Bend and Break* (2007). Available at www-irma.u-strasbg.fr/~debarre/Grenoble. pdf
42. L. Degoli, Sui sistemi lineari di quadriche riducibili ed irriducibili a Jacobiana identicamente nulla. Collect. Math. **35**, 131–148 (1984)
43. P. Deligne, *Letters to W. Fulton* (July/November 1979)
44. P. Deligne, *Le groupe fondamental du complément d'une courbe plane n'ayant que des points doubles ordinaires est abélien*. Sém. Bourbaki, n. 543, 1979/80. Lecture Notes in Mathematics, vol. 842 (Springer, Berlin, New York, 1981), pp. 1–10
45. P. Deligne, N. Katz, *Groupes de monodromie en géométrie algébrique* (SGA7). Lecture Notes in Mathematics, vol. 340 (Springer, Berlin, New York, 1973)

46. P. del Pezzo, Sulle superficie dell' n-esimo ordine immerse nello spazio a n dimensioni. Rend. Circ. Mat. Palermo **12**, 241–271 (1887)

47. V. Di Gennaro, Alcune osservazioni sulle singolarità dello schema di Hilbert che parametrizza le varietà lineari contenute in una varietà proiettiva. Ricerche Mat. **39**, 259–291 (1990)

48. I. Dolgachev, *Classical Algebraic Geometry—A Modern View* (Cambridge University Press, Cambridge, 2012)

49. W.L. Edge, The number of apparent double points of certain loci. Proc. Camb. Philos. Soc. **28**, 285–299 (1932)

50. B.A. Edwards, Algebraic loci of dimension $r - 3$ in space of r dimensions, of which a chord cannot be drawn through an arbitrary point of space. J. Lond. Math. Soc. **2**, 155–158 (1927)

51. L. Ein, Varieties with small dual variety I. Invent. Math. **86**, 63–74 (1986)

52. L. Ein, *Vanishing Theorems for Varieties of Low Codimension*. Lecture Notes in Mathematics, vol. 1311 (Springer, New York, 1988), pp. 71–75

53. L. Ein, N. Shepherd-Barron, Some special Cremona transformations. Am. J. Math. **111**, 783–800 (1989)

54. A. Elduque, S. Okubo, On algebras satisfying $x^2x^2 = N(x)x$. Math. Z. **235**, 275–314 (2000)

55. F. Enriques, Sulla massima dimensione dei sistemi lineari di dato genere appartenenti a una superficie algebrica. Atti Reale Acc. Scienze Torino **29**, 275–296 (1894)

56. G. Faltings, Formale Geometrie und homogene Räume. Invent. Math. **64**, 123–165 (1981)

57. G. Faltings, Ein Kriterium für vollständige Durchschnitte. Invent. Math. **62**, 393–402 (1981)

58. G. Fano, Sulle varietà algebriche che sono intersezioni complete di più forme. Atti Reale Acc. Torino **44**, 633–648 (1909)

59. J. Faraut, A. Korányi, *Analysis on Symmetric Cones* (Clarendon Press, Oxford, 1994)

60. G. Fischer, J. Piontkowski, *Ruled Varieties—An Introduction to Algebraic Differential Geometry*. Advanced Lectures in Mathematics (Friedr. Vieweg & Sohn Verlagsgesellschaft mbH, Braunschweig/Wiesbaden, 2001)

61. A. Franchetta, Forme algebriche sviluppabili e relative hessiane. Atti Acc. Lincei **10**, 1–4 (1951)

62. A. Franchetta, Sulle forme algebriche di S_4 aventi hessiana indeterminata. Rend. Mat. **13**, 1–6 (1954)

63. B. Fu, Inductive characterizations of hyperquadrics. Math. Ann. **340**, 185–194 (2008)

64. B. Fu, J.M. Hwang, Classification of non-degenerate projective varieties with non-zero prolongation and application to target rigidity. Invent. Math. **189**, 457–513 (2012)

65. T. Fujita, *Classification Theory of Polarized Varieties*. London Mathematical Society Lecture Note Series, vol. 155 (Cambridge University Press, Cambridge, 1990)

66. T. Fujita, J. Roberts, Varieties with small secant varieties: the extremal case. Am. J. Math. **103**, 953–976 (1981)

67. W. Fulton, *Intersection Theory*. Ergebnisse der Math. (Springer, Berlin, New York, 1984)

68. W. Fulton, On the topology of algebraic varieties. Proc. Symp. Pure Math. **46**, 15–46 (1987)

69. W. Fulton, J. Hansen, A connectedness theorem for projective varieties, with applications to intersections and singularities of mappings. Ann. Math. **110**, 159–166 (1979)

70. W. Fulton, R. Lazarsfeld, *Connectivity and Its Applications in Algebraic Geometry*. Lecture Notes in Mathematics, vol. 862 (Springer, New York, 1981), pp. 26–92

71. D. Gallarati, Sopra una particolare classe di varietà, ed una proprietà caratteristica delle V_r razionali rappresentabili sul sistema lineare di tutte le quadriche di S_r. Rend. Semin. Mat. Torino **15**, 267–280 (1955–1956)

72. A. Garbagnati, F. Repetto, A geometrical approach to Gordan–Noether's and Franchetta's contributions to a question posed by Hesse. Collect. Math. **60**, 27–41 (2009)

73. C.F. Gauss, *Disquisitiones generales circa superficies curvas*. Collected works, vol. 4, reprint of the 1873 original, (Georg Olms Verlag, Hildesheim, 1973)

74. A.V. Geramita, T. Harima, J.C. Migliore, Y.S. Shin, The Hilbert function of a level algebra. Mem. Am. Math. Soc. **186**(872), 139 pp. (2007)

75. N. Goldstein, Ampleness and connectedness in complex G/P. Trans. Am. Math. Soc. **274**, 361–373 (1982)

76. R. Gondim, On higher hessians and the Lefschetz properties. arXiv: 1506.06387
77. R. Gondim, F. Russo, Cubic hypersurfaces with vanishing hessian. J. Pure Appl. Algebra **219**, 779–806 (2015), see the corrected version as arXiv: 1312.1618
78. P. Gordan, M. Nöther, Ueber die algebraischen Formen, deren Hesse'sche Determinante identisch verschwindet. Math. Ann. **10**, 547–568 (1876)
79. P. Griffiths, J. Harris, Algebraic geometry and local differential geometry. Ann. Sci. Ecole Norm. Sup. **12**, 355–432 (1979)
80. A. Grothendieck, *Cohomologie locale des faisceaux cohérents et théorèmes de Lefschetz locaux et globaux*. Séminaire de Géométrie Algébrique (1962) (Publisher North-Holland Publishing Company, Amsterdam, 1968)
81. A. Grothendieck, *Techniques de construction et théorèmes d'existence en géométrie algébrique IV: les schemas de Hilbert*. Seminaire Bourbaki, 1960/61, Exp. 221, Asterisque hors serie 6, Soc. Math. Fr. (1997)
82. K. Han, Classification of secant defective manifolds near the extremal case. Proc. Am. Math. Soc. **142**, 39–46 (2014)
83. T. Harima, T. Maeno, H. Morita, Y. Numata, A. Wachi, J. Watanabe, *The Lefschetz Properties*. Lecture Notes in Mathematics, vol. 2080 (Springer, Heidelberg, 2013), xx + 250 pp.
84. J. Harris, A bound on the geometric genus of projective varieties. Ann. Scuola Norm. Sup. Pisa **8**, 35–68 (1981)
85. J. Harris, *Algebraic Geometry, A First Course*. Graduate Texts in Mathematics, vol. 133 (Springer, Berlin, 1992)
86. R. Hartshorne, *Ample Subvarieties of Algebraic Varieties*. Lecture Notes in Mathematics, vol. 156 (Springer, Berlin, 1970)
87. R. Hartshorne, Varieties of small codimension in projective space. Bull. Am. Math. Soc. **80**, 1017–1032 (1974)
88. R. Hartshorne, *Algebraic Geometry*. Graduate Texts in Mathematics, vol. 52 (Springer, New York, 1977)
89. R. Hartshorne, *Deformation Theory*, Graduate Texts in Mathematics, vol. 257 (Springer, New York, 2010)
90. O. Hesse, Über die Bedingung, unter welche eine homogene ganze Function von n unabhángigen Variabeln durch Lineäre Substitutionen von n andern unabhángigen Variabeln auf eine homogene Function sich zurück-führen lässt, die eine Variable weniger enthält. J. Reine Angew. Math. **42**, 117–124 (1851)
91. O. Hesse, Zur Theorie der ganzen homogenen Functionen. J. Reine Angew. Math. **56**, 263–269 (1859)
92. A. Holme, J. Roberts, Zak's theorem on superadditivity. Ark. Mat. **32**, 99–120 (1994)
93. J.M. Hwang, Geometry of minimal rational curves on fano manifolds, in *"Vanishing Theorems and Effective Results in Algebraic Geometry"*. I.C.T.P. Lecture Notes, vol. 6 (2001), pp. 335–393
94. J.M. Hwang, S. Kebekus, Geometry of chains of minimal rational curves. J. Reine Angew. Math. (Crelle's J.) **584**, 173–194 (2005)
95. J.M. Hwang, N. Mok, Rigidity of irreducible Hermitian symmetric spaces of the compact type under Kähler deformation. Invent. Math. **131**, 393–418 (1998)
96. J.M. Hwang, N. Mok, Birationality of the tangent map for minimal rational curves. Asian J. Math. **8**, 51–63 (2004)
97. J.M. Hwang, N. Mok, Prolongations of infinitesimal linear automorphisms of projective varieties and rigidity of rational homogeneous spaces of Picard number 1 under Kähler deformation. Invent. Math. **160**, 591–645 (2005)
98. J.M. Hwang, K. Yamaguchi, Characterization of Hermitian symmetric spaces by fundamental forms. Duke Math. J. **120**, 621–634 (2003)
99. P. Ionescu, *Embedded Projective Varieties of Small Invariants*. Lecture Notes in Mathematics, vol. 1056 (Springer, New York, 1984), pp. 142–186
100. P. Ionescu, Birational geometry of rationally connected manifolds via quasi-lines, in *Projective Varieties with Unexpected Properties (Siena, 2004)*, de Gruyter, 2005, pp. 317–335

101. P. Ionescu, On manifolds of small degree. Comment. Math. Helv. **83**, 927–940 (2008)
102. P. Ionescu, D. Naie, Rationality properties of manifolds containing quasi-lines. Int. J. Math. **14**, 1053–1080 (2003)
103. P. Ionescu, F. Russo, Varieties with quadratic entry locus, II. Compos. Math. **144**, 949–962 (2008)
104. P. Ionescu, F. Russo, Conic-connected manifolds. J. Reine Angew. Math. **644**, 145–158 (2010)
105. P. Ionescu, F. Russo, Manifolds covered by lines and the Hartshorne Conjecture for quadratic manifolds. Am. J. Math. **135**, 349–360 (2013)
106. P. Ionescu, F. Russo, On dual defective manifolds. Math. Res. Lett. **21**, 1137–1154 (2014)
107. T.A. Ivey, J.M. Landsberg, *Cartan for Beginners: Differential Geometry via Moving Frames and Exterior Differential Systems*. Graduate Studies in Mathematics, vol. 61 (American Mathematical Society, Providence, RI, 2003)
108. N. Jacobson, *Structure and Representations of Jordan Algebras*. American Mathematical Society Colloquium Publications, vol. XXXIX (American Mathematical Society, Providence, RI, 1968)
109. K. Johnson, Immersion and embedding of projective varieties. Acta Math. **140**, 49–74 (1978)
110. P. Jordan, J. von Neumann, E. Wigner, On an algebraic generalization of the quantum mechanical formalism. Ann. Math. **36**, 29–64 (1934)
111. J.P. Jouanolou, *Théorèmes de Bertini et applications*. Progress in Mathematics, vol. 42 (Birkäuser, Boston, 1983)
112. Y. Kachi, E. Sato, Segre's reflexivity and an inductive characterization of hyperquadrics. Mem. Am. Math. Soc. **160** (American Mathematical Society, Providence, RI, 2002)
113. G. Kempf, *Algebraic Varieties*. London Mathematical Society Lecture Note, vol. 172 (Cambridge University Press, Cambridge, 1993)
114. S.L. Kleiman, Concerning the dual variety, in *Proceedings 18th Scandinavian Congress of Mathematicians (Aarhus 1980)*. Progress in Mathematics, vol. 11 (Birkhäuser, Boston, 1981), pp. 386–396
115. S.L. Kleiman, Tangency and duality, in *Proceedings of the 1984 Vancouver Conference in Algebraic Geometry, Canadian Mathematical Society Conference Proceedings*, vol. 6 (American Mathematical Society, Providence, RI, 1986), pp. 163–225
116. J. Kollár, *Rational Curves on Algebraic Varieties*. Erg. der Math, vol. 32 (Springer, Berlin, 1996)
117. J. Kollár, Y. Miyaoka, S. Mori, Rationally connected varieties. J. Algebraic Geom. **1**, 429–448 (1992)
118. T.Y. Lam, *Introduction to Quadratic Forms over Fields*. Graduate Studies in Mathematics, vol. 67 (American Mathematical Society, Providence, RI, 2005)
119. J.M. Landsberg, On degenerate secant and tangential varieties and local differential geometry. Duke Math. J. **85**, 605–634 (1996)
120. J.M. Landsberg, On the infinitesimal rigidity of homogeneous varieties. Compos. Math. **118**, 189–201 (1999)
121. J.M. Landsberg, Griffiths–Harris rigidity of compact Hermitian symmetric spaces. J. Diff. Geom. **74**, 395–405 (2006)
122. J.M. Landsberg, C. Robbles, Fubini–Griffiths–Harris rigidity of homogeneous varieties. Int. Math. Res. Not. **7**, 1643–1664 (2012)
123. H. Lange, Higher secant varieties of curves and the theorem of Nagata on ruled surfaces. Manuscripta Math. **47**, 263–269 (1984)
124. R.K. Lazarsfeld, *Positivity in Algebraic Geometry I, II*. Erg. der Math. vols. 48 and 49 (Springer, Berlin, 2004)
125. R. Lazarsfeld, A. van de Ven, *Topics in the Geometry of Projective Space, Recent Work by F.L. Zak*. DMV Seminar 4 (Birkhäuser, Boston, 1984)
126. S. Lefschetz, On certain numerical invariants of algebraic varieties. Trans. Am. Math. Soc. **22**, 326–363 (1921)
127. C. Lossen, When does the Hessian determinant vanish identically? (On Gordan and Noether's Proof of Hesse's Claim). Bull. Braz. Math. Soc. **35**, 71–82 (2004)

128. T. Maeno, J. Watanabe, Lefschetz elements of Artinian Gorenstein algebras and hessians of homogeneous polynomials. Illinois J. Math. **53**, 591–603 (2009)

129. D. Martinelli, J.C. Naranjo, G.P. Pirola, Connectedness Bertini Theorem via numerical equivalence. arXiv: 1412.1978

130. H. Matsumura, *Commutative Algebra*, 2nd edn. (The Benjamin/Cummings Publishing Company, San Francisco, 1980)

131. K. McCrimmon, Axioms for inversion in Jordan algebras. J. Algebra **47**, 201–222 (1977)

132. K. McCrimmon, Jordan algebras and their applications. Bull. Am. Math. Soc. **84**, 612–627 (1978)

133. K. McCrimmon, Adjoints and Jordan algebras. Commun. Algebra **13**, 2567–2596 (1985)

134. K. McCrimmon, *A taste of Jordan algebras*. Universitext (Springer, New York, 2004)

135. N. Mok, Recognizing certain rational homogeneous manifolds of Picard number 1 from their varieties of minimal rational tangents, in *Third International Congress of Chinese Mathematicians. Part 1, 2*. AMS/IP Studies in Advanced Mathematics, vol. 42, pt.1, 2 (American Mathematical Society, Providence, RI, 2008), pp. 41–61

136. S. Mori, Projective manifolds with ample tangent bundle. Ann. Math. **110**, 593–606 (1979)

137. S. Mori, Threefolds whose canonical bundles are not numerically effective. Ann. Math. **116**, 133–176 (1982)

138. S. Mukai, Biregular classification of Fano threefolds and Fano manifolds of coindex 3. Proc. Natl. Acad. Sci. U. S. A. **86**, 3000–3002 (1989)

139. S. Mukai, Simple Lie algebra and Legendre variety. Preprint (1998), http://www.math.nagoya-u.ac.jp/~mukai

140. D. Mumford, Varieties defined by quadratic equations, in *Questions on Algebraic Varieties (C.I.M.E., III Ciclo, Varenna, 1969)* (Ed. Cremonese, Rome, 1970), pp. 2 9–100

141. D. Mumford, Some footnotes to the work of C.P. Ramanujam, in *C.P. Ramanujam—A Tribute* (Springer, New York, 1978), pp. 247–262

142. D. Mumford, *The Red Book of Varieties and Schemes*. Lecture Notes in Mathematics, vol. 1358 (Springer, Berlin, 1988)

143. O. Nash, K-Theory, $LQEL$-manifolds and Severi varieties. Geom. Topol. **18**, 1245–1260 (2014)

144. N.Y. Netsvetaev, *Projective Varieties Defined by a Small Number of Equations are Complete Intersections*. Lecture Notes in Mathematics, vol. 1346 (Springer, Berlin, 1988), pp. 433–453

145. M. Ohno, On degenerate secant varieties whose Gauss map have the largest image. Pac. J. Math. **187**, 151–175 (1999)

146. M. Oka, On the cohomology structure of projective varieties, in *Manifolds—Tokyo, 1973* (University of Tokyo Press, Tokyo, 1975), pp. 137–143

147. F. Palatini, Sulle superficie algebriche i cui S_h $(h+1)$-seganti non riempiono lo spazio ambiente. Atti Accad. Torino **41**, 634–640 (1906)

148. F. Palatini, Sulle varietá algebriche per le quali sono minori dell' ordinario, senza riempire lo spazio ambiente, una o alcune delle varietá formate da spazi seganti. Atti Accad. Torino **44**, 362–375 (1909)

149. I. Pan, F. Ronga, T. Vust, Transformations birationelles quadratiques de l'espace projectif complexe à trois dimensions. Ann. Inst. Fourier (Grenoble) **51**, 1153–1187 (2001)

150. R. Pardini, Some remarks on plane curves over fields of finite characteristic. Compos. Math. **60**, 3–17 (1986)

151. U. Perazzo, Sulle varietá cubiche la cui hessiana svanisce identicamente. G. Mat. Battaglini **38**, 337–354 (1900)

152. R. Permutti, Su certe forme a hessiana indeterminata. Ricerche di Mat. **6**, 3–10 (1957)

153. R. Permutti, Su certe classi di forme a hessiana indeterminata. Ricerche di Mat. **13**, 97–105 (1964)

154. L. Pirio, F. Russo, On projective varieties n-covered by irreducible curves of degree δ. Comment. Math. Helv. **88**, 715–756 (2013)

155. L. Pirio, F. Russo, Quadro-quadric Cremona maps and varieties 3-connected by cubics: semi-simple part and radical. Int. J. Math. **24**, 13, 33 (2013). doi: 10.1142/S0129167X1350105X 1350105

156. L. Pirio, F. Russo, The *XJC*-correspondence. J. Reine Angew. Math. (Crelle's J.) (to appear, DOI 10.1515/crelle-2014-0052)

157. L. Pirio, F. Russo, Quadro-quadric cremona transformations in low dimensions via the JC-correspondence. Ann. Inst. Fourier (Grenoble) **64**, 71–111 (2014)

158. L. Pirio, J.-M. Trépreau, Sur les variétés $X \subset \mathbb{P}^N$ telles que par n points passe une courbe de degré donné. Bull. Soc. Math. France **141**, 131–196 (2013)

159. F. Russo, On a theorem of Severi. Math. Ann. **316**, 1–17 (2000)

160. F. Russo, Varieties with quadratic entry locus, I. Math. Ann. **344**, 597–617 (2009)

161. F. Russo, Lines on projective varieties and applications. Rend. Circ. Mat. Palermo **61**, 47–64 (2012)

162. F. Russo, A. Simis, On birational maps and Jacobian matrices. Compos. Math. **126**, 335–358 (2001)

163. G. Scorza, Sulla determinazione delle varietá a tre dimensioni di S_r ($r \geq 7$) i cui S_3 tangenti si tagliano a due a due. Rend. Circ. Mat. Palermo **25**, 193–204 (1908)

164. G. Scorza, Un problema sui sistemi lineari di curve appartenenti a una superficie algebrica. Rend. Reale Ist. Lombardo Scienze e Lettere **41**, 913–920 (1908)

165. G. Scorza, Le varietá a curve sezioni ellittiche. Ann. Mat. Pura Appl. **15**, 217–273 (1908)

166. G. Scorza, Sulle varietá a quattro dimensioni di S_r ($r \geq 9$) i cui S_4 tangenti si tagliano a due a due. Rend. Circ. Mat. Palermo **27**, 148–178 (1909)

167. G. Scorza, *Corpi numerici e algebre* (Casa Editrice Giuseppe Principato, Messina, 1921)

168. G. Scorza, *Opere Scelte, I–IV* (Edizioni Cremonese, Roma, 1960)

169. C. Segre, Sulle varietà normali a tre dimensioni composte da serie semplici di piani. Atti della R. Acc. delle Scienze di Torino **21**, 95–115 (1885)

170. B. Segre, Bertini forms and Hessian matrices. J. Lond. Math. Soc. **26**, 164–176 (1951)

171. B. Segre, *Some Properties of Differentiable Varieties and Transformations*. Erg. Math. (Springer, Berlin, Heidelberg, 1957)

172. B. Segre, Sull' hessiano di taluni polinomi (determinanti, pfaffiani, discriminanti, risultanti, hessiani) I, II. Atti Acc. Lincei **37**, 109–117, 215–221 (1964)

173. J.G. Semple, On representations of the S_k's of S_n and of the Grassmann manifold $\mathbb{G}(k, n)$. Proc. Lond. Math. Soc. **32**, 200–221 (1931)

174. J.G. Semple, L. Roth, *Introduction to Algebraic Geometry* (Oxford University Press, Oxford, 1949 and 1986)

175. E. Sernesi, *Deformations of Algebraic Schemes*. Grund. der Math. Wiss., vol. 334 (Springer, Berlin, Heidelberg, 2006)

176. F. Severi, Intorno ai punti doppi impropri di una superficie generale dello spazio a quattro dimensioni e ai suoi punti tripli apparenti. Rend. Circ. Mat. Palermo **15**, 33–51 (1901)

177. F. Severi, Una proprietà delle forme algebriche prive di punti multipli. Rend. Acc. Lincei **15**, 691–696 (1906)

178. I.R. Shafarevich, *Basic Algebraic Geometry* (Springer, Berlin, Heidelberg, New York, 1974)

179. A. Simis, B. Ulrich, W.V. Vasconcelos, Tangent star cones. J. Reine Angew. Math. **483**, 23–59 (1997)

180. N. Spampinato, I gruppi di affinità e di trasformazioni quadratiche piane legati alle due algebre complesse doppie dotate di modulo. Boll. Accad. Gioenia Sci. Nat. Catania **67**, 80–86 (1935)

181. G. Staglianò, On special quadratic birational transformations of a projective space into a hypersurface. Rend. del Circolo Mat. Palermo **61**, 403–429 (2012)

182. G. Staglianò, On special quadratic birational transformations whose base locus has dimension at most three. Atti Accad. Naz. Lincei Cl. Sci. Fis. Mat. Nat. **24**, 409–436 (2013)

183. R.P. Stanley, Weyl groups, the hard Lefschetz theorem, and the Sperner property. SIAM J. Algebra. Discr. Methods **1**(2), 168–184 (1980)

184. E. Strickland, Lines in G/P. Math. Z. **242**, 227–240 (2002)

185. A. Terracini, Sulle V_k per cui la varieta' degli S_h $(h + 1)$-secanti ha dimensione minore dell' ordinario. Rend. Circ. Mat. Palermo **31**, 392–396 (1911)

186. A. Terracini, Su due problemi concernenti la determinazione di alcune classi di superficie, considerate da G. Scorza e F. Palatini. Atti Soc. Natur. e Matem. Modena **6**, 3–16 (1921–1922)

187. A. Terracini, Alcune questioni sugli spazi tangenti e osculatori ad una varietá, I, II, III, in *Selecta Alessandro Terracini*, vol. I (Edizioni Cremonese, Roma, 1968), pp. 24–90

188. P. Vermeire, Some results on secant varieties leading to a geometric flip construction. Compos. Math. **125**, 263–282 (2001)

189. G. Veronese, Behandlung der projectivischen Verhältnisse der Räume von verschiedenen Dimensionen durch das Princip des Prjjicirens und Schneidens. Math. Ann. **19**, 161–234 (1882)

190. A.H. Wallace, Tangency and duality over arbitrary fields. Proc. Lond. Math. Soc. **6**, 321–342 (1956)

191. A.H. Wallace, Tangency properties of algebraic varieties, Proc. Lond. Math. Soc. **7**, 549–567 (1957)

192. J. Watanabe, A remark on the Hessian of homogeneous polynomials, in *The Curves Seminar at Queen's*, vol. XIII. Queen's Papers in Pure and Applied Mathematics, vol. 119 (Queens Univ. Campus, Kingston, 2000), pp. 171–178

193. J. Watanabe, On the theory of Gordan–Noether on homogeneous forms with zero Hessian. Proc. Sch. Sci. Tokai Univ. **49**, 1–21 (2014)

194. H. Weyl, Emmy Noether. Scr. Math. **3**, 201–220 (1935)

195. J. Wiśniewski, On a conjecture of Mukai. Manuscripta Math. **68**, 135–141 (1990)

196. F.L. Zak, Severi varieties. Math. USSR Sbornik **54**, 113–127 (1986)

197. F.L. Zak, *Some Properties of Dual Varieties and Their Applications in Projective Geometry*. Lecture Notes in Mathematics, vol. 1479 (Springer, Berlin, Heidelberg, New York, 1991), pp. 273–280

198. F.L. Zak, *Tangents and secants of algebraic varieties*. Translations of Mathematical Monographs, vol. 127 (American Mathematical Society, Providence, RI, 1993)

Index

© Springer International Publishing Switzerland 2016
F. Russo, *On the Geometry of Some Special Projective Varieties*,
Lecture Notes of the Unione Matematica Italiana 18,
DOI 10.1007/978-3-319-26765-4

LECTURE NOTES OF THE UNIONE MATEMATICA ITALIANA

Editor in Chief: Ciro Ciliberto and Susanna Terracini

Editorial Policy

1. The UMI Lecture Notes aim to report new developments in all areas of mathematics and their applications - quickly, informally and at a high level. Mathematical texts analysing new developments in modelling and numerical simulation are also welcome.

2. Manuscripts should be submitted to
   Redazione Lecture Notes U.M.I.
   umi@dm.unibo.it
   and possibly to one of the editors of the Board informing, in this case, the Redazione about the submission. In general, manuscripts will be sent out to external referees for evaluation. If a decision cannot yet be reached on the basis of the first 2 reports, further referees may be contacted. The author will be informed of this. A final decision to publish can be made only on the basis of the complete manuscript, however a refereeing process leading to a preliminary decision can be based on a pre-final or incomplete manuscript. The strict minimum amount of material that will be considered should include a detailed outline describing the planned contents of each chapter, a bibliography and several sample chapters.

3. Manuscripts should in general be submitted in English. Final manuscripts should contain at least 100 pages of mathematical text and should always include

   - a table of contents;
   - an informative introduction, with adequate motivation and perhaps some historical remarks: it should be accessible to a
     reader not intimately familiar with the topic treated;
   - a subject index: as a rule this is genuinely helpful for the reader.

4. For evaluation purposes, please submit manuscripts in electronic form, preferably as pdf- or zipped ps-files. Authors are asked, if their manuscript is accepted for publication, to use the LaTeX2e style files available from Springer's web-server at
   ftp://ftp.springer.de/pub/tex/latex/svmonot1/ for monographs
   and at
   ftp://ftp.springer.de/pub/tex/latex/svmultt1/ for multi-authored volumes

5. Authors receive a total of 50 free copies of their volume, but no royalties. They are entitled to a discount of 33.3% on the price of Springer books purchased for their personal use, if ordering directly from Springer.

6. Commitment to publish is made by letter of intent rather than by signing a formal contract. Springer-Verlag secures the copyright for each volume. Authors are free to reuse material contained in their LNM volumes in later publications: A brief written (or e-mail) request for formal permission is sufficient.

Printed in the United States
By Bookmasters